50 YEARS OF SCIENCE
IN SINGAPORE

World Scientific Series on Singapore's 50 Years of Nation-Building

The complete list of titles in the series can be found at
http://www.worldscientific.com/series/wss50ynb

World Scientific Series on
Singapore's 50 Years of Nation-Building

50 YEARS OF SCIENCE IN SINGAPORE

Editors

B T G Tan
National University of Singapore

Hock Lim
National University of Singapore

K K Phua
Institute of Advanced Studies
Nanyang Technological University

World Scientific

NEW JERSEY · LONDON · SINGAPORE · BEIJING · SHANGHAI · HONG KONG · TAIPEI · CHENNAI · TOKYO

Published by

World Scientific Publishing Co. Pte. Ltd.

5 Toh Tuck Link, Singapore 596224

USA office: 27 Warren Street, Suite 401-402, Hackensack, NJ 07601

UK office: 57 Shelton Street, Covent Garden, London WC2H 9HE

Library of Congress Cataloging-in-Publication Data

Names: Tan, B. T. G., editor. | Lim, Hock, editor. | Phua, K. K., editor.

Title: 50 years of science in Singapore / editors, B.T.G. Tan, Hock Lim, K.K. Phua.

Other titles: Fifty years of science in Singapore

Description: New Jersey : World Scientific, [2017] | Series: World Scientific series on
    Singapore's 50 years of nation-building | Includes bibliographical references and index.

Identifiers: LCCN 2016047372| ISBN 9789813140882 (hardcover : alk. paper) |
    ISBN 9789813140899 (softcover : alk. paper)

Subjects: LCSH: Science--Singapore--History. | Science and state--Singapore--History.

Classification: LCC Q127.S55 A15 2017 | DDC 509.5957--dc23

LC record available at https://lccn.loc.gov/2016047372

British Library Cataloguing-in-Publication Data

A catalogue record for this book is available from the British Library.

Typeset by Stallion Press

Email: enquiries@stallionpress.com

Foreword

When Singapore became independent in 1965, the fledgling nation had little scientific resources or infrastructure. The most urgent matter with respect to Science was to build up an education system that could train and develop a pool of scientific and technical manpower to meet the needs of Singapore's industries.

Once this foundation had been firmly laid, both at the school and tertiary level, the next step was to raise our capabilities and to build up scientific and technological research infrastructure. This would enable Singapore and our industries to engage in knowledge creation and discovery, and for our economy to grow by moving towards high value-add sectors and activities.

In tandem with Singapore's monumental progress from third world to first, the past 50 years has seen major advances in Singapore's scientific infrastructure, both in research and education. In spite of our smallness, Singapore's achievements are remarkable and today we are recognised as a significant player in many key areas of scientific research.

To mark 50 years of Singapore's independence in 2015, World Scientific has embarked on the production of a series of volumes documenting Singapore's growth in various fields over this period. This volume tells the story of how Science in Singapore was seeded, nurtured and how the scientific landscape has flourished over the course of 50 years. Many of the articles are penned by authors who were personally and intimately involved in this journey. In the articles, they recount personal anecdotes and share candid stories of how they and their colleagues had contributed towards building up Singapore Science.

This volume is a valuable resource that chronicles the unique history, journey and development of Science in Singapore.

Professor Tan Eng Chye
Deputy President (Academic Affairs) and Provost
National University of Singapore

Preface

This book, *50 Years of Science in Singapore*, is being published as a volume in the World Scientific Series commemorating the 50th Anniversary of the Republic of Singapore.

In this volume, we have attempted to include as many accounts of the history and development of the various sectors of Singapore science in the last 50 years or so. The articles therefore cover the government agencies charged with the responsibility for science funding and research policy, the academic institutions and departments who have been in the forefront of the development of the nation's scientific manpower and research, the research centres and institutes which have been critical to the breaking of new ground in both basic and applied science research, science museums and education, and the academic and professional institutions which the scientific community itself has set up to enable Singapore scientists to serve the nation more effectively.

Wherever possible, we have invited those who were directly responsible and involved in the founding and development of the relevant bodies to author the articles. In a number of cases, we have modified already published material which may not have had a wide audience, and reprinted them in this volume where they will reach a larger and broader readership. We do not intend that the articles and essays in this volume be regarded as formal and meticulously researched historical documents. Our overall objective is to capture and record, in a more informal and personal but reasonably accurate manner, the historical recollections of those who were directly involved in the development of Singapore science. If this volume becomes a useful source for future historians of Singapore science, we will be well satisfied.

We, the Editors, would like to thank all our authors and contributors who took time off to pen their recollections and thoughts for this volume. Scientists are generally not ideal (or willing) writers of history, but we believe that the story of Singapore science should, as far as possible, be told by the scientists themselves who helped shape that history. We hope you will find this volume interesting and edifying.

B T G Tan, Hock Lim and K K Phua
Singapore, August 2016

Contents

About the Contributors

Antonio H Castro Neto got his PhD in Physics at University of Illinois at Urbana-Champaign in 1994. In 1994, he moved to the Institute for Theoretical Physics at the University of California at Santa Barbara as a postdoctoral fellow. In 1995, he became an Assistant Professor at University of California at Riverside. In 2000, he moved to Boston University as Professor of Physics. Since 2010, Prof Castro Neto is the Director of the Graphene Research Center and in 2014 he became Director of the Centre for Advanced 2D Materials funded by the National Research Foundation of Singapore. Prof Castro Neto is a Distinguished Professor in the Physics Department and Professor at the Department of Electrical and Computer Engineering and the Department of Materials Science Engineering at the National University of Singapore.

Chan Chui Theng is Associate Director, Admin at the Centre for Quantum Technologies (CQT) at NUS. She was one of the first members of staff to join CQT, recruited from the NUS Office of Alumni Relations in 2006. Her responsibilities include liaison with the Centre's stakeholders, assisting with grant applications and supporting management of the PhD programme. She graduated from the Faculty of Arts and Social Sciences, NUS.

Chan Sing Chai was lecturer in the Department of Physics, Nanyang University (1970–74). She was promoted to senior lecturer (1975) and was Head of the Computer Science Department in 1976–77. She was then posted to the Bukit Timah Campus (1978–80) to oversee the administrative work of the Department of Computer Science. Promoted to Associate Professor in 1983, she was tasked to handle students' internship posting and arrange career talks with the computer industries and related statutory boards. She was a committee member of the *International Chinese Computer Society* and the Chief Editor of the related *Chinese Computer Journal* (1985–1990).

Louis Chen received his PhD from Stanford University in 1971. He is currently Emeritus Professor at the National University of Singapore. He was Director of the Institute for Mathematical Sciences (2000–2012), Head of the Department of Mathematics (1996–2000) and Head of the Department of Statistics and Applied Probability (2002–2004). He was also President of the Bernoulli Society for Mathematical Statistics and Probability (1997–1999), President of the Institute of Mathematical Statistics (2004–2005), and Vice-President of the International Statistical Institute (2009–2011). His research interests are in probability and computational biology, focusing largely on Stein's method.

Chong Chi Tat is University Professor at the Department of Mathematics, National University of Singapore, and is Director of the Institute for Mathematical Sciences. At NUS, he has served as Head of the Department of Information Systems and Computer Science, Vice Dean of Science, Deputy Vice Chancellor, Deputy President and Provost, as well as Head of the Department of Mathematics. He received his PhD from Yale University and specialises in the research area of theory of computation.

Chou Loke Ming is Adjunct Research Professor at the Tropical Marine Science Institute of the National University of Singapore. He has 40 years of research experience on marine environment management and coral reef biology throughout Southeast Asia and is recognised for his leadership in the periodic compilation of Southeast Asia's coral reef condition for the global reef status reports published since 1998 by the Global Coral Reef Monitoring Network (GCRMN) of the International Coral Reef Initiative (ICRI). He is an Honorary Fellow of the Singapore Institute of Biology and a Fellow of the Singapore National Academy of Sciences.

Artur Ekert is the Director of the Centre for Quantum Technologies and Lee Kong Chian Centennial Professor at the National University of Singapore. He is also Professor of Quantum Physics, Mathematical Institute, University of Oxford. His invention of entanglement-based quantum cryptography in 1991 triggered an explosion of research efforts worldwide and continues to inspire new research directions. He has contributed to many important advances in the foundations and experimental realisations of quantum communication and computation. He was awarded the 1995 Maxwell Medal and Prize by the Institute of Physics and the 2007 Hughes Medal by the Royal Society. He was elected a Fellow of the Royal Society in 2016.

Feng Yuan Ping is a Professor in the Department of Physics, and past Head of Department (2007–2014). His research focuses on computational materials physics and materials prediction using first-principles approach. He is a fellow of American Physical Society (APS), Fellow of Institute of Physics, Singapore (IPS), and an academician of the Asia-Pacific Academy of Materials (APAM). He is currently serving as the Secretary of International Union of Materials Research Societies (IUMRS), and a Vice-President of Materials Research Society of Singapore (MRSS).

Go Mei Lin obtained her PhD from NUS in 1980. She is an associate professor in the Department of Pharmacy NUS and past Head of Department (1998–2000). She is a member of the Drug Development Unit at NUS which works on the preclinical characterisation of promising hits and leads. Her research interests lies in the area of design and synthesis of bioactive small drugs with drug-like properties, establishing structure-activity correlations and mode of action studies by biochemical and pharmacological approaches. She has authored more than 80 peer-reviewed publications and is co-inventor of three patents on biologically active entities.

Hew Choy Leong, an Emeritus Professor with NUS, is known for his research in fish anti-freeze proteins and fast growing transgenic salmon. Prof Hew received his PhD in Biochemistry from the University of British Columbia in 1970 and was a PDF at Yale till 1972. He has held several leadership roles, including Head of Biological Sciences at NUS from 1999–2008 and Deputy Director of the Mechanobiology Institute. He has also received many accolades, including *Most Outstanding Young Scientist in the Atlantic Provinces of Canada* in 1980, and winning the 1993 Award of Merit by the Federation of Chinese Canadian Professionals.

Andy Hor (BSc(Hon) Imperial College, DPhil(Oxon) Oxford, DSc London, Yale Postdoc) is the Vice-President and Pro-Vice-Chancellor (Research) and Chair Professor of the University of Hong Kong. He started his academic career in NUS in 1984. As Head of Chemistry for nearly a decade in 2000's, he was responsible for probably the most vigorous era of growth of the Department in its history. As Vice-Dean (Academic) of Science, he was involved in several landmark projects of NUS, such as the modular structure, mentorship, Special Programme in Science, Food Science and Technology, etc. He was seconded to A*STAR for 5 years as the Executive Director of the Institute of Materials Research & Engineering before the move to Hong Kong. He has published ~415 international papers with >880 citations in 2015.

Hsu Loke Soo was past Director of Computer Center, Nanyang University (1970–1974), Head of Department of Applied Computer Science, Nanyang University (1975–1978), Head of Department of Computer Science, Nanyang University (1978–1980), and Head of Department of Computer Science, NUS (1980–1989). He taught courses in Fortran, Cobol, Pascal and Systems Programming. He has been a member of the Editorial Board of the *International Journal on Education and Computing*. He was also an external examiner for Chinese University of Hong Kong, Central Lancashire University, UK and National Computer Centre.

Jenny Hogan is Associate Director for Outreach and Media Relations at the Centre for Quantum Technologies (CQT) at NUS. She graduated with a Master's Degree in Natural Sciences (Experimental Physics) from the University of Cambridge in 2002. Before moving to Singapore, she worked as a science journalist for *New Scientist* and *Nature* magazines. She has developed CQT's outreach programme to inspire local and international audiences about quantum physics.

Eugene Khor obtained his Bachelor of Science degree from Lakehead University in 1979, and went on to do his PhD at Virginia Tech where his research was on metal-containing polyamide materials. He joined the Chemistry Department of the National University of Singapore in 1984 where his principal research programme was on chitin materials, and retired from the University in 2011. He was the principal co-founder of BRASS, a medical technology tests services company, and guided the company from its founding till it became profitable in 2005 and its acquisition by Charles River Laboratories in 2013.

Kwek Leong Chuan is a Principal Investigator at the Centre for Quantum Technologies and holds two posts at Singapore's Nanyang Technological University: Deputy Director (Physical Sciences) of the Institute of Advanced Studies and Associate Professor at the National Institute of Education. He was a teacher for eight years before pursuing his doctoral degree at NUS, which he completed in 1999. His current research interests include the foundations of quantum theory and distributed quantum computing. He won Singapore's National Science Award in 2006. He is also an elected Fellow of the American Association for the Advancement of Science and the Institute of Physics, UK.

Lai Choy Heng is a Professor of Physics at the National University of Singapore (NUS) and the Deputy Director of the Centre for Quantum Technologies. He did his undergraduate and graduate education at The University of Chicago, and was a research fellow at the Niels Bohr Institute, Copenhagen, before joining NUS as a faculty member. Among the academic administration responsibilities, he was Head of the Department of Computational Science and Department of Physics, Dean of the Faculty of Science, Vice-Provost (Academic Personnel), and Executive Vice-President of the Yale-NUS College. His current research interests are in (quantum) information theory and dynamics on complex networks.

Lam Toong Jin, an Emeritus Professor with NUS, was the Head of Zoology Department (1981 to 1996), and then the Head of Biological Sciences Department (1996 to 1999) after the merger of Botany and Zoology Departments. He played key roles in the establishment of the Tropical Marine Science Institute, the preservation of the Zoological Reference Collection from the formal National Museum, and the founding of Institutional Animal Care & Use Committee. He has served as a member on the editorial boards of several international journals, and as consultant to the Food and Agriculture Organisation of the United Nations(FAO) and International Development Research Centre (IDRC). Prof Lam received the National Science & Technology Award in 1990 and the Distinguished Biologist Award (Singapore Institute of Biology) in 1999. He is still active in research on aquaculture, particularly on culturing Soon Hock with shrimps and plants in an integrated multi-trophic aquaculture system.

Lee Peng Yee is a mathematician, doing research in real analysis and functional analysis, in particular, the Henstock-Kurzweil integral. He has been teaching for 50 years at universities in Malawi (Africa), New Zealand, the Philippines, and Singapore. He had more than 100 research publications and over 20 PhD students. He was Head of the Mathematics Department at Nanyang University in the early '70s and at National Institute of Education and Nanyang Technological University in the late '90s. He served as Vice-President of the International Commission of Mathematical Instruction (1987–1990, 1991–1994). He retired at age 75 in 2013.

Leong Hon Wai is with the Department of Computer Science and the University Scholar's Programme (USP) at the National University of Singapore. He received his PhD in Computer Science from the University of Illinois at Urbana-Champaign. His research interest is in the design of efficient algorithms for problems from diverse application areas including VLSI-CAD, transportation logistics, multimedia systems, and computational biology. He has a passion for nurturing young talents and gives many outreach workshop on creative problem solving and computational thinking. He started the Singapore training program for the IOI (International Olympiad in Informatics) in 1992. He is a member of ACM, IEEE, ISCB, and a Fellow of the Singapore Computer Society.

Gloria Lim is a distinguished mycologist who is an Honours graduate of the Botany Department of the University of Malaya. She received her PhD from the University of London and served as Dean of the Faculty of Science as the Head of the Botany Department at the National University of Singapore (NUS). She was also the Foundation Director of the National Institute of Education at the Nanyang Technological Institute (NTU). She has served as a member of the Public Service Commission and on the Science Council, the Science Centre Board and the National Parks Board. She has received the Public Service Star, the Distinguished Science Alumni Award (NUS), and has been inducted into the Singapore Women's Hall of Fame.

Lim Hock is an Honours graduate of the Physics Department, NUS and received his PhD in meteorology from the University of Reading, England. He served in the Singapore Department of Meteorology before joining the faculty of the Physics Department. He was Founding Director of both the Centre for Remote Imaging, Sensing and Processing and the Temasek Laboratories at NUS. He is currently the Founding Director of the Singapore Nuclear Research and Safety Institute (SNRSI) and Director of Research Governance and Enablement at NUS. He has received the IPS President's Medal and is a Fellow of the Singapore National Academy of Science. He has also received the Public Service Medal and the Public Administration Medal (Silver).

Lim Tit Meng is the Chief Executive at the Science Centre Singapore. He is also an Associate Professor at the National University of Singapore. He is the President of the Singapore Association for the Advancement of Science and the First Vice-President of the Singapore National Academy of Science. He has been conferred Fellow of the Singapore National Academy of Science, and Fellow of the Singapore Institute of Biology. In the international science centre community, he serves as a Director in the Board of the Association of Science Technology Centres (ASTC), and the President of the Asia Pacific Science Centres Network (ASPAC).

Ling San is Professor of Mathematical Sciences in the School of Physical & Mathematical Sciences (SPMS), Nanyang Technological University (NTU), with a concurrent cross appointment in the School of Computer Science & Engineering. He has been Dean of the College of Science since August 2011, and had previously served as Chair of the SPMS and the Founding Head of the Division of Mathematical Sciences at NTU. Prior to joining NTU in April 2005, he was with the Department of Mathematics, National University of Singapore, for 13 years. He received his PhD from the University of California, Berkeley. His current research interests are coding and cryptography.

Loh Kian Ping is currently Provost's Chair Professor in NUS and a world leader in the field of graphene and 2D materials research. His team developed a strategy for the large area growth and transfer of graphene on silicon wafer (Nature 2014), a new strategy for the synthesis of graphene quantum dots from bucky balls (published in *Nature Nanotechnology*); developed an industrially scalable strategy to process solution processed graphene for green catalysis, leading to a spin-off company. His awards include the President's Science Award (2014), the University Outstanding Researcher award (2012), University's Young Scientist award (2008), and the American Chemical Society Nano Lectureship award (2013). He is currently the associate editor of Chemistry of Materials and also head of the 2D materials group in Centre for Advanced 2D Materials, NUS.

Low Boon Chuan is an Associate Professor at the Department of Biological Sciences, NUS, and was a co-founding member of the Mechanobiology Institute, where he is currently a Principal Investigator. His research focuses on the discovery of novel signalling proteins and protein domains that control cell morphology, motility, differentiation, cell growth and organ development. Previously a winner of the Singapore Young Scientist Award, he now serves in the judging panels for the Tan Kah Kee Young Inventors' Awards and the A*STAR Talent Search Awards. He is also the Deputy Director of the University Scholars Programme at NUS.

Low Teck Seng is currently the Chief Executive Officer of the National Research Foundation, Prime Minister's Office, Singapore. He graduated with a First Class Honours in Electrical & Electronic Engineering in 1978 from Southampton University and subsequently received his PhD from the same university in 1982. He joined the National University of Singapore in 1983 as an academic staff of the Department of Electrical Engineering. His research interests were in computational electromagnetics and spinelectronics. Low was awarded the National Science and Technology Medal in 2004, the highest honour bestowed on an individual who has made distinguished, sustained and exceptional contributions, and played a strategic role in Singapore's development through the promotion and management of R&D.

Paul Matsudaira is a Distinguished Professor and Head of the Department of Biological Sciences. He also directs the Centre for BioImaging Sciences and co-directs the MechanoBiology Institute. A cell biologist by training, Paul's research is focused on the mechanobiology and structure of cells and tissues. He received his PhD from Dartmouth College in 1980 and conducted postdoctoral research at the Max Planck for Biophysical Chemistry (Goettingen, Germany) and MRC Laboratory of Molecular Biology (Cambridge, UK). In 1985, he joined the Biology Department at the Massachusetts Institute of Technology and the Whitehead Institute for Biomedical Research. He left in 2008 as Professor of Biology and Bioengineering to join NUS.

Mok Kum Fun joined the Department of Chemistry, University of Singapore, as a lecturer in 1965 after obtaining his PhD from the University of Wellington, New Zealand under the Colombo Plan Scholarship scheme. He retired from the Department of Chemistry, National University of Singapore in 2001.

Oh Choo Hiap was Head (July 2000–June 2006), Physics Department, National University of Singapore. He is a Senior Fellow at the Institute of Advanced Studies, Nanyang Technological University, and also Fellow of the American Physical Society (APS). Prof Oh won the National Science Award 2006 (Singapore). His main research areas include Yang-Mills gauge field theories and high energy physics, quantum algebra and Yang-Baxter equations, and quantum information and computation.

Ong Chong Kim is a recipient of the prestigious NUS Outstanding Researcher Award (2010). He is a Fellow of the Institute of Physics (IOP) UK and the Institute of Physics, Singapore. He served as Vice-President of the Singapore Academy of Science (1998–2000), President of the Institute of Physics, Singapore (1996–2000), Founding Director of the Centre for Superconducting and Magnetic Materials and was Deputy Head of the Department of Physics, NUS (2006–2008). He sat on the editorial board of three international journals and co-authored more than 566 papers with a H-index of 41. He also co-authored a textbook on *Microwave Electronics* (Wiley) and holds a patent in America.

Phua Kok Khoo obtained his PhD in Mathematical Physics from the University of Birmingham in 1970. He was awarded the Institute of Physics Singapore (IPS) President's Award by the IPS Council in 2006. He is a Fellow at the American Physical Society, the Founding Director of the Institute of Advanced Studies at Nanyang Technological University, an Adjunct Professor at the National University of Singapore, and Honorary Professor in many universities in China. He is also Chairman of World Scientific Publishing Company. Professor Phua's research interests are in theoretical high energy physics, science education and science policies, and has published many scientific papers in internationally refereed journals. For nearly 40 years, Professor Phua continues on strengthening scientific research in Asia and promoting physics education, higher education and scholarly exchanges at the international level.

Seeram Ramakrishna, *FREng* is the Director of Center for Nanofibers & Nanotechnology at the National University of Singapore (NUS). He is a Highly Cited Researcher in Materials Science. Thomson Reuters identified him among the World's most influential scientific minds. He is an elected Fellow of Royal Academy of Engineering, UK, Academy of Engineering, Singapore, NAE and ISTE, India, AAET, FBSE, IES, AAAS, ASM International, ASME, AIMBE, IMechE, and IOM3. His university leadership includes NUS's Vice-President (Research Strategy); Dean of Faculty of Engineering; Director of NUS Enterprise; Director of ILO; Founding Director of NUS Bioengineering; Founding Co-Director of NUSNNI; Founding Chairman of SERIS, and Founding Chair of Global Engineering Deans Council.

Sim Keng Yeow is a graduate of the NUS Chemistry Department and was a Commonwealth and Fulbright Scholar. He was a long-serving and distinguished Head of the NUS Chemistry Department from 1988 to 1996. He has served as President of the Singapore National Institute of Chemistry (SNIC), Vice-President of the Singapore National Academy of Science (SNAS) and Secretary-General of the Federation of Asian Chemical Societies (FACS). He is a Hon Fellow of SNIC and has received the SNIC Distinguished Service Award, and has also been made a Fellow of SNAS. He has received the FACS Citations Award for the development of chemistry in Asia.

Kuldip Singh has worked at NUS since 1985. He is currently an Associate Professor in the NUS Department of Physics, Admin Director for the Centre for Quantum Technologies and a Residential Fellow at Cinnamon College at NUS UTown. He also has a teaching appointment with the University Scholars Programme (USP). He has been recognised for his teaching in USP and the Faculty of Science with five Faculty Teaching Excellence Awards. In 2011, he received a Commendation Medal at Singapore's National Day Awards. Having helped to initiate research in quantum information in Singapore, his research interests continue in quantum computation, quantum algebras and mathematical physics.

Sow Chorng Haur is currently the Head of the Department of Physics at the National University of Singapore (NUS). He is also the President of the Institute of Physics Singapore. Previously he has served as Vice Dean (Research) and Assistant Dean (Outreach) for the Faculty of Science NUS. His research interests include the studies of laser-materials interaction of a wide variety of nanostructured systems. These include 2D materials, carbon nanotubes and nanoparticle systems.

R Subramaniam has a PhD in physical chemistry. His current research interests are mainly in the areas of science education, science communication, and science & technology developments in Singapore. He is an associate professor at the National Institute of Education in Nanyang Technological University. He also serves as honorary secretary of the Singapore National Academy of Science.

Bernard Tan Tiong Gie attended the University of Singapore (Bachelor of Science with Honours in Physics, 1965) and Oxford University (Doctor of Philosophy in Engineering Science, 1968). He joined the then University of Singapore (now NUS, the National University of Singapore) in 1968, where he has served as Dean of Science, Head of Physics and Dean of Students. He is currently a Professor of Physics at NUS, where he is also Director of the Centre for Maritime Studies. He is a Fellow of the Singapore National Academy of Science, Fellow of the Institute of Physics (UK), Fellow of the Institute of Physics, Singapore and Member of the Institution of Engineering and Technology (UK).

Evon Tan is one of the founding members of the Centre for Quantum Technologies' (CQT) administrative team at NUS. Now a Senior Admin Manager, she has over 10 years' experience in events management, web development and secretarial support. Her responsibilities at CQT have included the administration of the PhD programme, the organisation of local and international conferences and coordination of the centre's outreach activities.

Leo Tan Wee Hin is a marine biologist by training, with research interests also in science education and science & technology developments in Singapore. He is currently Director of Special Projects in the Faculty of Science at the National University of Singapore, in which role he was instrumental in raising $46 million for the setting up of the first natural history museum in Singapore — the Lee Kong Chian Natural History Museum. His previous key appointments included the following: President of the Singapore National Academy of Science, Director of the National Institute of Education, Professor of Biological Sciences at the Nanyang Technological University, Director of the Singapore Science Centre, and Chairman of the National Parks Board.

Kevin Y L Tan is a legal academic and historian. A graduate of NUS and the Yale Law School, he taught full-time at the Faculty of Law, National University of Singapore from 1986 to 2000 when he founded Equilibrium Consulting Pte Ltd, a boutique consultancy focused on history, heritage and publishing. He is active in many civic organisations and was previously President of the Singapore Heritage Society (2001–2011) and is currently President of ICOMOS Singapore. He has edited and written some 40 books on the law, history and politics of Singapore. He is currently Adjunct Professor at the Faculty of Law, National University of Singapore as well as at the S Rajaratnam School of International Studies, Nanyang Technological University. His particular areas of research and teaching are: Constitutional and Administrative Law; the Singapore Legal System: International Human Rights; and Legal History.

Tang Seung Mun obtained his BS, MS and PhD in physics from Indiana University and his MSc in computing science from the University of Newcastle-Upon-Tyne. He joined the Singapore University in 1968, served as the Head of Physics Department, National University of Singapore (NUS) from 1988 to 1995, and retired from NUS in 2005. His research interests encompass a number of areas, including beta & neutron spectroscopies, applications of X-ray fluorescence & particle-induced X-ray emission in porcelain/gem-stone authentication & aerosol/sea-sediment analysis, and Monte Carlo simulations. He has published over 100 research papers.

Elizabeth Taylor has a degree in Pharmacy, and a PhD in Clinical Pharmacology from the University of London, UK. She was appointed one of the first Research Scientists in the Department of Medicine, NUS in 1982 after working at the University of Oxford, UK on cardiovascular research, and after which she spent a sabbatical year at the University of Cambridge, UK. She became one of the main drivers behind the formation of the Tropical Marine Science Institute at NUS and was appointed Deputy Director in 1998. This visionary step led to the first multi-faculty NUS research entity with government partners, and an offshore research laboratory that is now a National Research Facility. Currently, she heads the Marine Mammal Research Laboratory.

Raj Thampuran has been Managing Director of A*STAR since July 2012. He was responsible for spearheading the efforts to develop the science and technology research blueprint for the Science and Technology 2015 Plan. Since joining A*STAR in 2000, Dr Thampuran has held various senior leadership positions including Executive Director of the Science and Engineering Research Council, as well as the Executive Director of the Institute of High Performance Computing, a research organisation of over 200 staff and one of the 14 Research Institutes supported by A*STAR, where he spearheaded its R&D portfolio and industry development efforts until June 2010.

Thio Hoe Tong held the Computer Centre directorship from three universities for a period of 25 years; they included the former Nanyang University, the former University of Singapore and the National University of Singapore (NUS). He was concurrently Head of the Department of Computer Science in NUS from November 1983 till October 1985. He obtained his PhD from the University of Rochester in the States in 1970. He has spearheaded the development of the ISO 9001-certified IT infrastructure and services in NUS for more than two decades. He was instrumental in bringing the Internet technology and services to Singapore in the late eighties.

T Venkatesan is currently the Director of the Nano Institute at the National University of Singapore (NUSNNI) where he is a Provost Chair Professor of ECE, Physics, MSE and NGS. He wore various hats at Bell Labs and Bellcore for about 17 years before becoming a Professor at University of Maryland for another 17 years. He is the inventor of the pulsed laser deposition (PLD) process, a Fellow of the APS, winner of the Bellcore Award of excellence, was a Guest Professor at Tsinghua University and Winner of the George E Pake Prize awarded by American Physical Society (2012).

Andrew T S Wee is a Provost's Chair of Physics, and Vice President (University and Global Relations) at the National University of Singapore (NUS). He is President of the Singapore National Academy of Science (SNAS), Fellow of the Institute of Physics UK (IoP), Institute of Physics Singapore (IPS), and an academician of the Asia-Pacific Academy of Materials (APAM). His research interests include scanning tunneling microscopy (STM) and synchrotron radiation studies of the molecule-substrate interface, graphene and 2D materials, and related device studies. He is an Associate Editor of *ACS Nano*, and on the Editorial Boards of several other journals.

Steven John Wolf is Head of the Science Communications Unit at the Mechanobiology Institute, National University of Singapore. Steven graduated in 2004 with a Bachelor of Science (Hons) from the University of New South Wales, Australia, and a PhD in cancer pharmacology from the University of Sydney, Australia, in 2009. Since 2011, Steven has pursued a career in Science Communications. The MBI Science Communications Unit develops an educational website on mechanobiology called MBInfo (mechanobio.info) and seeks to introduce the next generation of scientists to the interdisciplinary field of mechanobiology through outreach programmes, video and animation productions, and media publications.

Richard M W Wong joined NUS in 1997 and is currently Head and Professor of the Department of Chemistry. After his PhD from the Australian National University in 1989, he held postdoctoral positions at IBM Kingston and Yale University, and a prestigious Australian Research Fellowship at University of Queensland. He has authored and co-authored about 200 scientific publications, with nearly 8,600 citations. He received the NUS Outstanding Research Award in 2002. He is on the international advisory board of *Asian Journal of Organic Chemistry* and is Associate Editor of the *Australian Journal of Chemistry*. His research interest involves the application of computational quantum chemistry to a wide range of chemical problems such as organocatalysis, weak intermolecular interactions, hydrogen storage materials, and drug design.

Francis Yeoh is Professorial Fellow for Entrepreneurship at the School of Computing, National University of Singapore (NUS), overseeing the school's entrepreneurship efforts, teaching entrepreneurship and mentoring university startup companies. He is also Executive Director for the Mediapreneur Incubator, a corporate incubation programme for media startups under Singapore's Mediacorp Pte Ltd. In a career spanning 30 years, he had been a research scientist, research institute director, internet startup CEO, venture investor, professor and government policy maker. He stepped down as the first CEO of the National Research Foundation (NRF) of Singapore in 2012 after more than six years.

Chapter 1

The Science Council and Singapore science in the '60s and '70s

B T G Tan

When Singapore gained its independence from Malaysia on 9 August 1965 rather suddenly and dramatically, the immediate focus of our leaders was our survival as a new nation with no natural resources. One of the prime tools which could enable us to leverage our way into economic survival would be science, which was quickly seen by them as a necessary prerequisite for industrial and technological progress.

The Science Council of Singapore

In February 1966, Deputy Prime Minister Toh Chin Chye, himself a physiologist and a member of the Faculty of the University of Singapore (the predecessor of the National University of Singapore or NUS), announced a proposal in Parliament for the formation of a new statutory board, the Science Council of Singapore. On 24 May 1967, the Bill for the formation of a Science Council of Singapore had its first reading in Parliament.[1] The functions of the new body would be to make reports and recommendations on:

- Scientific and technological research and development.
- The effective training and utilisation of scientific and technological manpower in Singapore.
- The establishment of official relations with other scientific organisations.

The Straits Times of 14 November 1967 announced that Lee Kum Tatt had been appointed Chairman of the new Science Council.[2] He was a leading bio-chemist in Singapore and Head of the Department of Scientific Services, and he would set the Council's initial directions and became prominent in Singapore's scientific and technological policy making. Apart from him, ten other members of the Council were appointed, including physicist Hon Yung Sen and chemist Kiang Ai Kim, both faculty members of the University of Singapore's Science Faculty.

The new Science Council, temporarily housed in Fullerton Building, immediately swung into action and organised the National Conference on Scientific and Technological Cooperation between Industries and Government Bodies in October 1968, with the objective of discussing how science and technology could work hand-in-hand with industry to boost economic development.[3] Lee Kum Tatt on announcing the National Conference declared that "It is now recognised more than ever before that knowledge and application of science and technology is the key to economic development."

The first Annual Report of the Council for 1967 was presented by Lee Kum Tatt to Toh Chin Chye on 30 March 1968, and reported on the Council's relations with international bodies.[4] In particular, the Council appointed a subcommittee to look into matters related to the International Atomic Energy Agency or IAEA and organised an IAEA First Regional Research Coordination Meeting in Singapore in November 1967. A UNESCO mission to Singapore was asked by the Council to look into the possibity of establishing a Technical University in Singapore. Relations with the Commonwealth Scientific Committee were also discussed.

In 1969, the year of the 150th Anniversary of the founding of Singapore by Stamford Raffles, the Science Council organised an exhibition entitled "Science in the Service of Man" at the Victoria Memorial Hall. The co-organiser of this exhibition was the new Ministry of Science and Technology, whose founding Minister was Toh Chin Chye. The exhibition ran from 15 to 28 October, and culminated in a gala event — the first ever "Science Ball" organised by the Council. The highlight of the Ball was the awarding of the Council's first Gold Medal for Applied Research, which was won by TG Ling, an industrial chemist who had made significant contributions in the field of animal nutrition and feeds.

The Science Centre

Perhaps the most influential activities of the Science Council in its first decade were in the promotion of science. In late 1968, directed by the Minister for Science and Technology, the Science Council appointed a Special Committee to look into the setting up of a Science Centre in Singapore.[5] I had joined the Physics Department of the University of Singapore on 8 November 1968 as a Lecturer as the most junior member of the Department and had met Toh Chin Chye, the Vice-Chancellor, soon after joining.

I was quite surprised to be named as a member of the Special Committee, alongside such senior persons as Ronald Sng, the Chairman of the Special Committee and a founding member of the Science Council.[6] The other members were Sng Yew Chong, the Director of Technical Education at the Ministry of Education and the father of technical education in Singapore, and Rex Shelley, a senior engineer at Hume Industries who later became a longstanding member of the Public Service Commission.

The most immediate task of the Special Committee was to gather information on Science Centres and Science Museums around the world, and I was asked to attend a meeting in India of ICOM, the International Council of Museums. The subject of this ICOM meeting was Science Museums, and this served as my crash course on science museums. I, with the other participants of the meeting, visited the leading Science Museums in India, including the Birla Industrial and Technological Museum in Calcutta and the Visvesvaraya Industrial and Technological Museum in Bangalore.

I was also fortunate to meet well-known science museum directors, including the director of the famous Deutsches Museum in Munich, arguably the most famous science museum in the world alongside the renowned London Science Museum. I did visit the Deutsches Museum in 1999 and it was indeed a magnificent institution with many exhibits of great historical significance.

The Special Committee then asked UNESCO in early 1969 for an expert on science museums who could draw up a proposal for a science museum for Singapore. UNESCO selected Margaret Weston, then a senior curator at the famed London Science Museum in South Kensington, who was in Singapore from 27 September to 30 November 1969. I was asked to assist Margaret, a delightful lady who took a considered and rational approach to her mission, who made sure that all aspects of the issue were looked after. Her report became our blueprint for a modern Science Centre which would cover the aspects of physical science and engineering of relevance to Singapore's development.[7] The Science Centre would not house exhibits of historical interest but focus on up-to-date science and technology. (Ironically, the London Science Museum also started as a museum of modern science and technology, but over the years the "modern" exhibits became historical exhibits.)

At around the same time, consideration was being seriously given to the redevelopment of the National Museum (formerly the Raffles Museum). The National Museum at that time had been housing, in addition to the exhibits on the history and anthropology of our region, the famed Raffles Collection, which consisted of thousands of zoological specimens collected by Stamford Raffles and others over many years. This had become an invaluable reference collection of immense value to biologists interested in the fauna of the region, but unfortunately the collection's value was then not well understood by policy makers. The National Museum was to focus purely on the history, culture and anthropology of the region, which meant the Raffles Collection exhibits had to find a new home.

A decision was then made to include biological sciences as a second theme to the Science Centre, in addition to the original theme of physical sciences and engineering. With the inclusion of biological sciences, the Special Committee was enlarged into a Joint Committee in March 1970 to "formulate proposals for the integration of the natural history component of the National Museum into the Science Centre".[5]

Some of the mounted specimens of the Raffles Collection did go to the Science Centre, but most of the Collection eventually found a home in the NUS Faculty of Science, properly cared for and housed as the Zoological Reference Collection (ZRC), accessible to researchers but not displayed in all its glory. Today, the ZRC together with the important Botanical Collection from the NUS Herbarium is housed and displayed in the new Lee Kong Chian Museum of Natural History under magnificent display conditions.

Eric Alfred, a noted zoologist who was then Acting Director of the National Museum, was originally supposed to be transferred to the Science Centre to head its biological sciences division, and there was a notion that I might join the Centre to head its physical sciences/engineering division. As it turned out, neither of us went to the Science Centre, as Eric stayed on with the National Museum, and I stayed on with the NUS Physics Department.

Ms Weston's proposal, with the addition of the biological sciences, was accepted and work started on the design and construction of the new Centre. As I recall, the current site at Jurong East was not the only site mooted, one of the several others suggested being at Kallang where the Sports Hub is now located. Kenneth V Jackman from the Lawrence Hall of Science, Berkeley was appointed as Director of the Science Centre, and I found him to be dedicated and knowledgeable. However, he vacated the Director's position before the Science Centre Building (imaginatively designed by noted Singapore architect Raymond Woo) was completed, and was succeeded by my Physics Department colleague RS Bhathal. The Science Centre was officially opened on 10 December 1977 by Toh Chin Chye and has become a key institution in Singapore for the promotion of science and technology.

The Science and Industry Quiz

Perhaps the other most well-known of the Science Council's activities in its first decade was the Science and Industry Quiz, known to its many fans as the S and

I Quiz. It was believed that an effective method of popularizing science and technology over television would be a quiz show which could combine education with entertainment. Preparations started for the quiz which was to be contested by student teams from secondary schools. Each team would field four members for the actual quiz programme with two reserve members standing by.

The format for the quiz consisted of two separate parts. The first part was a Preliminary Round in which all competing schools had to answer two sets of question papers. The first set, Set A, consisted of 100 questions and the second set, Set B, consisted of 50 questions. Each team was given just 15 minutes to answer each set of questions. The Preliminary Round for the very first S and I Quiz was held at the Raffles Institution Hall on 28 July 1972, with a total of 54 teams competing.

From the results of the Preliminary Round, 12 school teams were selected for the second and more publicly visible part of the Quiz, the Televised Series. This series was run in three stages: four Quarter Finals, two Semi Finals and one Finals, all of which were prerecorded television programmes (in subsequent years, the Finals would be televised "live"). The 12 teams took part in the Quarter Finals three teams at a time, with the winning team going on to the Semi Finals in which two teams competed at a time. The two Semi Finals winners would then go on to compete in the grand Finals for the top position.

Each programme of the televised series had a similar format consisting of four rounds. The first round was a team round in which each team would be given a fixed time limit (60 seconds in the Quarter Finals and 90 seconds in the Semi Finals and Finals) to answer as many questions as they could. Two points were awarded for a correct answer and there was no deduction of points for a wrong answer. Any member of the team could answer and the teams could confer amongst themselves.

The second round was an individual round in which each member of a team would be asked a question which had to be answered within 60 seconds without help from the other members. Two points would be awarded for a correct answer with no deduction of points for a wrong answer. If the team member could not answer the question, another member of the team could answer within the time limit and would be awarded one point if the answer was correct. In this round, most of the questions featured an illustration or diagram which had to be identified correctly. Each team would have to answer eight questions for the Quarter Finals and 12 questions for the Semi Finals and Finals.

Round three was a team round in which the same question was posed to all the teams and each team had to answer as a team. The six questions in this round allowed teams to exercise their problem-solving and teamwork skills, as the questions would require some calculation or deduction beyond just simple recall or memory work. For a correct answer, two points were awarded with no points deducted for an incorrect answer.

The fourth and last round was usually the most exciting round of all. In this round, each question was posed to all the teams simultaneously, and a member of any team who wanted to answer could press the team's button to sound a buzzer. The buzzer system ensured that only the first button pressed for that question would activate the buzzer. For each correct answer, four points would be awarded but unlike the other rounds, two points would be deducted for an incorrect answer.

This final round often saw dramatic changes in the teams' fortunes, with a team which was in the lead losing points for wrong answers, thus allowing a competing team to catch up or even overtake them. The team members thus had to be swift and accurate, and the attempts to answer questions became a matter of who could press the buzzer first. Incidentally, the technical support for the quiz was quite rudimentary (in those early days of television), so the electronic buzzer system was designed by me and constructed in the workshop of the NUS Physics Department.

This Organising Committee for this first S and I Quiz (and subsequent Quizzes) was Ang How Ghee, and the Questions Subcommittee and Panel of Assessors during the televised programmes were chaired by Mok Kum Fun of the NUS Chemistry Department.[64] The winner of the first S and I Quiz was Anglo-Chinese School (Secondary) and they and their main rivals, Raffles Institution, would be the teams to beat throughout the series.

The Quizmaster for the this first S and I Quiz was Tay Eng Soon of the NUS Electrical Engineering Department who became a Singapore celebrity as a result. (Eng Soon later entered politics and rose to the position of Senior Minister of State for Education, becoming the guiding force behind many key educational initiatives such as the development of the ITE and Polytechnics, as well as of the schools' Choral Excellence programme, before his untimely death in 1993).

Eng Soon was the Quizmaster for all the programmes in the six years of the S and I Quiz which ran from 1972 to 1977, except for 1976 when I took over as Quizmaster when he was unavailable.[8] I also took over for one programme in the series for 1975,[9] and was the standby Quizmaster for 1977.[10] After the first Quiz, I took over as Chairman of the Questions Subcommittee and of the Panel of Assessors.[11]

As virtually the first quiz programme on Singapore television organised by RTS (Radio Television Singapura), the popularity of the S and I Quiz quickly grew to phenomenal proportions well beyond our most optimistic projections.

The huge viewership was glued to their TV sets during each episode, and particularly for the Finals programme. Fierce battles developed between leading schools vying for the top prize in the Quiz, and many schools selected their team members and trained them well ahead of the coming year's Quiz.

By 1977, RTS had started to bring up the question of a fee for broadcasting the Quiz. This became a contentious issue, particularly as all the people involved in the quiz and the preparation of questions (a very onerous and tedious task) were working *pro bono*. The Quiz was carried on for one more year in 1978 when Patrick Pestana was the Quizmaster, and I chaired the Organising Committee. The format was changed by replacing the first round with a practical assignment section, with the teams carrying our simple projects like the construction of working models. The practical assignments were actually done before the televised programmes, as the students had been given the topics a few days beforehand.

In 1979, the S and I Quiz was replaced by a completely new television programme, The Innovators. This was not a Quiz programme but a series of six programmes in which teams of students at the Junior College level would work with RTS producers to produce television programmes, each of which was based on a particular scientific or technological theme but with the focus on innovation. The three best programmes were chosen by a Panel of Judges and awarded prizes. The winners were Anglo-Chinese Junior College whose programme was entitled "... And Life Goes On ...", Raffles Institution ("Food Encounters") and Temasek Junior College ("The Miracle Gene").

To complement the S and I Quiz and to give the students in the vocational and technical institutes a chance to show their skills to a wider public, the Science Council collaborated with the Vocational and Industrial Training Board (VITB) to launch a new television quiz programme, "Top of the Trade", pitting students of different vocational institutes against each other. This programme highlighted trades like woodworking, building construction, applied arts, machine tools and the automotive trades. The programme did not run every year, but the first series appears to have been launched at the end of 1977, followed by two other series at the end of 1978 and 1979/1980. I believe I was the Quizmaster for some of these programmes, certainly the one in 1978.[12] After a lapse of a few years, another series of Top of the Trade appears to have been produced in 1983.

In 1980, The Innovators was replaced by the Science Council with a new television programme for secondary schools called Science Challenge, which returned to the Quiz format and whose Organising Committee I chaired. The four rounds of the S and I Quiz were now modified, such that in the first round, the teams would have to solve problems within a time limit of three minutes, instead of being asked questions. The second round was essentially unchanged, but in the third round, each team would give a presentation on a project which they had been working on for the last two weeks, and marks would be given for their presentation. The final round remained the same — the exciting speed round with the buzzer. The Quizmaster for this revamped programme was Lawrence Chia of the NUS Chemistry Department.

Science Challenge seems not have been run again until 1989, when it was revived just for one year as Science Challenge '89. The first round was, as in the original S and I Quiz, a team round but now a picture could be used to illustrate a question. In round two, the teams each had to solve a practical science problem on the spot using the given apparatus, such as the identification of a set of unknown liquids, within a fixed time limit. Next, the judges gave their evaluation of a specific project on which each of the teams had been working for the last few weeks. (During the programme, video clips were shown of each team working on this project.) The final round was the same as the exciting final round of the S and I Quiz, i.e. the buzzer speed round.

The lapse of 8 years after the last Quiz series allowed Science Challenge '89 to become more modern and slicker than the previous programmes, with the copious use of computer graphics and more polished production values. Another innovation was the introduction of two Quizmasters instead of just one — Lai Choy Heng of the NUS Physics Department, and Karina Gin of the NUS Civil Engineering Department, which certainly lent a touch of glamour to the programme!

Science Challenge '89 was the last in the series which started with the S and I Quiz in 1972, and new Science-based quiz programmes such as the National Science Challenge have carried on this tradition. The S and I Quiz garnered national attention and interest well beyond the reach of today's programmes, for three reasons: it was the first local quiz programme on television; viewers were transfixed by the fierce competition between the top schools participating; and there was very little competing local content on television at that time. The S and I Quiz also seems to have defined the format for many of today's local quiz programmes on television.

The Science Council's Other Activities

The Science Centre and the S and I Quiz were undoubtedly the most visible of the Science Council's activities during the 1970s. Indeed, the Quiz programmes organised by the Council helped the Council to win the Guinness Award for Scientific Achievement in 1980, the first time a Singapore entry had won this award.[23] Much of the Council's other activities continued to be mainly concerned with Singapore's relations with international bodies, the organisation of surveys, seminars and conferences in various scientific and technological fields relevant to Singapore's development, and the general promotion of science and science education.

In the Council's Annual Report for 1968,[13] the following standing committees of the Council reported on their work for the year:

- International Atomic Agency Committee chaired by Lee Kum Tatt and then by Tay Sin Yan.
- Biological Sciences Committee chaired by Cheng Tong Fatt.
- Engineering Committee chaired by A. Robert Edis and then by Hiew Siew Nam.
- Natural Resources and Their Utilization Committee chaired by Lee Kum Tatt and then by Tan Eng Liang.
- Physical Sciences Committee chaired by Hon Yung Sen and then by Kiang Ai Kim.
- Social Science Committee chaired by You Poh Seng.

After just a year or so of operation, it is clear that the Science Council had firmly established itself as the national body on science and technology policy.[14]

The 1971–1972 Annual Report of the Council records the holding of a Regional Workshop on Water Resources Environment and National Development jointly sponsored by the US National Academy of Sciences.[15] There was also a report on the Central Agency for Industrial and Business Orientation which placed tertiary students on attachments with various industrial and commercial companies. The Applied Research Fellowships were started in 1971 to encourage scientists and engineers from industry to undertake applied research in academic institutions.

An important development was the Research Grants scheme which channeled selected research proposals to the Ministry of Science and Technology for funding. Nine projects were recommended requiring a grand total of $76,867. In the field of education the Science Council initiated the biennial Creative Science Teachers' Awards, with the aim of giving recognition to science teachers for their outstanding contributions to science education in both primary and secondary schools. In 1971, two primary school teachers and two secondary school teachers were honoured with these awards, which were each worth a sum of $1,000.

The next volume of the Council's Annual Reports covered the period 1972–1975 and recorded the success of the Science and Industry Quiz.[11]

In 1972, the Council collaborated with the Ministry of Education on a project to produce a comprehensive range of science teaching aids for primary schools.[11,16] By 1979, this project had borne fruit, as is stated by its Annual report for 1977/78.[17] A Workshop on Corrosion and a study on the Evolution of Singapore's Technological Institutions with aid from the Canadian International Development Research Agency were also reported. The 1975/76 Annual Report also mentioned the continued success of the S and I Quiz, as well as continuing schemes like the Applied Research Fellowship and the Research Grants Scheme.[19]

In 1975 Lee Kum Tatt, who had so solidly laid down the Science Council's foundations, was succeeded as Chairman by Choo Seok Cheow of the NUS Electrical Engineering Department.[20,21] When Choo Seok Cheow became Deputy Vice-Chancellor of NUS in 1977, he was in turn succeeded at the end of 1977 by Ang How Ghee of the NUS Chemistry Department.[17,22] I joined as a Council member in 1977, and served till the Council's transition into the National Science and Technology Board in 1990/91.

In 1976, the Council jointly organised, with the United States Information Service, a Seminar on Transfer of Technology which was chaired by Tay Eng Soon as reported in the Council's Annual Report for 1976/77.[21] The same Annual Report stated that the Council's Applied Research Fellowship would be discontinued, and that the Council had also assisted the Ministry of Science and Technology in reviewing the Research Grants Scheme.

The Council also organised a five-day international symposium on science and technology for development in January 1979, in collaboration with ICSU, the International Council of Scientific Unions.[25,26] The first issue of a new science publication, the Science and Technology Quarterly, was published by the Council in July 1980.[27,28] The aim of this new publication was to feature programmes and policies on science and technology for development at both national and international levels.

The next three volumes of the Council's Annual Reports show the steady progress made by the Council in promoting and chronicling the growth of science and technology in Singapore, as well as our relations with international scientific bodies.[23,29-30] One key project was the national Survey of R&D Activities undertaken jointly in 1982–83 with the Ministry of Trade and Industry and the Economic Development Board.

In 1983, Choo Seok Cheow returned as Chairman of the Science Council.[31] In the Council's Annual Report for 1984/85, the Council's research funding scheme which was started in 1981, the Research and Development Assistance Scheme or RDAS, was reported to have disbursed in the previous year a total sum of $7.36 million for 14 research projects — nine in the biomedical area and five in the physical science engineering area.[18] This brought the total value of RDAS grants so far disbursed to $23.3 million.

There was also a report on the development of the Singapore Science Park, for which the Council assumed responsibility in April 1984. It was designed to serve as a focal point for industrial R&D facilities and related "brain services" in Singapore, and to foster collaboration between industry and academia.

By the time of the next Annual Report, the RDAS had been established as the most important grant scheme for applied research, and two RDAS Advisory Committees had been established.[32] Choo Seok Cheow himself chaired the RDAS Advisory Committee on Science and Technology (on which I served), and Edward Tock, Dean of the NUS Medical Faculty, chaired the RDAS Advisory Committee on Biomedical Research. For FY 1985, a total of $4,023,928 was approved for six projects.

National Surveys of R&D were now being regularly conducted by the Council, and the 1985 Survey revealed that the gross national spending on R&D in 1984/85 was $214.3 million. The private sector contributed 45% of this figure with the government and higher education contributing 18% and 33% respectively. The Council had also appointed two International Panels, one on Biological Sciences and the other on Engineering/Physical Sciences.

Several International Conferences organised by the Council were also reported, including the International Conference on Artificial Intelligence (AI) on 23–27 March 1986, whose Organising Committee I chaired and which was the first international meeting on AI in Singapore. The theme of the Conference was "Artificial Intelligence and its Applications — A State of Art View", and we were fortunate to have Edward Feigenbaum from Stanford as our Keynote Speaker (who was also the NUS Lee Kuan Yew Distinguished Speaker). We also had several other distinguished AI researchers from overseas, such as Randall Davis of MIT, Raj Reddy of Carnegie-Mellon and Don Walker of Bell Communications Research.

Two other notable Conferences that year were the Biotech '85 Asia Conference on 27–29 November 1985, and the Modern Engineering Technology Seminar (better known as METS) on 19–21 August 1986. Biotech '85 Asia marked Singapore's move into the biotechnology field, and had as Keynote Speaker Ronald Cape, Chairman of Cetus Corporation.

The 86/87 Annual Report of the Council reported that the RDAS scheme disbursed a sum of $8.06 million in FY86,[33] and an almost identical sum of $8.02 million the next year.[34] The National Science and Technology Awards, which succeeded the Council's Gold Medal awards were presented at the culminating dinner of Technology Month, an event inaugurated in 1987 by the Council. The first two recipients of these Awards in 1987 were Lee Seng Lip and Wong Hock Boon. The following year in 1988, the Council initiated the National Young Scientist and Engineer Awards to give recognition to young researchers.[35]

The Council's 88/89 Annual Report reported that the initial block vote of $50 million for RDAS from the Ministry of Trade and Industry in 1981 was being followed by a new block vote of $50 million in 1988.[24] Over the years from 1981 to 1988, $43 million had been disbursed to 72 projects, and the Principal Investigators of two of the completed projects, Lee Seng Lip and SS Ratnam, had won National Science and Technology Awards, which was testimony to the quality of the research done under RDAS. The progress of the Science Park from its establishment in 1982 with just two tenants, to 50 tenants in 1988, was another demonstration of the Council's credibility.

The last two reports of the Science Council, for 1989/90 and 1990, continued to highlight the progress of RDAS and the Science Park. The 1989/90 report gave the number of RDAS projects approved during the year as 11, for a total sum of $12.3 million.[36] It also listed the ASEAN R&D projects awarded to Singapore by the ASEAN Committee on Science and Technology (COST). The 1990 report highlighted 14 new RDAS projects, with sums granted ranging from $197,000 to $759,400.[37] A graph shows that the cumulative sum granted since the inception of RDAS was around $65 million for over 90 projects. The number of tenants in the Science Park had risen to 74, and in April 1990 the Council handed over the operation of the Science Park to Technology Parks Pte Ltd.

The transition from Science Council of Singapore to National Science and Technology Board in 1990 was accomplished smoothly, with Choo Seok Cheow still Chairman at the transition to NSTB in January 1991, having brought the Council to a point when it was ready to oversee the next stage of Singapore's scientific and technological development. We should also recognize the contribution of the administrative Heads of the Council. Ng Chon Choo, Khoo Hun Hock, Siew Weng Hin, Siew Hing Yun, Vincent Yip and Chou Siaw Kiang have held the positions of Secretary, Chief Administrative Officer and Executive Director at various times in the Council's existence, and have ably implemented the policies and decisions of the Council.

The Ministry of Science and Technology

Following the establishment of the Science Council in 1967, the Cabinet agreed to set up the Ministry of Science and Technology in 1968 with Toh Chin Chye as its founding Minister. In most of its early years, its most senior civil servant was Au Yee Pun, who held the rank of Principal Assistant Secretary. The most important item in its portfolio was the Science Council, through which most of its important initiatives like the Science Centre were executed. The other statutory board under the Ministry was the Metrication Board under the Chairmanship of Baey Lian Peck.[38] The Straits Times of 28 February 1973 reported that a nuclear energy unit would be formed in the Ministry with a budget of $120,000 for research.

The Ministry undertook a National Survey of Scientists in 1971,[39] produced a Directory of Scientific and Technical Research and Consultancy Establishments in 1973,[40] and undertook two other National Surveys, one of Scientific Manpower and the other of Engineering Manpower, in 1974.[41] The Ministry also carried out Surveys on the Use of Computers in Singapore in 1973, 1976 and 1980.[42] The education of radiographers also seems to have been another task allocated to the Ministry.[43]

Toh Chin Chye served as Minister from 1968 to 1975 when he was offered the Ministry of Education, but instead became Minster for Health. He was succeeded in June 1975 by the then Minister for Education Lee Chiaw Meng who served till December 1976. EW Barker took on the Science and Technology portfolio in addition to his duties as Minister for Law until the Ministry was dissolved on 1 April 1981 because of "the difficulty in coordinating its multiple functions". Its functions were transferred to the Ministries of Education, Trade and Industry, and Health.[44]

The Singapore National Academy of Science

In 1967, scientists in Singapore came together for the first time to form the Singapore National Academy of Science or SNAS. This created an organisation which would represent them collectively as a scholarly as well as a professional body. The intention clearly was that SNAS should become recognised as Singapore's equivalent of the UK's Royal Society or the USA's National Academy of Sciences.

SNAS was inaugurated on 31 July 1967 with the following objectives[45]:

- To promote the advancement of science and technology in the Republic of Singapore.
- To discuss scientific, technological and macroeconomic problems, in particular those of national interest.
- To represent the scientific opinion of the members and fellows of the academy.

The membership of SNAS was made up of individual Singapore scientists who applied for membership and who were admitted on the basis of their scientific credentials. The first President of SNAS was Tom Elliott of the University of Singapore Pharmacy Department, who was also closely associated with the founding of the Trades Union movement in Singapore.[46]

The new Academy immediately set to work, organising its First Science Congress on the theme "New Frontiers in Science"[47] from 7 to 14 August 1968. SNAS also produced a new Singapore-based journal, the *Journal of the Singapore National Academy of Science*, starting in 1969. Volume 1 No 1 of the new Journal was entirely devoted to the Proceedings of the First Science Congress. The Organising Committee of the Congress was chaired by Kiang Ai Kim who revealed in his Introductory Address at the Opening of the Congress that the idea for the formation of the Academy was conceived by Toh Chin Chye.[48] The Congress Proceedings as published in the Journal were compiled by the Congress Editorial Committee chaired by Tham Ah Kow of the University of Singapore Zoology Department.

The first issue of the Journal carried the Opening Address of the First Science Congress by the Minister for Science and Technology, Toh Chin Chye.[49] He concentrated on the role of science in industry and commerce, and in the final words of his address asked "How can science and technology be applied towards our economic development? That is the challenge we face today." Also in the Journal was the Address at the Conference Dinner by Goh Keng Swee, Minister for Finance.[50] Unsurprisingly, he concentrated on the economic aspects of science and technology and declared, "If we keep both feet on the ground, forgo grandiose schemes and sterile generalisations, I believe that the yield from sensibly directed generously-financed research ...... will yield very substantial dividends."

SNAS held its Second Science Congress in November 1971, on the theme of Science and the Urban Environment in the Tropics.[51,52] The nine-day Congress was opened by Toh Chin Chye, and attracted 300 participants and 140 papers. At the Congress banquet, distinguished clinician ES Monteiro was elected as the first Honorary Member of SNAS, and Kiang Ai Kim was elected as an Honorary Fellow. A novel feature of this Congress was the presentation of papers by pre-university students.[53] Subsequent Science Congresses were organised by SNAS in 1977 and 1984.[54,55]

In March 1971, the then Minister of State for Education, Lee Chiaw Meng, who was also a well-established civil engineer, was elected as President of the Singapore National Academy of Science. He held office as President until 1978, when he was succeeded by Ang Kok Peng of the University of Singapore

Chemistry Department, who became President from 1978 to 1992. Lee Chiaw Meng was a dynamic personality who made his mark in Singapore, not only with SNAS which underwent momentous changes under his Presidency, but also as Minister for Education and Minister for Science and Technology. His untimely death in 2001 robbed the nation of a highly talented individual who could have contributed so much more to Singapore.

The Institute of Physics, Singapore

A parallel development to the formation of SNAS during the 1970s was the move by Singapore scientists in specific scientific disciplines to organise themselves into their own scholarly and professional bodies. The physicists in the two physics departments, i.e. at the University of Singapore and Nanyang University, decided in 1972 to form the Institute of Physics, Singapore. The initiative came from the Physics Department of the University of Singapore, and a strong and early advocate of this move was my colleague, Wong Kwei Cheong (who later went into politics).

I joined him in reaching out first to our colleagues in our Department and then to those at Nanyang University (Physics Department). The mutual respect which the physicists in both departments had for each other was a major factor in our quickly reaching consensus and registering the new body with the Registrar of Societies. I remember putting together the constitution of the new Institute rather hastily with the aid of the constitutions of existing professional bodies.

The formation of the Institute of Physics, Singapore (IPS) was duly reported in the Straits Times of 3 October 1972.[56] The main aims of the Institute were to carry out an active programme of activities for its members to help upgrade their professional abilities, to seek a public voice for physicists, and to disseminate information about physics and its applications. The first President of IPS was Hsu Loke Soo of Nanyang University and the Vice-President was Lim Yung Kuo of the University of Singapore. I was the Honorary Secretary and Ng Ser Choon of Nanyang University was the Honorary Treasurer. The fine balance in the first IPS Council between the two universities showed the desire to demonstrate our high mutual respect for each other.

To mark its founding, the IPS held its Inaugural Dinner on 11 January 1973 at the Shangri-La Hotel, with Minister for Communications Yong Nyuk Lin as Guest of Honour. To demonstrate the broad support which the IPS had from the local scientific and engineering community, the President of SNAS, the Chairman of the Science Council, the President of the Science

Teachers' Association of Singapore and the President of the Institution of Engineers Singapore were honoured guests at this dinner.[57] In his speech, the Minister declared, ". . . many of the industries already here, such as the electronics and camera industries, are dependent on physics research for their very survival."

The IPS also immediately launched a new publication, the *IPS Bulletin*, to serve as its official mouthpiece, as well as to disseminate news items about its activities and about physics and its applications. The *IPS Bulletin* was published about four times a year and I was its Editor for most of the years of its existence. The Annual Report of the IPS was usually published in the pages of the *IPS Bulletin*, and in the second issue of the Bulletin, the first Annual Report of the IPS announced that the Institute had, amongst other activities, donated two medals for the best Honours Physics Students at each of the two universities. An essay contest on Physics for pre-university students was also established.[58]

The IPS held its First Symposium on Physics from 19–20 March 1976 at Nanyang University. The distinguished physicist and Nobel Laureate, Chen Ning Yang, presented an Invited Paper "Gauge Fields, Magnetic Monopoles and Fibre Bundles". 42 papers were accepted for presentation at the Symposium, and a Symposium Proceedings was published with myself and Ong Phee Poh as Co-Editors.[59]

The Restructuring of SNAS

Singapore mathematicians had long had their own society, the Singapore Mathematical Society (SMS) even before the founding of SNAS in 1967. The chemists had formed the Singapore National Institute of Chemistry (SNIC) a short time after SNAS was formed, and as related above, the IPS was formed in 1972. The Singapore Institute of Biology (SIBiol) was formed a few years later, while the science teachers had also formed the Science Teachers' Association of Singapore (STAS), some years ago (John Yip being one of its key founders).

Sometime in 1975/76, it was realised by SNAS and the specialist science bodies that a better form of organisation which could unite all Singapore scientists in a single body would give us a stronger national voice. Consequently, we started discussing the formation of just one national body, and it was decided that all scientists in Singapore would unite under one banner — SNAS. This meant that the separate societies for each discipline i.e. SMS, IPS, SNIC and SIBiol would be dissolved, and their members would automatically become members of SNAS, with separate sections for each discipline.

However, many members of the separate specialist societies were uneasy about their dissolution and the disappearance of their traditions, short though some of their histories might have been. As one of the members of IPS who would have been very sorry to see it go out of existence, I pondered about an alternative arrangement where we could still unite under SNAS. I then wrote to Lee Chiaw Meng about this alternative arrangement which would give us unity under SNAS but which would preserve the separate societies.

Lee Chiaw Meng was now Minister for Education, and immediately after receiving my note outlining my proposal, he called me up for a meeting with him at his Ministry office. When I had fully explained my proposal, which called for SNAS to become a federation of constituent societies such as SMS, IPS, SNIC and SIBiol, he quickly saw its merits and directed SNAS to implement it instead of the previously agreed plan.

First, the name of the current SNAS was changed to the Singapore Association for the Advancement of Science or SAAS. A new society was then registered as a federation of societies, bearing the same name as the old SNAS i.e. the Singapore National Academy of Science. This new SNAS would not directly admit individual members , but would have SAAS, SMS, IPS, SNIC, SIBiol, and STAS as its constituent bodies. The individual members of these six societies were automatically members of the new SNAS. Unity under the new SNAS was therefore achieved, but with the retention of the individual specialist societies and their traditions.

The new SNAS was launched in 1976 with Lee Chiaw Meng as its President.[60] The six societies were designated as the Founder Members of SNAS, but other societies representing other scientific societies could join SNAS as Affiliate Members. Firms and other organisations interested in the promotion of science and technology could join as Associate Members. The new Academy had just two objectives:

- To promote the advance of science and technology in the Republic.
- To represent the scientific opinion of the Founder/Affiliate Members of the Academy.

Sometime in the mid-1990s, the National Science and Technology Board discontinued the National Young Scientist and Engineer Award. I then wrote to someone in authority (I cannot quite remember who but it could have been the Chairman of the NSTB) to express my disagreement with this move, and to argue strongly for the reinstatement of the Award. I cited the National Arts Council's Young Artists' Award which had done so much to encourage young artists to excel in their work, and pointed out that the discontinued award had done much to similarly encourage young scientists and engineers. The National Young Scientist and Engineer Award was, thankfully, subsequently reinstated and its selection process brought under the charge of SNAS where it remains today.[61]

The University of Singapore

When I was an undergraduate in the Science Faculty of the University of Singapore (from 1962 to 1965), the Faculty was then situated at the Bukit Timah campus and the Dean of the Faculty was HB Gilliland who was also Head of the Department of Botany. The other Departments in the Faculty were Mathematics, Physics, Chemistry, Botany, Zoology and Pharmacy. I was admitted in the 1962/63 academic year to the second year of the Faculty where I read Pure Mathematics, Physics and Chemistry and went on to read Honours in Physics. (I was closely involved with the Science Society, co-editing its Science Journal together with Lui Pao Chuen in 1963.[62]) The science degree structure was such that you read four subjects in your first year, three in your second year, then two in your third year after which you could graduate with a Pass degree. If you qualified, you could be admitted for the fourth (Honours) year to concentrate on just one subject for your Honours degree.

HB Gilliland was succeeded as Dean by Hon Yung Sen of Physics, who served in this position from 1965 to 1968 (during which time I was in the UK for my postgraduate degree). When I came back and joined the University in November 1968, Toh Chin Chye had become Vice-Chancellor and the Dean was Kiang Ai Kim. The Head of Physics was Hon Yung Sen. The Deanship later passed on to Ang Kok Peng of the Chemistry Department, Gloria Lim of the Botany Department and then Koh Lip Lin of the Chemistry Department after the establishment of NUS in 1980.

The University was largely a teaching university at that time, and there was very little funding for research which members of the teaching staff could apply for, possibly only in the region of a few hundred thousand dollars in all. Consequently, staff members did not do too much research, and whatever was done was often done with teaching equipment. Two of my colleagues — Tan Kuang Lee and Ong Phee Poh — and I thought that we might try to design and construct a working carbon dioxide laser, just to while the time away. Though such lasers were relatively new then, I have to confess that what we did was not really of a high research level, since even enterprising schoolchildren in the US had managed to build a carbon dioxide laser.

Nevertheless with our very meager resources, building such a laser was still a daunting task. With very little funds, we used materials which were as cheap as possible. The transformer for the laser was an inexpensive (of the order of $1,000) neon sign transformer. The flat transparent ends of the laser tube ideally should have been made of quartz, in order to allow the infra-red emission from the laser to pass through. We couldn't afford quartz, but bought plates made of sodium chloride (common salt) instead, which were also transparent to infra-red light. The drawback was that these plates deteriorated rapidly in our humid atmosphere.

We certainly couldn't afford to buy an infra-red radiation detector, so when we had constructed the laser and switched it on, we had no idea whether it was working properly, as the infra-red light which the laser was supposed to emit was totally invisible. After running the laser several times it was only when I noticed that a piece of perspex plastic in the path of the laser had deformed through being heated by the infra-red light, that we had any indication that the laser was indeed working! (It was fortunate that none of us suffered any permanent eye damage as we did not use any eye protection).

When the IPS held its First Symposium on Physics in March 1976, we submitted a short paper on our carbon dioxide laser which was accepted for presentation, and which was published in the Symposium Proceedings.[63] We still claim to this day that our laser was the first ever laser constructed in Singapore, and possibly in the region.

Arthur Rajaratnam became Head of Physics in 1969 and stayed on as Head till 1982. In 1975, Toh Chin Chye stepped down as Vice-Chancellor and was succeeded by Kwan Sai Kheong until the formation of the National University of Singapore in 1980. Tony Tan Keng Yam, a graduate of the NUS Physics Department, was the first Vice-Chancellor of NUS and was followed by Lim Pin from 1981 to 2000. From 1980 onwards, much improved research funding led to the remarkable development of NUS as a research university. With the parallel developments in the governance of science and research, and the growing confidence of the community of Singapore scientists, the stage was now set for a rapid expansion of scientific and engineering research in the nation, both in quantity and quality.

References

[1] "A science council to be set up in Singapore," *The Straits Times*, May 28, 1967.

[2] "New science council," *The Straits Times*, November 14, 1967.

[3] "A conference to link science and industry," *The Straits Times*, September 7, 1968.

4   Science Council of Singapore, "Annual Report of the Science Council of Singapore 1967," 1969.

5   "When and how it all began," *The Straits Times*, December 10, 1977.

6   Science Council of Singapore, "Annual Report of the Science Council of Singapore Jan 1 1969 — March 31 1970," 1971.

7   Weston, M.K. "Proposals for the setting up of a science centre in Singapore", 1969.

8   Science Council of Singapore, "Report on Science and Industry Quiz '76," 1976.

9   Science Council of Singapore, "Report on Science and Industry Quiz '75," 1975.

10 Science Council of Singapore, "Report on Science and Industry Quiz '77," 1977.

11 Science Council of Singapore, "Science Council of Singapore Annual Reports for 1972–1975," 1976.

12 "Pick of the Week," *Sunday Nation*, October 22, 1978.

13 Science Council of Singapore, "Annual Report of the Science Council of Singapore 1968," 1971.

14 "Science Council of Singapore, "Annual Report of the Science Council of Singapore April 1 1970 — March 31 1971," 1972.

15 Science Council of Singapore, "Science Council of Singapore Annual Report 1971–1972," 1973.

16 Rav Dhaliwal, "Making science a simple subject," *The Straits Times*, December 21, 1979.

17 Science Council of Singapore, "Science Council of Singapore Annual Report 1977/78," 1979.

18 Science Council of Singapore, "Science Council Annual Report 84/85," 1986.

19 Science Council of Singapore, "Science Council of Singapore Annual Report 1975/76," 1976.

20 "Council Head," *The Straits Times*, August 27, 1976.

21 Science Council of Singapore, "Science Council of Singapore Annual Report 1976/77," 1978.

22 "New council head," *The Straits Times*, December 26, 1977.

23 Science Council of Singapore, "Science Council of Singapore Annual Report 80/81," 1982.

24 Science Council of Singapore, "Annual Report 1988/89," 1990.

25 "Science and technology meeting," *The Straits Times*, December 4, 1978.

26 Science Council of Singapore, "Science Council of Singapore Annual Report 1978/79," 1980.

27 "First science quarterly," *The Straits Times*, June 30, 1980.

[28] Science Council of Singapore, "Science Council of Singapore Annual Report 79/80," 1981.

[29] Science Council of Singapore, "Science Council of Singapore Annual Report 1981/1982," 1982.

[30] Science Council of Singapore, "Science Council of Singapore Annual Report 82/83," 1984.

[31] Science Council of Singapore, "Science Council of Singapore Annual Report 1983/84," 1985.

[32] Science Council of Singapore, "Science Council of Singapore Annual Report 85/86," 1987.

[33] Science Council of Singapore, "Science Council of Singapore Annual Report 86/87," 1988.

[34] Science Council of Singapore, "Science Council of Singapore Annual Report 87/88," 1989.

[35] "New awards for outstanding young scientists and engineers," The Straits Times, June 11, 1988.

[36] Science Council of Singapore, "Science Council of Singapore Annual Report 1989/90," 1991.

[37] Science Council of Singapore, "Science Council Annual Report 1990," 1992.

[38] "The industry goes metric next February," The Straits Times, October 3, 1972.

[39] "Good response to scientist survey," The Straits Times, September 15, 1971.

[40] "Scientific guide," The StraitsTimes, July 23, 1973.

[41] "Survey on jobs," The Straits Times, February 11, 1974.

[42] "Computer survey," The Business Times, March 22, 1980.

[43] "Thirty pass technician course," The Straits Times, November 11, 1972.

[44] "Science ministry to be dissolved," The Straits Times, February 18, 1981.

[45] Habibul Haque Khondker, "Science and Technology Policies for Development: The Case of Singapore," in Handbook of Development Policy Studies. New York: Marcel Dekker, 2004, pp. 331–343.

[46] "Don: why we must strive to produce our own scientists," The Straits Times, October 25, 1967.

[47] Lee Chiaw Meng, Presidential Address to the Second Science Congress of the Singapore National Academy of Science, November 19, 1971.

[48] Kiang Ai Kim, "Introductory Address," Journal of the Singapore National Academy of Science, vol. 1, no. 1, p. 1, 1969.

[49] Toh Chin Chye, "Opening Address," Journal of the Singapore National Academy of Science, vol. 1, no. 1, pp. 2–5, 1969.

[50] Goh Keng Swee, "Address at the Congress Dinner," *Journal of the Singapore National Academy of Science*, vol. 1, no. 1, pp. 6–9, 1969.

[51] "300 Asian delegates for science meet," *The Straits Times*, November 15, 1971.

[52] P.N. Avadhani, "Proceedings of the Second Congress of the Singapore National Academy of Science," in *Science and the Urban Envinronment in the Tropics*, 1973.

[53] "Student papers at SNAS congress," *The New Nation*, November 15, 1971.

[54] "Jek to open congress tomorrow," *The Straits Times*, July 13, 1977.

[55] "Distinguished scientist to give keynote talks," *The Straits Times*, May 22, 1984.

[56] "Physics institute to upgrade local skills," *The Straits Times*, October 3, 1972.

[57] "Yong: Republic needs brain power of physicists," *IPS Bulletin*, vol. 1, no. 1, pp. 1–3, January/March 1973.

[58] "The Institute of Physics Singapore Annual Report 1972/73," *IPS Bulletin*, vol. 1, no. 3, July/September 1973.

[59] B.T.G. Tan and P.P. Ong, *Proceedings of the IPS Symposium 1976.*: The Institute of Physics Singapore, 1976.

[60] New SNAS is formed IPS Bulletin, Vol. 4 no. 3, July/September 1976.

[61] "Scientist award," *The Straits Times*, June 12, 1997.

[62] L.P.C. Lui and B.T.G. Tan, *The Science Journal*, vol II, The Science Society, University of Singapore, 1963.

[63] P.P. Ong, K.L. Tan, and B.T.G. Tan, "An inexpensive CO_2 laser," in *Proceedings of the IPS Symposium 1976*, 1976, pp. 70–74.

[64] Science Council of Singapore, "Report on the Science Quiz '72," 1972.

Chapter 2

R&D in Singapore — The Early Years of NSTB

Francis Yeoh

Preamble

I was part of the National Science & Technology Board (NSTB) on two separate occasions — the first was at its founding in 1991, where I served 18 months as Director Planning, and then Director, Technology Division. My second stint was from 1998 to 2000, as Assistant Chief Executive, with primary responsibility for funding research at the universities and public research institutes & centres, but with later involvement in the Technopreneurship21 initiative to develop technology entrepreneurship and the venture capital industry in Singapore. This is my recollection of NSTB's first 18 months.

Background

When I returned to Singapore as a fresh engineering PhD from the UK in 1984, I was interviewed and promptly offered a job by Mr Lim Swee Say,[1] who was then head of a unit called JSEP (Joint Software Engineering Programme) straddling the System & Computer (S&C) organisation at the Ministry of Defence (MINDEF) and the newly formed National Computer Board (NCB) under the Ministry of Finance (MOF). Swee Say wanted to build up a core team of technical experts to support a massive new initiative underway to computerise the civil service. The NCB, set up only a few years earlier, had the mission 'to drive Singapore into the information age' and was hurriedly hiring staff for its civil service computerisation programme (CSCP). S&C was the MINDEF unit with a mandate to look at new technologies for potential defence use.

Not long after I joined JSEP however, Swee Say, ever the visionary, saw a greater need beyond technical support, of building up research and innovation capability in IT (information technology) and sought to transform JSEP into an IT R&D entity. His perseverance paid off when he managed to convince the

MOF to allocate funds to move the 30-strong JSEP team into a new R&D arm for the NCB to be called the Information Technology Institute (ITI). ITI was thus formed in 1986 to carry out R&D in three areas — software engineering, computer & communications, and knowledge systems.

R&D?

With hardly any R&D activity in Singapore in the 1980s, being a researcher was something of a novelty. Apart from some classified defence research, there were only modest research efforts in agencies such as the Public Utilities Board (PUB) and the Meteorological Service unit. Swee Say used to tell us of the great challenges he faced, trying to convince MOF about the importance of investing in R&D for the future. When he first presented his plans for ITI, he was 'reminded' by MOF that 'money does not grow on trees'. MOF's stance soon softened however, and within a few years, several more R&D entities were established. The Institute of Systems Science (ISS), a training institution at the National University of Singapore (NUS), set up in partnership with IBM to produce system analysts for IT jobs, also started moving into research. In 1987, as part of an investment incentive package offered by the Economic Development Board (EDB) to pharmaceutical giant Glaxo, a sizeable life science research institute was created at the NUS, with the mission of conducting world-class research in molecular and cell biology under the guidance of eminent biologist (now Nobel Laureate) Dr Sydney Brenner. The Institute of Molecular & Cell Biology (IMCB) was by far the largest research institute at that time, with a planned strength of several hundred scientists. Over at the Nanyang Technological University (NTU), an institute for computer integrated manufacturing, called GINTIC, was also set up in partnership with German firm, Grumman International.

The '80s were the early years of R&D in Singapore. Apart from ITI, the other research institutes were all established in partnership with foreign companies. The reasoning was that expertise was lacking locally and technology transfer from foreign partners was necessary for Singapore to build up its own capability.

The government did not play an active role in national R&D planning then. There was a small unit called the Singapore Science Council[2] with about a dozen staff dispensing modest research grants, mainly to university researchers. The four institutes — ITI, ISS, IMCB and GINTIC all came about independently through different bottom up initiatives supported by different ministries.

Moving Up the Value Chain

The years following Singapore's independence in 1965 were a period of rapid economic development. As Singapore continued to grow economically, the government recognised the need for the country to move up the economic value chain into higher value-added activities. During the recession of 1985–86, the Economic Committee that was set up to chart new directions for Singapore, reiterated the need for Singapore to move away from its low cost advantage of a developing country into activities offering higher value and skills content. Chaired by then Minister of State for Trade & Industry Lee Hsien Loong,[3] the Committee recommended that Singapore should aim to be an international business centre and an exporter of high value goods and services. Various other committees and workgroups at that time echoed the need to develop capabilities in research and innovation to support this ambition.

It was against this backdrop of events that the government decided to take a more proactive approach to invest in building up the R&D capability. The plan was to create a new agency that would have the mandate to develop national policies for R&D in a coordinated manner and be given substantial resources to fund research activities.

Birth of NSTB

Thus the National Science and Technology Board (NSTB)[4] came into being in January 1991, through a parliamentary act, as a statutory board under the Ministry of Trade & Industry (MTI). The NSTB was entrusted with the mission to develop Singapore into a centre of excellence in selected fields of science and technology so as to enhance national competitiveness in the industrial and services sectors.

This mission meant that NSTB's focus was to promote industry-driven R&D, defined as such R&D that would contribute to current and future national economic competitiveness.

Mr Lam Chuan Leong, the permanent secretary for trade and industry at that time was appointed the founding chairman of NSTB.

The dozen or so staff of the Science Council were transferred to NSTB. In addition, Dr S K Chou, an engineering lecturer from NUS was seconded as the first Executive Director. Because of my research background at ITI, I was seconded from the NCB to be Director Planning while another EDB manager came on board as Director Industry. NSTB took up office at the Pasteur, one of the early blocks of the Singapore Science Park, located near the entrance to the park.

Starting Almost from Scratch

Starting up a new organisation involves a lot of hard work, but is always exciting. As the government had indicated its willingness to set aside a substantial budget to the tune of S$2 billion for five years to support R&D, NSTB was expected to make a significant impact to the R&D landscape of Singapore. S$2 billion was a huge allocation at that time — easily an order of magnitude more than existing funding levels!

The core NSTB management team did not have much operational experience in managing a national R&D agency, so there were many questions about what and how things should be done. There was no Google to search for information online, so we depended on books and published articles, the advice of visiting experts and of course, large doses of common sense. The operating environment was highly dynamic, much like that in a startup!

MTI was actively involved during the first year of NSTB's operation. It was not uncommon for us to be summoned multiple times each day to MTI's office at the old Treasury Tower in Shenton Way. Indeed, for a few weeks during the first six months of NSTB, my entire Planning Team was relocated to MTI, so that we could work directly with NSTB's Chairman and MTI officials.

The advantage of starting with an almost clean slate was that there were no legacy issues to worry about. However, as a new entity with a large fund to manage, there was considerable pressure to show progress quickly. The upside of this was that we were able to put through new ideas and programmes for approval without layers of bureaucracy and multiple rounds of iterations. There was simply no resources or time for this! Indeed, for much of the first year, there were more board members than staff in the organisation!

With a tentative allocation of S$2 billion as a national R&D budget for five years, much of NSTB's early effort was focused on developing a national R&D strategic plan to propose the allocation of funds and identify major R&D initiatives to support. One of the first steps taken was to form committees to study various technology areas in depth and recommend good initiatives to support. Expert Committees were thus formed in nine areas deemed to be important and relevant to Singapore: Information Technology, Microelectronics, Electronic Systems, Manufacturing, Materials, Biotechnology, Food & Agrotechnology, Medical Sciences and Energy, Water & Resources. These committees, with representation from academia, government and industry, were tasked to analyse Singapore's needs and capabilities in their specific areas and recommend new R&D initiatives for NSTB's consideration. The Planning

Department, with a handful of officers under my charge, acted as the secretariat for these committees. We were thus involved in endless rounds of expert committee meetings! Thankfully, additional support was provided by statutory boards within the MTI family — the EDB and SISIR[5] both played a big role in helping to identify relevant technology areas to study as well as identifying experts for the committees.

In parallel, MTI's strategic planning group would work with NSTB to address non-technology specific topics such as infrastructure and manpower development. After much effort and many rounds of iterations, the final report entitled *National Technology Plan — Window of Opportunities* was ready to be launched.

National Technology Plan

In September 1991, then Deputy Prime Minister and Minister for Trade & Industry, Lee Hsien Loong unveiled the National Technology Plan (NTP) at the first Technology Month dinner (see Box Story: Technology Month). The media headline was on the S$2 billion R&D Fund, plus the target set to achieve a gross expenditure on R&D of 2% of GDP and have 40 RSEs (research scientists and engineers) per 10,000 labour force by 1995. The expectation was for at least half of this R&D expenditure to come from the private sector. The emphasis of the NTP was on research directed towards economic upgrading, i.e., with results that would eventually contribute to the nation's economic competitiveness.

Key Thrusts of NTP

The NTP outlined five key thrusts that the government, through the NSTB, would undertake:

- Sets aside a S$2 billion Research and Development Fund (RDF) to support industry-driven R&D over the period 1991–95.
- Provides grants and fiscal incentives to encourage more R&D in the industry.
- Assists in developing and recruiting R&D manpower.
- Supports and funds research centres and institutes that can train manpower or provide the technological support to enable companies to undertake R&D.
- Provides assistance for R&D commercialisation and develops technology infrastructure.

GERD

One of the new terms I learnt moving into a R&D planning role was the acronym GERD — gross expenditure on R&D. This is used in measuring national R&D activity in terms of spending. In comparing across countries, GERD is often expressed as a percentage of GDP. GERD/GDP is probably the single most used indicator to represent a country's level of R&D. A complementary measure is the number of research scientists or engineers (RSE) per 10,000 labour force, which gives an indication of the concentration of R&D jobs in the workforce.

Singapore's expenditure on R&D had grown from $38m or a miniscule 0.2% of GDP in 1978 to $572m or 0.9% of GDP in 1990. R&D manpower over the same period grew from 1,672 to 7,094. There were 4,329 graduate RSEs in 1990 or 29 per 10,000 labour force. Compared to the developed countries and other newly industrialised economies such as Taiwan and South Korea, Singapore was a long way behind (see Figure 1 for the comparison of GERD/GDP and RSE/10,000 workforce among countries).

There was probably an element of competitive pressure as well. South Korea had unveiled in 1989 an aggressive plan to reach an expenditure target on technology of 5% of Gross National Product by the year 2000 from 1.8% then. Likewise, Taiwan's plan in 1988 was to reach 2.2% of GDP by 1996, from a level of 1.4%.

In reality, GERD/GDP, as an input measure, is not the best indicator of R&D capability. Nevertheless, it continues to be widely used because it is simple and measurable and allows international comparisons. To be sure, there has been reasonable correlation between GERD/GDP and R&D outcomes over the years. Countries known for scientific depth and innovation, such as Sweden, Switzerland, Finland and Israel score high in this measure. Singapore

(Continued)

Fiscal Incentives and Grants

The use of tax incentives to encourage R&D activities was a widespread practice in many industrialised countries. To stimulate greater industry R&D, grants were offered to companies embarking on research projects as a way for the government to share the risk. The plan was to encourage both MNCs and local corporations to develop new technologies and products through R&D in Singapore.

One popular grant was the Research and Development Assistance Scheme (RDAS), carried over from the Science Council. One of the first tasks I did at

(Continued)

has continued to use this metric even today, but supplemented by numerous other KPIs (key performance indicators).

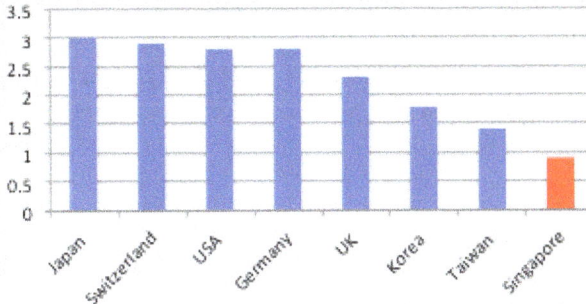

GERD as a % of GDP (1990)
(World Competitiveness Report 1992)

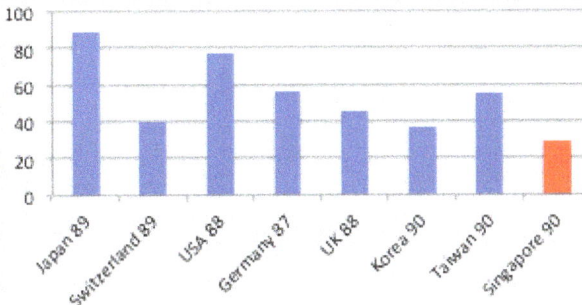

Research Scientists/Engineers per 10,000 labour
force (OECD and national sources)

Fig. 1. Comparison of R&D indicators across different countries.

NSTB was to simplify the application process for the grant while increasing its scope of coverage, making it more attractive to companies. The RDAS funded up to 70% of the direct cost of a project, such as research manpower, equipment and material, equivalent to a support of about 50% of total project cost. As a grant, no repayment was required, but the terms stipulated that royalties up to twice the amount of grant received would be payable if the research led to a commercial product. In practice, this was difficult to track or enforce and in all probability, royalties were never collected.

For the larger corporations and MNCs, more attractive incentives were necessary beyond supporting direct project costs. NSTB built on the myriad incentive schemes administered by EDB, such as pioneer status, double deduction of R&D expenses, accelerated depreciation for capital expenditure and concessionary tax rates, to create a package of R&D support known as Research Incentive Scheme for Companies (RISC). This proved to be very popular and was taken up by many MNCs and large local companies alike, including HP, Philips, Motorola, ST Electronics, etc. The RISC scheme was eventually transferred to EDB.

R&D Manpower Development

With the target announced of increasing RSE manpower from 31 to 40 per 10,000 labour force in five years, a slew of schemes was created to attract bright graduates into R&D careers as well as to step up the recruitment of foreign scientific talent.

Technology Month

The idea of setting aside a period of time each year with a high concentration of science-related activities was started by NSTB's predecessor, the Singapore Science Council in 1986. Each year, a week was designated as 'Science & Technology Week' during which the institutes of higher learning and the schools were asked to organise activities related to science and technology. These could be public lectures by scientists, informal talks, science quizzes, conducted tours of labs etc. Given the increased emphasis on R&D with NTP, NSTB decided to expand the period to a full month and designated September each year as TechMonth. Efforts were stepped up to fill the calendar with activities to increase awareness of and appreciation for science and technology. The anchor event of TechMonth was the S&T gala dinner organized by NSTB — it was at the inaugural gala dinner that the first NTP was announced. The TechMonth dinner eventually became the signature event for major announcements on S&T matters.

To increase the visibility of R&D and the take-up of R&D grants, companies receiving RDAS and RISC grants were invited on stage during the TechMonth dinner to receive award certificates given out by the officiating minister. Companies receiving grants would buy tables for their staff and guests to attend in celebration. Smaller tenants of the Science Park would use the occasion as their company annual dinner! The first TechMonth dinner in September 1991 turned out to be a roaring success, with the entire Shangri-La Hotel ballroom filled to capacity, with close to 100 tables.

Among these were:

- Joint industrial fellowships with companies for Masters and PhD studies, in which NSTB would cover 70% of the cost.
- Scholarships to encourage fresh graduates to pursue technical post-graduate courses.
- Programmes allowing R&D engineers to earn degree credits while working on company research.
- Funding of a pool of foreign RSEs to act as a "float" from which organisations carrying out R&D could recruit.
- Improving the terms of employment for foreign R&D talent, such as providing housing assistance and allowances.
- Setting up of an office to provide career guidance and placement services for graduates considering a career in R&D.
- Giving recognition for R&D contribution through various awards.

The Joint Industry Masters or JIM program was especially popular as it allowed R&D personnel in industry to pursue a Masters degree at very little cost to the company. Recipients of the award had to work at least two years in R&D after graduation.

The R&D talent pool scheme came in at a time when China, under Deng Xiaoping, was just beginning to open up to the world. It attracted many high quality applicants from China (graduates of top Chinese universities) who saw the programme as an opportunity to travel out of the country. Many were deployed in public research institutes (RIs) such as ISS and ITI. Indeed, some of them took the opportunity to escape to the West! I recall that from the first group of five Chinese scientists deployed at the ITI, two suddenly left Singapore under the pretext of returning to China to deal with family emergencies. They were later found to have made their way to the West. The remaining three settled in Singapore and became citizens.

Research Institutes and Centres

An important role that NSTB was tasked to undertake was to be the funding agency for the growing number of public research institutes and centres (RICs) being set up. The intent was for NSTB to provide coordination and oversee policy guidance to all such institutes in Singapore.

Four institutes — ISS, ITI, IMCB and GINTIC, were already in operation prior to the establishment of NSTB but were funded and driven by different ministries. Two others — the Institute of Microelectronics (IME) and the

Institute of Manufacturing Technology (IMT) were just starting up. Both were established as part of the plan to develop capabilities required to support new industry sectors targeted by EDB. In IME's case, it was in semi-conductor and micro-electronics, where VLSI (very large scale integration) technology was the state of the art. EDB was then aggressively courting MNCs to set up wafer fabs in Singapore, as they had very high value-added contribution to GDP. Singapore's sovereign fund Temasek had also ventured into the industry, investing heavily to build up a homegrown microelectronics company in Chartered Semiconductor Manufacturing Ltd[6] to compete globally. IMT like-wise, was created in response to the promise and potential for high-value jobs offered by intelligent manufacturing technologies. IME and IMT, funded by EDB at the start, were transferred to NSTB to consolidate the funding of public research institutes in a single agency. Table 1 shows the state of the six research institutes in 1992.

In addition to RIs, some smaller research centres were also set up, again mainly in response to a perceived need to build capability in a new technology area that was, or would become important to Singapore. Two such centres set up around that time were the Magnetics Technology Centre (MTC) and the Centre for Wireless Technology (CWC), both driven by NUS's Engineering Department. The MTC was focused to serve the needs of the large and growing disk drive industry. Singapore was then the world leader in the manufacture of hard disk drives, with many top players such as Seagate, Maxtor, Micropolis and Western Digital all having large production facilities here. The CWC was similarly targeted at the emerging wireless communication sector, and sought to partner with world leaders Motorola, Ericsson and NTT Docomo in R&D. MTC was upgraded to become the Data Storage Institute (DSI) while CWC was eventually merged with ITI and ISS to form the present Institute for Infocomm Research (I²R).

NSTB recognised that the public research institutes and centres (RICs) provide the enabling resources of manpower, skills, technology, knowl-edge, products and processes needed by the private sector. They were intended to complement the R&D activities undertaken by companies in Singapore.

Commercialisation of Research

Even as it pushed the message of industry-driven R&D, NSTB recognised that successful commercialisation of research required more than good

Table 1. Public R&D Institutes in Singapore in 1992.

Research Institute	Year Formed	Parent Organisation	Size in 1992	Research Areas
Institute of Systems Science (ISS)	1986	National University of Singapore	115	Neural networks, virtual reality, image processing, multimedia, Chinese computing
Information Technology Institute (ITI)	1986	National Computer Board	130	Software engineering, computer & communications, knowledge systems
GINTIC Institute of Computer-integrated Manufacturing	1985	Nanyang Technological University	90	Computer-integrated manufacturing, precision machining, PCB assembly & testing
Institute of Molecular & Cell Biology (IMCB)	1987	National University of Singapore	120	Rational drug design, study of human diseases, transgenic animals, genetically engineered tropical plans, aquaculture
Institute of Microelectronics (IME)	1990	National University of Singapore	40	Microelectronics system applications, VLSI/CAD, silicon processing technology, reliability & failure analysis
Institute of Manufacturing Technology (IMT)	1990	Nanyang Technological University	30	Flexible manufacturing systems, simulation, concurrent engineering, computer-aided design, networking, intelligent manufacturing systems

technology — other factors such as manufacturing, marketing and distribution had to be in place. Nevertheless, at that time, more emphasis was placed on growing the supply of technology. The number of patents registered in Singapore over the 1986–89 period was very low. Singapore had only 0.2 patents per 100,000 population, way below the numbers in other countries — 43.7 in Japan, 42 in Switzerland, 31.3 in Taiwan and 18 in the USA.

In NTP, NSTB had considered setting up a one-stop centre to aid in technology commercialisation, providing guidance in product market research, feasibility studies, application for patents and even product marketing. The centre could even match-make technology suppliers with users, entrepreneurs with venture capitalists and investors. However, this was not pursued in the end. Instead, NSTB left it to the universities and RICs to develop and implement their own technology transfer and commercialisation programmes.

Technology Infrastructure

The government recognised that as the country entered into a more innovation-driven phase of its economic journey, suitable infrastructure to support knowledge-intensive activities had to keep pace. The accepted wisdom at that time was that knowledge-based activities flourish best in a localised area with a good mix of complementary players — high tech companies, research institutes and centres, universities, startups and professional services — all integrated within an attractive living environment with social and recreational amenities. Such an area was known by various terms, such as a 'science habitat' or a 'technopolis'. Examples being studied then included Silicon Valley, Route 128, North Carolina's Research Triangle Park, Sofia Antipolis in France, Taiwan's Hsinchu Science Park and Japan's Tsukuba Science City.

The closest that we had as a 'technopolis' in the early '90s was the Ayer Rajah area near NUS, where the 30 hectare Singapore Science Park was built in 1984. This was where NSTB was located together with government organisations dealing with technology, such as SISIR, NCB and some 80 tenants, which included the technology centres of Xerox, Fujitsu, Sony and Tata.

In its plan to grow the technopolis, NSTB entered into a partnership with Jurong Town Corporation (which was responsible for developing industrial parks and science parks) to pursue adding three more phases of science park development to increase the total area to 120 hectares. The land parcels for phases two, three and four of the science park development were all in parkland around the South Buona Vista area. Science park development was then undertaken by JTC's subsidiary, Technology Parks Pte Ltd (TPPL). The NSTB set up a new corporation jointly owned with JTC called SSPPL (Singapore Science Parks Pte Ltd) to undertake the development of subsequent phases of the science park, but contracted TPPL to do the work. Perhaps as a consequence of this expanded scope, TPPL grew rapidly, was eventually renamed as

Ascendas,[7] and became a leading developer of technology parks, undertaking huge billion dollar infrastructure projects all over the world.

An interesting term that was used in the NTP to refer to the technopolis being envisioned was the 'Technology Corridor'. This referred to a belt of land in the South-Western part of Singapore, stretching from NTU in the far West to the science park at Kent Ridge. Along this stretch was a concentration of tertiary educational institutions (NUS, NTU, Singapore Polytechnic, Ngee Ann Polytechnic), hi-tech businesses at the science park and Ayer Rajah Industrial Park, good quality housing, recreational amenities (Temasek Club, Holland Village) and good access to major highways, such as the Ayer Rajah Expressway (AYE). Figure 2 shows a schematic representation of the Technology Corridor drawn up by NSTB in NTP.

NSTB had plans to develop this corridor, in cooperation with other organisations to create an attractive living environment that would attract and retain scientific talent and foster innovation and creativity.

Fast forward a quarter century later, the technopolis envisioned in 1991 has grown way beyond what was originally envisioned and is continuing to expand. The One-North area is not only home to education and research institutes (NUS University Town, Insead Campus, Biopolis, Fusionopolis, etc), it has also developed into a thriving eco-system for startups, housing hundreds of technology startups in the several blocks of flats around Block 71, Ayer Rajah Crescent.

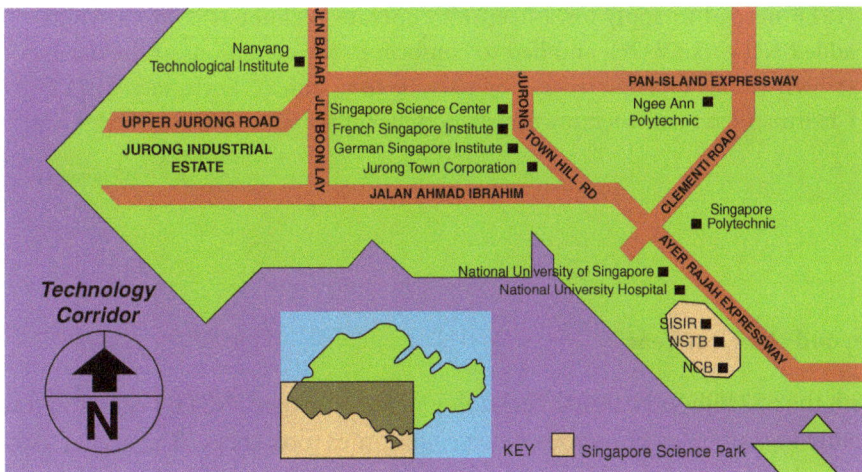

Fig. 2. Technology Corridor in the National Technology Plan 1991.

Technet — Singapore's First Internet Services Provider

A major highlight of NSTB's first year was the launch of Technet. During the early '90s, the Internet was still an unknown network, originally built by US Defense Agency ARPA to test the robustness of electronic communication for the military. It was then used mainly by US research scientists to run programmes on remote computers. Some academics at NUS were using it to access information and data in the US, connecting via very low speed international leased lines. This was a time when a T1 leased line (1.544 Mbps) cost a few hundred thousand dollars a month. NSTB saw an opportunity to play a value adding role by improving access to the Internet for the universities and making it available to the wider R&D community in Singapore, including private companies. After several rounds of discussion with Dr Thio Hoe Tong, then Director of the NUS Computer Centre and his deputy Dr Tommi Chen, we put up a proposal to build an Internet network for Singapore and named it Technet (short for Technology Network). The idea we had was for NSTB to fund the expensive leased line to the US Internet landing point and allow companies to share the line, thereby greatly reducing their cost of connection while improving access speed. Since NSTB did not have computing resources, the operation of Technet was to be outsourced to NUS Computer Centre, a proposition gladly accepted by Dr Thio.

There was strong objection to the project, however, from Singapore Telecoms, which then had a monopoly in the local telecommunication market and a lucrative business in long distance leased lines. Telecoms (as SingTel was then known) feared a loss of revenue from this arrangement and said NSTB needed to apply for a license to operate Technet as it was a 'Value-added Network (VAN)' and had to conform to the conditions of the license (which would disallow access by private companies). Fortunately, NSTB Chairman Lam had been previously permanent secretary for the Ministry of

(Continued)

Beyond the First Year

With the S$2 billion National Technology Plan launched, NSTB began to grow rapidly in staffing as numerous new programmes took shape. The end of 1992 saw a change in leadership as then permanent secretary for Transport and Communications Teo Ming Kian succeeded Lam Chuan Leong as chairman. Ming Kian defined the role of NSTB more sharply as the agency to bring about

(*Continued*)

Communications and had served a stint as general manager (CEO) of Telecoms. Supportive of our proposal, he summoned Telecoms' senior management to his office and 'arm-twisted' them into grudging agreement.

With the obstacle cleared, the Technet proposal was presented to the Minister of Trade and Industry (Lee Hsien Loong) and was approved with a S$3 million budget over five years. The plan was for Technet to charge users a fixed subscription charge for access and become self-sufficient eventually as the numbers picked up. As it turned out, our projections for take-up were grossly under-estimated. Consequently, the upgrading of the leased line (we started with 128 kbps) was moved forward to cope with the surge in demand.

Technet was launched in NUS by the late Dr Tay Eng Soon, who was then Senior Minister of State for Education. I chaired a Steering Committee to decide on usage and charging policies. The R&D community responded warmly to Technet and applications for access picked up quickly.

When the World Wide Web and Mosaic browser came out in 1994, it took the world by storm and transformed the Internet from being a tool for geeks in research into something anyone could use. Every organization wanted to connect to the Internet! Technet was actually playing the role of an ISP (Internet Services Provider) before the term was coined and had a rapidly growing number of subscribers by then.

It did not take long for Sembawang Media (a new business unit started by EDB and Sembawang Corporation Chairman Philip Yeo to enter the new Internet business) to recognise Technet's commercial value. Technet was acquired by Sembawang Media for a sum that was more than its accumulated cost and was expanded and renamed Pacific Internet, eventually listing in NASDAQ. Thus Technet was a rare example where a government initiated developmental project actually returned more than its investment! I was no longer with NSTB then, having left for the US in June 1992 to pursue a programme in the management of technology.

the creation and application of knowledge and innovation to sharpen Singapore's competitive edge and maintain its relevance. R&D was to become an increasingly critical factor for Singapore's continuing competitiveness as the country seeks to evolve into an innovation-driven economy.

When I returned to NSTB a second time in 1998, the organisation was highly energised, having taken on an exciting challenge to promote technology entrepreneurship ('technopreneurship') in a big way. Spearheaded by Teo Ming

Kian, and strongly supported by then Deputy Prime Minister Dr Tony Tan, the Technopreneurship21 (T21) programme was given a US$1b budget to develop entrepreneurship and the venture capital industry in Singapore.

I was fortunate to have another opportunity to work for Ming Kian and Dr Tan, when the National Research Foundation (NRF) was formed in 2006 with a S$5 billion budget to take Singapore significantly further in research, innovation and enterprise. In addition to supporting high impact research at the universities, NRF put in place a national framework for innovation and enterprise to develop the entrepreneurial eco-system and make Singapore a global hub for technology entrepreneurship.

Conclusion

Singapore has come a long way since the 1980s when a career in R&D was uncommon and someone with a PhD had few options outside academia. We now rank highly in international comparisons on academic research, scientific excellence, patent filing, tech-savviness, innovation and entrepreneurship. These have come about through substantial and sustained investments in science and technology over the years. Government spending on R&D has grown eight times in 20 years! The first significant step in this national R&D journey was taken almost a quarter of a century ago with the establishment of NSTB and the launch of the first National Technology Plan.

References

[1] Swee Say has been a minister in the Singapore Government since 2001.

[2] The Science Council was set up under the now defunct Ministry of Science & Technology in 1968. The Ministry was closed down after 12 years in 1981.

[3] Lee Hsien Loong became Prime Minister of Singapore in 2004.

[4] NSTB was renamed as the Agency for Science, Technology and Research (A*STAR) in 2002.

[5] Singapore Institute for Standards & Industrial Research, a statutory board under MTI, now part of SPRING Singapore.

[6] Chartered Semiconductor Manufacturing was eventually acquired by GlobalFoundries in 2009.

[7] TPPL was rebranded as Arcasia in 1997 and renamed Ascendas in 2001.

Chapter 3

Science, Technology and Open Innovation — The A*STAR Journey[†]

Raj Thampuran and Colleagues at A*STAR

BUILDING AN ASIAN INNOVATION CAPITAL

R&D Amidst a Changing Economic Landscape

Up until the 1980s, Singapore's economic growth was driven mainly by foreign direct investments, and the country's Gross Expenditure on Research and Development (GERD) never exceeded 0.5 per cent of its gross domestic product (GDP).

However, as regional competition in traditional manufacturing and services sectors intensified, Singapore recognised the need to broaden and diversify its economic base to new clusters and activities. It had to move up the value chain and transform itself into a knowledge-based economy. Investment in R&D became essential to country's long-term competitiveness.

Some of Singapore's early R&D efforts originated in the institutes of higher learning; others were carried out by the research arms of various public agencies or in collaboration with private sector organisations.

In 1985, an R&D Group was set up within the Institute of Systems Science (ISS) to drive applied research programmes. The same year, GINTIC was formed through a collaboration between the then-Nanyang Technological Institute (now NTU) and Grumman International to address the R&D needs of the manufacturing industry.

New institutes were formed in the ensuing years, focusing on different areas of research. They included the Information Technology Institute which was set up in 1986 as part of the then-National Computer Board, the Institute of Microelectronics (IME) which was established in 1991, the Magnetics Technology Centre and the Centre for Wireless Communications which were

[†] Adapted from *Science, Technology and Open Innovation — The A*STAR Journey*. Published by A*STAR (Agency for Science, Technology and Research), Copyright © 2015 A*STAR.

set up in 1992, the Centre for Signal Processing in 1996, the Institute of Materials Research and Engineering (IMRE) in 1997 and the Institute of High Performance Computing (IHPC) in 1998. Some of these research institutes were later merged and their research programmes consolidated.

Speaking at the opening of Fusionopolis One in 2008, Prime Minister Lee Hsien Loong described Singapore's public research institutes (RIs) as the bridge that translated R&D into useful outcomes for the economy and society. "Many research institutes have grown alongside the industrial sectors they support. By intertwining their research programmes with the needs of industry, our research institutes provide companies access to new knowledge and innovative technologies that help to sustain their competitive edge."

And so, from labour-intensive manufacturing in the 1960s and capital- and skill-intensive manufacturing in the 1970s-1980s, Singapore went on to develop new capabilities in areas such as data storage, wireless communications, semiconductor, chemicals and materials research.

The data storage industry was an example of a sector which benefited from public sector R&D. With the support of R&D from the Data Storage Institute (DSI), it progressively shifted out of low-end disk drive assembly activities into complex manufacturing and storage-related services. By 2008, Singapore had close to 80 per cent share of the world market for high-end enterprise disk drives, and 40 per cent share of the world market for hard disk media. The move up the value chain was made possible through DSI's extensive collaboration with key industry players on next-generation product developments, as well as its own research which spawned new high value-added activities.

Another striking illustration of the industry relevance of A*STAR's RIs is the Institute of Chemical and Engineering Sciences (ICES), whose roots can be traced back to the late 1990s when it was set up as a research centre in the Faculty of Engineering of the National University of Singapore (NUS).

When Singapore began building a petrochemical ecosystem on Jurong Island to tap on the opportunities presented by the chemicals industry, the NUS centre was upgraded to a research institute. It was subsequently moved to Jurong Island as ICES, to help move the industry up the value chain by fostering research and development in new products and processes.

ICES' custom-designed facilities were opened on Jurong Island in 2004. Since then, it has hosted numerous corporate laboratories for companies such as Mitsui Chemicals, Dystar, Syngenta and BASF and allowed these companies to share laboratory facilities, state-of-the-art analytical equipment and research expertise to jumpstart their R&D activities. These collaborations paid off, as many of these companies have since established their own research presence in Singapore.

The Fourth Pillar

R&D was also instrumental in creating the fourth pillar of Singapore's economy — the biomedical sciences — alongside electronics, engineering and chemicals.

Up to the 1990s, the Institute of Molecular and Cell Biology (IMCB) was the only full-fledged RI focusing on biomedical sciences (BMS). Pockets of biomedical research were located in the Singapore Science Park and NUS, but there was no critical mass of BMS research activities in either the public or private sector.

During the 1997 Asian Financial Crisis, it became apparent that new engines of growth had to be created to diversify the economy, generate jobs and provide the impetus for the next wave of economic prosperity.

BMS was identified as a prime candidate to fulfil this role.

Singapore was already a trusted healthcare centre which served a much wider region. It was also a natural base for the pharmaceutical industry offering excellent intellectual property protection, and A*STAR's IMCB, which was established in 1987, was gaining international recognition for its work.

But with BMS, Singapore would have to chart a slightly different course from its traditional manufacturing-led approach to economic development.

At the time, the Economic Development Board (EDB) was starting to promote pharmaceuticals for manufacturing and production, Mr Philip Yeo, then-Chairman of EDB was concerned that if Singapore focused only on production, there would be no intellectual capital created for the industry.

A study group was formed to learn about the complexities in establishing a BMS research infrastructure. Comprising Mr Yeo, then-Deputy Prime Minister Dr Tony Tan and Nobel Laureate Dr Sydney Brenner, the group visited top biomedical research laboratories in the United Kingdom and Europe. Its members were quick to realise that in order to position itself as an R&D powerhouse for BMS, Singapore would have to raise its game.

Birth of the Biomedical Sciences Initiative

To develop a BMS hub, Singapore would need a whole spectrum of resources that could support medical research effectively. An infrastructure had to be established, from building state-of-the-art research facilities to attract top-notch researchers and scientists to work here, to implementing corporate R&D activities to create a buzz around BMS.

The Government wasted no time. In 2000, it launched a BMS Initiative to look into the development of a full spectrum of capabilities, from bringing together top scientists to building state-of-the art facilities for research in biomedical sciences. A budget of S$1.48 billion was set aside to support these efforts.

The Biomedical Research Council (BMRC) was formed under A*STAR (then-NSTB) to develop public sector research and build up Singapore's talent pool in the biomedical sciences.

During Phase 1 of Singapore's BMS Initiative from 2001 to 2005, new research institutes were created to jumpstart the BMS industry. The Bioinformatics Institute (BII) was formed in 2001 to drive the use of computing systems in biomedical research. The Singapore Genomics Programme, which was introduced in 2000, was renamed the Genome Institute of Singapore (GIS) to reflect its status as a research centre. In 2003, the Bioprocessing Technology Centre, which started off as a Bioprocessing Technology Unit under an EDB programme in 1990, was renamed the Bioprocessing Technology Institute (BTI). The same year, the Institute of Bioengineering and Nanotechnology (IBN) was also formed.

In Phase 2 of the BMS Initiative from 2006 to 2010, the focus was extended to building up a strong Translational & Clinical Research (TCR) capability to facilitate the translation of scientific discoveries into new treatments and diagnostics.

The focus on public-private sector collaboration intensified as Singapore entered the next stage of capability development under the BMS Initiative, which was to go beyond basic science to work with industry on translational and applied research.

New Translational Clinical Research (TCR) Flagship programmes led by clinician-scientists at National University Health System (NUHS), Tan Tock Seng Hospital (TTSH) and Singapore Eye Research Institute (SERI) were established to bring together the best complementary research strengths in hospitals, national disease centres, universities and A*STAR research institutes (RIs) to focus on disease or research themes of strategic importance. Some of the A*STAR RIs involved in these projects included the Singapore Bioimaging Consortium (SBIC), the Singapore Immunology Network (SIgN), Institute of Medical Biology (IMB), Singapore Institute of Clinical Sciences (SICS) and Experimental Therapeutics Centre (ETC). The Singapore Stem Cell Consortium (SSCC) was also formed during this time to provide a boost to stem cell research.

Changes in the Public Sector R&D Landscape

In January 2002, the National Science and Technology Board (NSTB) was restructured and re-named the Agency for Science, Technology and Research (A*STAR). In parallel with BMRC, the Science and Engineering Research Council (SERC) was formed as the overarching group for the physical sciences and engineering institutes. The changes were aimed at fostering greater collaboration and reducing duplication in Singapore's R&D efforts. Rather than grow out of university research and disparate research institutes, Singapore's R&D agenda was now clearly set out and driven by these two councils under a single agency.

Providing the infrastructure backdrop for these developments was one-north, an integrated focal point for R&D in Singapore.

ONE-NORTH

The Masterplan is Unveiled

In 2001, the Government unveiled the one-north masterplan, with a focus to promote R&D in the biomedical sciences (BMS) and in infocomm and media (ICM). Public research institutes would be located in close proximity to each other and to private sector organisations, promoting collaboration and facilitating the exchange of ideas. The research institutes under BMRC were to be housed in Biopolis — the BMS research hub, and their SERC counterparts in Fusionopolis — the Science and Engineering research hub. There would also be residential and recreational facilities aimed at creating a sense of community.

JTC Corporation (JTC) was appointed the lead agency for planning and development, with private developers invited to participate in the project. London-based Zaha Hadid Architects was appointed as the masterplan consultant for the project.

BIOPOLIS

Within 30 months, the first R&D cluster was ready. Opened in 2003, Phase 1 of Biopolis was a 185,000 sq m research complex comprising seven high-rise buildings linked by sky bridges to encourage and facilitate multi-disciplinary collaborations. It also promoted the concept of co-location of public-private research institutes and shared facilities, and provided companies with a plug-and-play infrastructure to help them jump-start their operations.

BII was the first A*STAR research institute to move in when Phase 1 was completed. Progressively, the other research institutes — GIS, BTI, IBN and IMCB — moved in to occupy five of the seven buildings — Centros, Genome, Matrix, Nanos and Proteos. The other two buildings, Chromos and Helios, were earmarked for the private sector.

Biopolis was subsequently expanded through five phases to 13 buildings with a total floor area of more than 340,000 sq m, housing A*STAR's research institutes and consortia, corporate laboratories, as well as local and foreign private research agencies and entities.

Building the BMS Community

In the beginning, it was difficult to attract companies to invest in research activities in Biopolis as Singapore did not have a strong track record in BMS. One of the first projects embarked on by EDB was to attract Novartis to set up the Novartis Institute of Tropical Diseases (NITD) there. To lay the groundwork, a team of experienced scientists from A*STAR's research institutes was seconded to Novartis to embark on collaborative research. This paved the way for NITD to become the first corporate laboratory to be established in Biopolis.

Singapore stepped up its efforts to attract multinational companies to set up their R&D divisions in Biopolis, to capitalise on the knowledge and skills of the local scientists and the spectrum of capabilities that was available, especially in TCR.

Facilities built in Phase 2, comprising the buildings Neuros and Immunos and developed by Ascendas (Tuas) Pte Ltd, were marketed to the private research institutes.

The co-location of companies' research operations and public research institutes paved the way for many successful partnerships. For example, in the area of genome technology, United States-based Fluidigm Corporation collaborated with the Genome Institute of Singapore (GIS) to set up its Single Cell Omics Centre in Biopolis, which is dedicated to accelerating the understanding of how individual cells work, and how diagnosis and treatment might be enhanced through insight derived from single cells.

Another company which moved in during Biopolis' Phase 2 development was ARKRAY, a Japanese manufacturer of clinical tests and in vitro diagnostics devices. In 2013, IBN finalised a memorandum of understanding (MOU) to establish an R&D partnership with ARKRAY. Under the MOU, ARKRAY would invest S$9.1 million to establish a new R&D centre in Biopolis, co-located with IBN, and hire 21 research staff over five years.

Phase 3 of Biopolis' development saw the completion of Synapse and Amnios by Crescendas Bionics Pte Ltd, which boosted Biopolis' research space by another 42,000 sqm. These are purpose-built multi-tenanted biomedical research facilities which extend basic research activities into TCR as well as medical technology research. One of the anchor tenants for Phase 3 was Chugai Pharmabody Research (CPR).

CPR involved an investment of S$200 million spread over five years. Its 60-man R&D unit focused on antibody engineering. It was the company's second facility here and its only research centre outside of Japan at the time.

In 2015, Chugai Pharmaceutical announced another S$476 million investment over seven years to expand its research institute here. The investment will make CPR one of the biggest pharmaceutical research and development operations in Singapore.

Diversifying into Food, Nutrition and Personal Care

Meanwhile, the organic infrastructure of Biopolis has allowed the R&D metropolis to remain adaptive to changes required of a dynamic industry. Over the last few years, it has been diversifying beyond pharmaceuticals, biologics and medical technology to attract multinational corporations from the food and nutrition, as well as the personal care industry. The move has paid off handsomely.

Leading medical nutrition company Danone set up Nutricia Research (then known as Danone Research Centre for Specialised Nutrition) in Biopolis in 2011. It was the company's first research centre in Asia to focus fully on science for child and maternal health. In December 2012, L'Oréal, a worldwide leader in cosmetics, opened an Advanced Research Centre right next to IMB, a move that exemplifies IMB's expertise in skin biology research. Under the collaboration, IMB will help L'Oréal better understand Asian skin types and the effects of the tropics on skin.

Phase 4 of Biopolis' development saw Procter and Gamble Company (P&G) investing S$250 million to build a mega innovation centre in Singapore, the second of two such facilities that the company has in Asia. The 32,000 sq m centre was officially opened by Deputy Prime Minister Tharman Shanmugaratnam in 2014. It houses P&G's research for beauty care, home care and personal health care, as well as pilot plant functions and support offices, and is the largest private research facility in Singapore. It is the largest P&G life sciences research facility outside of Mason, Ohio, and the largest in Asia.

P&G has also signed a Master Research Collaboration Agreement (MRCA) with A*STAR to promote knowledge sharing between the Singapore Innovation Centre and the research community in Singapore. The agreement is one of P&G's largest public-sector research collaborations, and will enable P&G to partner with Singapore research, medical and educational institutions for five years and generate up to S$60 million in joint funding.

Capitalising on the interest of companies such as P&G and L'Oréal, A*STAR has developed an integrated and internationally competitive research programme on skin biology in Singapore, in partnership with the National Skin Centre and Nanyang Technological University (NTU). The new Skin Research Institute of Singapore (SRIS) will focus on epithelial biology, allergy, pigmentary disorders and wound healing, encompassing needs for both the pharmaceutical and consumer care industries.

In a boost for the nutritional sciences, the new Singapore Centre for Nutritional Sciences, Metabolic Disease and Human Development (SiNMeD) was established by A*STAR and NUS to bring together scientific research, translational initiatives and clinical practices in this field. The collaboration between the NUS Yong Loo Lin School of Medicine and A*STAR's Singapore Institute for Clinical Sciences (SICS) focuses on fundamental, clinical and translational research to understand the role of nutrition and early development in the onset and progression of obesity and metabolic diseases like diabetes.

With Phase 5 of Biopolis completed in 2013 by Ascendas Venture Pte Ltd, the concept of ready fitted-out laboratory space was introduced. The 46,000 sq m development known as Nucleos has been earmarked for private sector R&D units.

FUSIONPOLIS

Fusionopolis One Opens

Fusionopolis One, the first phase of development in the Fusionpolis precinct, was designed by the late Dr Kisho Kurokawa, a renowned Japanese architect. The two-tower podium complex comprised the buildings Connexis and Symbiosis, and housed research institutes as well as shops and apartments, providing researchers and scientists with an environment where they could exercise their creativity, engage in experimentation and bring ideas from mind to market.

The strategic intent was for the Fusionopolis precinct to become a hub for public-private sector collaboration. Two A*STAR research institutes, Institute for Infocomm Research (I²R) and Institute of High Performance Computing (IHPC), were identified as anchor tenants in Fusionopolis One. They would act as "queen bees" to attract commercial research partners and other ICT companies with shared facilities and expertise. I²R undertakes research on data mining, internet, mobile and wireless technologies, media processing and human-machine interfaces. IHPC's expertise lies in modelling, simulation and computational chemistry.

The institutes brought with them accompanying infrastructure and high-end equipment, offering prospective tenants a one-stop shop for high-end computing and IT services.

As a further incentive, Fusionopolis One's branding as a prototyping facil-ity and its serviced apartments would provide valuable test-bedding opportuni-ties for commercial firms looking to trial products and services in a contained yet flexible environment.

To capitalise on the institutes' presence and fully exploit their facilities, A*STAR mapped out a target tenant mix of firms that would benefit from the synergy. Potential tenants included communications manufacturers (especially wireless products), those requiring supercomputing facilities to model their product designs and simulations, and also software developers and engineer-ing companies such as large engineering companies like Rolls-Royce.

The strategy paid off. When Fusionopolis One was launched, 13 companies moved in and public-private sector collaborations involving the two anchor research institutes in Fusionopolis One continued to gain traction.

Lloyd's Register (LR) and IHPC opened a joint laboratory to leverage IHPC's capabilities in computational fluid dynamics and engineering mechan-ics, and deliver innovative technological solutions to address the challenges faced by the marine, energy and offshore sectors. In other collaborations, Baidu, the leading Chinese language internet search provider, established a joint laboratory with I²R to leverage the research institute's unique capabilities in Southeast Asian Language Resources, Natural Language Processing, Information Retrieval and Extraction, and Speaker Verification technology.

Bringing SERC Research Institutes Under One Roof

In October 2015, Fusionopolis Two was completed — bringing together SERC and the rest of A*STAR's research institutes under one roof. These included the Data Storage Institute (DSI), Institute of Materials Research and Engineering (IMRE), Institute of Microelectronics (IME) and the Singapore Institute of Manufacturing Technology (SIMTech). The seventh SERC research institute,

ICES, continues to be located on Jurong Island to be in proximity to the chemicals, biomedical and process engineering industries that it supports.

Fusionopolis Two comprises three new buildings — Innovis, Kinesis and Synthesis that provide 104,000 sq m of business park and R&D space. The complex offers a massive test-bed for new technologies, housing Singapore's largest R&D clean room, dry and wet laboratories as well as vibration sensitive test-bedding facilities.

With the research institutes under one roof, the Fusionopolis precinct offers a concentration of diverse capabilities in areas such as materials science and engineering, data storage, microelectronics, manufacturing technology, in addition to high performance computing and infocomm research.

Fusionopolis' proximity to Biopolis, which is located just 600 metres away, also paves the way for the two research hubs to meld their R&D efforts, to create exciting new outcomes that combine the physical sciences, engineering and the biomedical sciences.

NEW CHALLENGES AND OPPORTUNITIES

In May 2009, the Economic Strategies Committee (ESC) was formed to develop strategies for Singapore to build capabilities and maximise opportunities in order to achieve sustained and inclusive growth. The ESC noted that "with the shift of markets to Asia in the post-crisis world, it was essential for Singapore to seize the 'window of opportunity to create a strong presence in Asia over the next five to 10 years'".

Innovation was identified as one of three bases for sustaining Singapore's economic growth, together with skills development and productivity. "Singapore is in an excellent position to become a key global R&D hub and Asia's Innovation Capital — the home for private sector R&D activities and innovation, in partnerships and collaborations with world-class public sector R&D institutes, a hub for innovation and enterprise, and a location of choice for commercialisation," said the ESC.

The ESC also recommended that foundational research should continue to be nurtured especially within the universities. At the same time, more resources and effort should be given to translate the knowledge created into impact. Undergirding all these, Singapore should continue to invest in attracting and retaining the best talent both from within and beyond Singapore.

Following ESC recommendations, the Government announced that it would invest S$16.1 billion — or 1 per cent of Singapore's estimated GDP — in R&D from 2011 to 2015, and target to reach a GERD of 3.5 per cent by 2015.

Against the backdrop of the ESC's recommendations, planning got under-way on Research, Innovation and Enterprise (RIE) 2015, to chart the direction of Singapore's R&D efforts for 2011–2015. Under RIE2015, key strategies that were identified included continued investment in basic science and knowl-edge to seed intellectual capital for future innovations, a stronger focus on economic outcomes and greater emphasis on competitive funding to spur innovation.

At the Crossroads

2010 presented a critical juncture in Singapore's aspiration to become an innovation-driven economy as global companies looked to the country as a preferred location to launch into Asian markets, and Asian companies increas-ingly chose to be based here to globalise their products and services.

The A*STAR research institutes were thus a significant enabler in making Singapore an Innovation Capital of Asia. They provided innovative solutions, attracted top talent and were flexible and responsive to various needs of enter-prises such as multinational corporations, globally competitive companies, local companies and high-tech start-ups.

Under the A*STAR Science, Technology & Enterprise Plan (STEP) 2015, the agency aligned its R&D agenda with Singapore's economic priorities and the needs of the industry.

A*STAR continued to work with the wider research community and engage in public-public and public-private collaboration to drive R&D in sup-port of Singapore's key economic clusters and to capture growth through emerging industries. "Enterprise" would be a key thrust in A*STAR's STEP plan with enhanced and concerted efforts to engage and anchor MNCs in the country, seed innovative capacities and gear local enterprises for growth.

THE FUTURE

In January 2014, Nestlé signed a three-year Framework Research Agreement with A*STAR, which involves the participation of all 18 A*STAR research institutes. A similar model of partnership has also been established with other leading companies, for example, Merck, General Electric and The Coca Cola Company.

Going forward, the co-location of public and private sector research, which is a unique feature of Biopolis and Fusionopolis, will continue to be a major strength, enabling scientists from A*STAR and the industry to exchange ideas and explore collaborations, thus facilitating the commercialisation of Singapore's research outputs. It would also enable private sector laboratories to make use of shared infrastructure such as research equipment, core scientific services and conference facilities that support the public sector institutes.

The research ecosystem is set to grow as ties are forged and networks strengthened amongst the community of multinational corporations, small and medium enterprises, start-ups and public research institutes and private sector research operations at one-north. At the same time, the infrastructure will continue to develop and evolve to meet these needs.

From this research oasis, the work that is being carried out will have a significant impact on the wider society and economy. The BMS research institutes, for example, are working closely with clinicians to translate their research into tangible benefits for the clinical community and the masses, while the physical sciences and engineering research institutes are collaborating with private sector partners and other stakeholders to define the future of manufacturing, maritime transport, aviation, infocomm technology and other key sectors of the economy.

The integration between biomedical sciences and physical sciences and engineering will also continue to take on added significance as Singapore — and many other countries around the world — grapple with new, multi-faceted challenges on the horizon such as climate change; the need for renewable clean energy; the need for quality and cost effective healthcare services in the face of a rapidly ageing population; and the need to enhance the liveability of cities, given rapid urbanisation.

Chapter 4

National Research Foundation

Low Teck Seng and the National Research Foundation

About the National Research Foundation

Research and Development (R&D) is a cornerstone of Singapore's national strategy to develop a knowledge-based, innovation-driven economy. In 2006, the National Research Foundation (NRF) was established to set the national direction for R&D by developing policies, plans and strategies for research, innovation and enterprise.

Over the years, NRF has funded strategic initiatives and built up R&D capabilities in our research ecosystem. In partnership with our key stakeholders, it has built up a strong R&D base at the universities, research institutes and hospitals. It has also catalysed stronger linkages between research and industry, which strengthens the innovative capacity of our companies. NRF has also supported the scaling up of our startups so that technologies developed in Singapore can achieve worldwide impact. We attract top scientists to conduct R&D in Singapore, and they now recognise Singapore as a major R&D hub.

In January 2016, under the latest Research, Innovation and Enterprise (RIE) 2020 Plan, the Singapore government committed a further S$19 billion to support Singapore's RIE efforts over the next five years. RIE2020 will focus on translating R&D in key technology domains into solutions that have direct relevance in addressing our national needs, encouraging technology adoption in companies to strengthen competitiveness, while supporting these efforts by growing a strong research and innovation community. There will also be continued focus in funding basic research in the physical sciences, engineering, and mathematical areas that will provide the foundation to our scientific base and research expertise.

NRF's Beginnings: Building Our RIE Capabilities

Identification of Strategic Research Programmes
to Create New Industries and High Growth

On 1 January 2006, the Singapore Government set up NRF as a department under the Prime Minister's Office. An initial funding of S$5 billion, set aside in the National Research Fund, was to be administered by NRF from 2006 to 2010. At that time, the NRF Board identified three strategic areas of research, namely, Biomedical Sciences, Environmental and Water Technologies, and Interactive and Digital Media, as these were areas deemed to be poised for rapid growth in Asia and the world.

The Research, Innovation and Enterprise Council (RIEC), chaired by Prime Minister Lee Hsien Loong and comprising both public and private sector members, held its inaugural meeting in Singapore in 2006. The RIEC was tasked to advise the Singapore Cabinet on national research and innovation policies and strategies to drive Singapore's transformation into a knowledge-based economy, built upon a strong foundation of capabilities in R&D. The RIEC also proposes measures to encourage new initiatives in science and technology and catalyse new areas of economic growth through research, innovation and enterprise. At its inaugural meeting, the RIEC approved programmes under the three strategic areas of research, to be implemented over a five-year period from 2006 to 2010, with an allocated funding of S$1.4 billion.

The Biomedical Sciences programme would focus on deepening basic research, building drug discovery capabilities, and strengthening Singapore's strong foundation in healthcare services delivery. In particular, the translation of highly promising scientific research into clinical application and use would be supported. The Environmental and Water Technologies programme would take a leading role in developing new technological solutions for managing the water life cycle, building on Singapore's strong foundation in water technologies and management. Lastly, the Interactive and Digital Media programme would leverage Singapore's foundation in infocommunication infrastructure to create new innovative niches in the fast-paced sector.

International Collaborations

Establishment of CREATE

The concept for Campus for Research Excellence and Technological Enterprise (CREATE) was approved by RIEC in 2006. An international collaborator established in Singapore, CREATE houses research centres set up by top universities internationally. Its modern laboratory design has also won CREATE a Laboratory of the Year award.

At CREATE, researchers from diverse disciplines and backgrounds work closely together to perform cutting-edge research in strategic areas of interest, for translation into practical applications that can lead to positive economic and societal outcomes for Singapore. They carry out research projects in partnership with Singapore universities and research institutes, bringing significant intellectual strength to topics of relevance to Singapore.

The research centres in CREATE focus on four interdisciplinary thematic areas of research: human systems, energy systems, environmental systems and urban systems, unified by an emphasis on studying the interaction across systems and adopting a system of systems approach. Research projects are carried out in partnership with Singapore universities and research institutes, bringing significant intellectual strength to topics of relevance to Singapore.

CREATE has strengthened Singapore's position as a nexus for international R&D collaborations. Since its inception, CREATE has established 15 joint research programmes between our local universities and nine top overseas institutions. As of 2015, CREATE laboratories have collectively produced over 2,350 publications in leading academic journals and worked with more than 100 companies. The research outcomes have also led to 16 spin-off companies.

The research centres in CREATE are:

- Berkeley Education Alliance for Research in Singapore (BEARS).
- Cambridge Centre for Advanced Research in Energy Efficiency in Singapore (CARES).
- Energy and Environmental Sustainability Solutions for Megacities (E2S2).
- Nanomaterials For Energy and Water Management (NEW).
- Cellular and Molecular Mechanisms of Inflammation (NUS-HUJ-CREATE).
- Singapore-ETH Centre for Global Environmental Sustainability (SEC).
- Singapore-MIT Alliance for Research and Technology (SMART).
- Singapore-Peking University Research Centre (SPURc).
- Singapore-Technion Alliance for Research and Technology (START).
- Technische Universität München (TUM) CREATE.

With the overseas institutions co-located at CREATE, Singapore is able to leverage the aggregation of expertise to generate thought leadership in areas of Singapore's national priorities. This would help to shape conversations and future strategies. One such example is in Urban Mobility where SMART FM, SEC FCL and TUM CREATE contributed to a comprehensive set of tools that allow policy makers to generate options for urban mobility. The three programmes have also jointly organised an Urban Mobility Symposium to showcase the cutting-edge research in Urban Mobility, which has attracted interest from international researchers and the local transport agencies and ministry, and contributed to the conversations for Singapore's Urban Transport Vision. There are ongoing plans to hold similar symposiums on Environmental Challenges and Building Energy Efficiency.

Strengthening Research Capabilities in Singapore

NRF is committed to enhance Singapore's research capacity by growing a pool of top research talent and developing the platforms on which they could achieve breakthroughs in their research in Singapore. Scientists and researchers based in publicly-funded institutes of higher learning as well as research centres in Singapore also have access to national research programmes and awards. The following schemes have been established to strengthen research capabilities in Singapore.

- *NRF Fellowship Scheme*

The NRF Fellowship Scheme is an early career award that provides opportunities for outstanding young scientists from all over the world to lead impactful research in Singapore. It is open to PhD holders with research experience in any discipline of science and technology. Each Fellow is provided with a research grant over five years, to support projects that exhibit high likelihood of a research breakthrough. The award enables the recipient to assemble and lead a small research team at a Singapore-based research organisation and can also be used to cover equipment and consumables costs. Examples of research organisations hosting NRF Research Fellows include A*STAR, NUS, NTU, Singapore Management University (SMU), Singapore University of Technology and Design (SUTD), Duke-NUS Medical School (Duke-NUS), and Temasek Life Sciences Laboratory (TLL). Since the inception of the scheme in 2007, NRF has launched 9 calls and awarded 75 NRF Fellows.

- *NRF Investigatorship*

The NRF Investigatorship provides opportunities for established, innovative and active mid-career researchers to pursue ground-breaking, high-risk research in Singapore. It is designed to support a small number of excellent Principal Investigators who have a track record of research achievements that identify them as leaders in their respective field(s) of research. As of 2016, we have 15 NRF Investigators carrying out research in chemistry, physics, engineering, cancer research, etc.

- *Competitive Research Programme*

Funding by NRF through the Competitive Research Programme (CRP) seeks to foster the formation of multi-disciplinary teams to conduct cutting-edge research that are of relevance to Singapore, our economy and/or our society. This allows a coordinated, integrated and sustained way of bringing together complementary research groups in Singapore to conduct high-impact research.

The CRP scheme selects the best use-inspired basic research proposals, on the basis of scientific excellence and relevance to Singapore, through a scientific merit review process. The proposed research programme must, in the first instance, be motivated by an important need or problem to be solved. Since the inception of CRP in 2007, NRF has launched 17 calls and awarded 68 projects.

- *Research Centres of Excellence in our local universities*

Singapore has a comparative advantage in education. We have an internationally well-regarded education system — our students are internationally regarded as being among the top in math and science, while our universities do cutting edge research.

In 2007, NRF and the Ministry of Education in Singapore proposed to establish a small number of Research Centres of Excellence (RCEs) within universities in Singapore to step up our research efforts. The objectives of RCEs are to attract, retain and support world-class academic investigators; catalyse the development of Singapore's universities into research-intensive universities and strengthen their prestige globally; enhance graduate education in our universities and train quality research manpower; engender interest in research among students in Singapore and encourage them to pursue research careers; and create new knowledge in selected areas of focus.

Today, NRF has set up five RCEs. These include the Centre for Quantum Technologies (CQT Singapore) at NUS — the first of Singapore's RCEs, established in 2007 — which focuses on quantum physics and computer science in information processing; the Earth Observatory of Singapore (EOS) at NTU which focuses on earth sciences; the Cancer Science Institute of Singapore (CSI Singapore) at NUS, which develops new approaches to treat cancer; the

Mechanobiology Institute (MBI Singapore) at NUS which focuses on cell and tissue mechanics by studying living systems and diseases; and the Singapore Centre on Environmental Life Sciences Engineering (SCELSE), a joint venture between NTU and NUS.

The RCEs have built strong teams around areas of cutting edge research, and are now internationally regarded as being among the top centres in their respective fields.

Encouraging Innovation and Enterprise in Singapore

National Framework for Innovation and Enterprise

In 2008, RIEC approved a S$350 million National Framework for Innovation and Enterprise (NFIE) to support the government's efforts to make innovation and enterprise pervasive in Singapore. The framework encourages universities and polytechnics to engage actively in innovation and academic entrepreneurship to bring their R&D results from lab to market.

Initiatives under the NFIE included the University Innovation Fund, the Translational R&D Grant, an Innovation Vouchers Scheme for SMEs to procure R&D and other services from IHLs and public research institutes, and an Innovation Policy Centre to carry out studies in innovation for the government and industry. There is also an articulation of a national framework of intellectual property (IP) principles for publicly-funded R&D to support the process of technology transfer, as well as enterprise support structures, such as proof-of-concept studies and technology incubator programmes, that aim to bridge the gap between university research and market needs.

Early stage venture capital (VC) funds, managed by professional VCs, were also catalysed through matching investments from NRF. The funds were to invest in technology-based startup companies in Singapore, particularly those generated from R&D in Singapore. More details on the key initiatives under NFIE are provided. The following schemes were implemented under NFIE:

* *Translational Research and Development grant for Polytechnics*

In 2009, NRF provided S$25 million to fund translational research and development at the five polytechnics in Singapore. The Translational Research and Development (TRD) grant provides the polytechnics with resources to build on the research results from universities and research institutes, to turn them into products and applications that can be commercialised by industry.

- *University Innovation Fund*

In 2009, NRF also established the University Innovation Fund (UIF) to support academic entrepreneurship efforts in three universities — NTU, NUS and SMU. The $50 million UIF aims to catalyse innovation and facilitate the creation of high tech startup companies to bring R&D results from the lab to the market. Supported activities related to innovation and enterprise include establishing entrepreneurship education programmes, setting up technology incubators, and having experienced entrepreneurs-in-residence on campus to interact with students and faculty.

- *Early Stage Venture Fund*

The Early Stage Venture Fund (ESVF) scheme was launched in 2008. ESVF seeks to catalyse venture capital (VC) for the hi-tech startup sector. NRF invests on a 1:1 matching basis, for up to S$10 million per fund, to seed the formation of VC funds managed by professional VCs that invest in early-stage Singapore-based high-tech startups. Three instalments of ESVF had been announced as of May 2016.

In the first round of ESVF in 2008, NRF seeded five VC funds with a total commitment of S$50 million, to co-invest in early-stage Singapore-based high-tech startups on a dollar-to-dollar matching basis. The first batch of five VC funds selected as ESVF partners were Bioveda Capital, Extream Ventures, New Asia Investments, Raffles Venture Partners and Walden International. To-date, these five VC funds have collectively invested S$42 million in 28 startups from various technology sectors.

In 2014, NRF selected five more VC funds in the second tranche of ESVF. The five were: Golden Gate Ventures, Jungle Ventures, Monk's Hill Ventures, Tembusu ICT Fund I 2, and Walden International. NRF committed another S$50 million in total, and capitalised another S$100 million for early-stage investment in Singapore-based high-tech startups.

In 2016, NRF announced ESVFIII, which seeks to catalyse the growth of early-stage Singapore-based high-tech startups through government co-investments in corporate venture funds set up by Large Local Enterprises (LLEs). This facilitates the deepening of our technology ecosystem, where smaller companies can grow around a core of larger companies, contributing to value creation and growth of the industry cluster. It serves as a way for LLEs to renew their technology base through the innovative technology startups they invest in. In return, investee startups can leverage the resources of its large corporate partner to more easily overcome barriers to commercialisation, and access networks and expertise to scale-up and go global. In this third round of investment, NRF again invested S$10 million in each fund on a matching basis. The four LLEs selected are: CapitaLand Limited, DeClout Limited, Wilmar International Limited, and YCH Group Pte Ltd.

- *Proof-of-Concept Grant Scheme*

In 2008, NRF launched its first proof-of-concept (POC) grants known as the POC scheme. POC demonstrates that a new process or technique is feasible for use in commercial applications. The POC scheme provides funding to researchers from public hospitals and institutes of higher learning to enable them to carry out further research on their inventions or ideas. The resulting product or application could then be licensed to interested companies or be marketed by a new startup company. Up to $250,000 is awarded to each project.

- *Technology Incubation Scheme*

The Technology Incubation Scheme (TIS) is an initiative announced in 2008 for technology incubators willing to invest in early-stage high-tech startup companies and nurture their growth in Singapore. The $50 million fund under TIS invites technology incubator investors, local or foreign to set up in Singapore. The scheme is modelled after the successful Technological Incubator Programme of Israel.

NRF's Continued Mission: Ramping up Efforts in RIE

Building on Singapore's earlier efforts and successes in R&D, S$16.1 billion was allocated under the Research, Innovation and Enterprise (RIE) 2015 Plan. This was a substantial increase in funding compared with previous science and technology plans, demonstrating Singapore's strong commitment to both basic and mission-oriented research in our public sector research institutions. In the implementation of the RIE2015 Plan, NRF continued efforts to support Singapore's long-term vision to be a research-intensive, innovative and entrepreneurial economy.

NRF directed our efforts in the RIE 2015 plan towards three focus areas: growing future capabilities by establishing a strong science base, supporting future growth and ensuring economic relevance of Singapore's research, and meeting future challenges by addressing national needs.

Establishing A Strong Science Base

Through initiatives that develop a strong science and talent base, NRF helped to build future science and research capabilities that are of strategic importance to Singapore. The key RIE 2015 initiatives to establish a strong science base in Singapore include:

- *Medium-Sized Research Centre*

The Medium-Sized Research Centre funding scheme was launched in 2014 to consolidate research activities across departments, faculties and universities to create a critical mass of leading researchers in areas of research that are of strategic importance for Singapore. These centres have helped to spin out innovative technologies that are relevant to the needs of our industries and create value for the economy. Two research centres have been set up under this scheme — the Centre for Advanced 2D Materials (CA2DM) and Singapore Centre for 3D Printing (SC3DP):

CA2DM seeks to become the world-leading research centre in the area of two-dimensional material science and technology. The centre is led by Prof Antonio Castro Neto from the National University of Singapore's Department of Physics, a leading theorist and one of the most cited researchers in the field of graphene research. Expertise in the synthesis and applications of two-dimensional materials could have direct impact on multiple industry sectors including aerospace, biomedical, data storage and energy.

SC3DP seeks to be a global 3D printing centre of excellence with a focus on manufacturing of high quality components, parts and products via research, education and applications. It will deepen capabilities in the area of future manufacturing to explore novel processes in the aerospace, building and marine industries. SC3DP is led by Prof Chua Chee Kai from the Nanyang Technological University's School of Mechanical & Aerospace Engineering, one of the most cited researchers globally in the field of 3D printing.

- *Returning Singaporean Scientists Scheme*

The Returning Singaporean Scientists Scheme attracts top overseas-based Singaporean scientists back home to lead research in areas important to the growth of Singapore. This grows our local pool of scientists and researchers, strengthens the Singaporean R&D talent core, and further builds R&D capabilities in Singapore. To date, four scientists have returned to continue their research in Singapore. They are:

- Prof Ho Teck Hua, an award-winning behavioural scientist, who returned to take up the position of Deputy President (Research & Technology) at NUS.

- Prof Andrew Lim, who returned to Singapore to head the Department of Industrial and Systems Engineering at NUS' Faculty of Engineering.
- Prof Chua Nam Hai, a world-renowned biotechnology expert, who returned to Singapore to focus on his research at the Temasek Life Sciences Laboratory (TLL). He is also deputy chairman of TLL's management board and chairman of its research strategy committee.
- Prof Aaron Thean, a well-respected opinion leader in the microelectronics community at IMEC, Belgium, who returned to lead a center at NUS with a focus on translating basic device research and creating industry collaboration.
- Prof Peh Li-Shiuan, an expert in low-power interconnection networks, on-chip network and parallel computer architectures, will return to lead Systems research at the NUS Computer Science department.

Human Frontier Science Program

The Human Frontier Science Program (HFSP) is an international program of research support for life sciences. Research is funded at all levels, from individual researchers to program grants. Singapore became a Management Supporting Partner (MSP) of HFSP in 2014. One of the criteria to be an approved MSP is the ability to demonstrate scientific excellence in life sciences.

As an MSP, Singapore researchers can compete for funding support under the various HFSP funding programmes. These include research grants for ground-breaking collaborations entailing extensive collaboration among teams of scientists working in different countries and in different disciplines; and postdoctoral fellowships awarded to scientists who wish to work on foreign laboratories. HFSP awardees from other countries are also able to work in labs in Singapore.

Since the first HFSP grants and fellowships were awarded in 1990, 26 grant awardees have gone on to win Nobel Prizes in the fields of Physiology or Medicine, Chemistry and Physics. Singapore's participation as a MSP strengthens the opportunity for local scientists and researchers to be part of a leading international community in life sciences.

Ensuring Economic Relevance of Singapore's Research

To create value from technology advancements and innovation, NRF sought to support research that is of economic relevance to Singapore as part of the RIE 2015 plan, hence ensuring our future economic growth. This is achieved through stronger public-private R&D collaboration and the translation of good research into solutions that have direct relevance to industry needs or in addressing national challenges.

- *Corporate Laboratory@University Scheme*

The Corporate Laboratory@University Scheme was launched by NRF in March 2013 to support the establishment of key laboratories by industries in our universities. It seeks to strengthen Singapore's innovation system by encouraging public-private R&D collaboration between universities and companies. It ensures ideas and innovations from our universities and research institutes are translated into products, services and businesses that benefit industries and Singaporeans.

The scheme also creates employment opportunities for Singaporeans in technology-driven industries and trains a pool of industry-ready research manpower who can push the technology boundaries for these industries. Currently, nine corporate laboratories have been launched in a wide range of areas: oil and gas exploration, advanced robotics and autonomous systems, energy and water, cyber security and urban rail systems, etc.

- *Research Consortia*

NRF facilitated the establishment of two consortia in the areas of spintronics and photonics to strengthen collaborative research partnerships between Institutes of Higher Learning and industry. They act as a platform to encourage greater interaction amongst academia and industry to realise innovation from the knowledge generated by our research efforts, which in turn attracts more companies to conduct related research activities in Singapore.

In December 2014, the Singapore Spintronics Consortium (SG-SPIN) was established with founding members that included NUS, NTU, Applied Materials, Inc., Delta Electronics and Globalfoundries. Spintronics is an emerging technology that utilises the intrinsic spin of electrons and its associated magnetic moment, in addition to their electronic charge that is exclusively used in existing electronic devices. This area of research can potentially lead to more energy-efficient, larger capacity and faster devices compared to current technology.

In September 2015, NRF announced the launch of the LUX Photonics Consortium. Its founding members included NUS, NTU, Agency for Science, Technology and Research (A*STAR), and seven industry members Technolite (Singapore) Pte Ltd, Denselight Semiconductors Pte Ltd, II-VI Singapore Pte Ltd, Finisar Singapore Pte Ltd, STELOP Pte Ltd, DSO National Laboratories and Coherent Singapore. Photonics research includes developing the next generation ultra-fast Internet through new fibre-optic cables or ground-breaking electronic circuits powered by light instead of electricity. Membership of the consortium has since grown and the consortium is also expanding its linkages with the international community, e.g. European Photonics Industry Consortium (EPIC).

- *Innovation Cluster Programme*

The Innovation Cluster Programme was announced in October 2013 to encourage technology organisations and economic agencies to work with industry to form innovation clusters. This strengthens partnerships across companies, universities, research institutes and government agencies to bring technology ideas quickly to market, raise productivity, create jobs and grow the sector.

It is a key platform for Singapore-based enterprises, including startups and small-and-medium-sized enterprises, to translate and capture the value of research from our publicly-funded research performers. Enterprises can also use the intellectual property to build up their innovation capacity, shorten the time needed to market new products and create new growth opportunities. Four innovation clusters were identified in the areas of diagnostics, additive manufacturing, speech and language technologies and membranes, of which two have been launched:

- <u>Diagnostics</u>: The Diagnostics Development Hub was launched in November 2014 and is led by Exploit Technologies Pte Ltd. It leverages Singapore's strengths and leading clinicians/medical consortiums in areas such as oncology, ophthalmology, infectious and cardiac diseases, as well as ready access to Asian patient samples, to develop diagnostic solutions tailored to diseases predominantly found in Asia.
- <u>Additive Manufacturing</u>: The National Additive Manufacturing Innovation Cluster (NAMIC) was launched in September 2015, led by NTU in partnership with NUS and the SUTD. NAMIC will focus on sectors where Singapore has developed capabilities with competitive advantages, including aerospace, marine and offshore, precision engineering, medical technology, building and construction, and design.

Meeting Future Challenges by Addressing National Needs

National Innovation Challenge

Under RIE2015, a new National Innovation Challenge (NIC) was launched to harness Singapore's R&D capabilities to tackle large, complex problems facing

cities. The NIC seeks to harness Singapore's multi-disciplinary research capabilities to develop practical, impactful solutions to national challenges in areas such
as energy resilience, environmental sustainability and urban systems. In addition
to improving the lives of Singaporeans, these solutions also carry the potential
of commercial spinoffs both at home and abroad.

- *Energy National Innovation Challenge*

The Energy National Innovation Challenge (ENIC) was established in February
2011 to develop and deploy cost-competitive energy solutions within 20 years to
improve energy efficiency, reduce carbon emissions and broaden Singapore's energy
options. A funding of S$300 million was allocated from 2011 to 2015, covering
areas including Green Buildings, Green Data Centres, Energy Storage, Waste to
Energy and Smart Multi Energy Systems (SMES). With these, we are better-placed
to prioritise energy R&D funding to address existing gaps and strengthen efforts
in areas with greater potential to meet the ENIC's objectives. In particular, SMES
aim to accelerate the translation of energy R&D projects and ideas into deployable
solutions with a systems approach, provide a platform for integrated test-bedding
to optimize energy usage, and build up capabilities in system-level integration.

- *Land and Liveability National Innovation Challenge*

The Land and Liveability National Innovation Challenge (L2 NIC) is a multi-
agency effort led by the Ministry of National Development and NRF. It leverages
research and development to create and optimise Singapore's space capacity for
future sustained growth, while supporting a highly-liveable environment for
future generations. The first phase of L2 NIC takes place from 2013 to 2018 with
a funding of S$135 million.

Two R&D thrusts have been identified to support the Land and Liveability
NIC objectives. The first thrust focuses on creating new space in a cost-effective
way. This includes research into creating useable underground spaces in caverns and areas of poor soil, new land reclamation methods that require less
sand, and developing alternatives to land reclamation. The second R&D thrust
focuses on optimising the use of space while maintaining high liveability.
Potential areas of research include intensification of low-density land uses like
industry and infrastructure; achieving a good quality living environment even
with intensified developments; reducing space demands for infrastructure,
such as transport; co-locating urban infrastructure, utilities and services with
other spaces; and reducing unproductive use of land.

Ten research projects in areas such as sensor-enabled homes, personalised
care for seniors, and mega underground caverns in Singapore were awarded

under L2 NIC's first call for proposals in 2013. Another five research projects to improve the cost-effectiveness of underground developments, as well as human comfort and well-being were awarded in May 2016 under the second call for proposals.

- *National Innovation Challenge on Active and Confident Ageing*

The National Innovation Challenge on Active and Confident Ageing, spearheaded by the Ministry of Health, seeks to catalyse innovative ideas and research in Singapore that can transform the experience of ageing in Singapore. It comprises three key research thrusts: lengthening health spans; productive longevity; and ageing in place.

The NIC on Active and Confident Ageing has opened three calls for proposals. The Care-At-Home Innovation Grant, launched in 2015, encourages greater innovation and adoption of technology by providers to serve home care clients in more productive and cost-effective ways. The Grant Call on Cognition, which concluded in early 2016, focused on lengthening health span by improving cognitive functioning and delaying the onset of dementia in older adults. The Ageless Workplaces Innovation Grant, which will conclude in 2016, aimed to improve the productivity and health of older workers, as well as enable older workers to be employable beyond 65 years of age.

Strategic Research Programmes

More Strategic Research Programmes (SRPs) were identified to support investments in areas of research to create new industries and enable high growth:

- *Biomedical Sciences Translational & Clinical Research*

Singapore's Biomedical Sciences Translational & Clinical Research programme focuses on bench-to-bedside translation of basic biomedical discoveries into better medicines and therapies for improved healthcare. Multi-institutional flagship programmes have been established in gastric cancer, eye disease, infectious diseases, metabolic diseases and schizophrenia.

The strategic competitive grant programmes for healthcare are driven by the National Medical Research Council, Ministry of Health and Biomedical Research Council, as well as the Agency for Science, Technology and Research (A*STAR).

- *Environment & Water Technologies*

In 2006, the Environment & Water Industry Programme Office (EWI) was set up as a whole-of-government effort to coordinate the Environment & Water

Technology Strategic Research Programme, to grow Singapore into a 'Global Hydrohub' for leading edge technologies in the water industry. With an initial allocated funding of S$470 million, EWI accelerates the formation and growth of startups through financial incentives and mentoring. Funding opportunities are made available for both basic and applied R&D projects as well as programmes for expediting commercialisation of R&D outcomes. In 2011, an additional S$140 million was allocated to EWI under RIE2015. Today, Singapore has attracted water companies from across the world to be based in Singapore.

- *Energy*

The S$365 million Clean Energy SRP was established in 2007 with an emphasis on clean energy, and later renamed the Energy SRP in 2013 to cover a wider scope of energy technologies. The Energy SRP has established research centres in key energy technology areas, including the Solar Energy Research Institute of Singapore (SERIS) and Energy Research Institute at NTU (ERI@N).

SERIS has been playing an instrumental role to spur solar adoption and catalyse the growth of the solar industry in Singapore. Leveraging its R&D capabilities in industrial silicon wafer solar cell fabrication, SERIS has collaborated with leading solar companies such as REC Solar and Trina Solar to develop new products and processes with higher efficiencies and lower costs.

ERI@N has a comprehensive range of capabilities across key energy research domains. ERI@N has been successful in forming research partnerships with companies such as Bosch, BMW, Rolls-Royce and Johnson Matthey, many of which have gone on to set up R&D centres in Singapore. ERI@N is also spearheading two large-scale, integrated piloting and demonstration programmes to support the translation and commercialisation of new technologies. These are the EcoCampus project, which aims to meet NTU's vision to become the greenest university campus in the world, and the Renewable Energy Integration Demonstrator — Singapore (REIDS) project, a micro-grid test-bed on Pulau Semakau landfill.

- *Interactive & Digital Media*

Overseeing the efforts in the Interactive Digital Media area is the multi-agency IDM Programme Office (IDMPO) hosted by the Media Development Authority which coordinates efforts among agencies such as the Agency for Science, Technology and Research (A*STAR), Defence Science and Technology Agency (DSTA), Economic Development Board (EDB), Infocomm Development Authority (IDA), International Enterprise (IE) Singapore, Ministry of Communications and Information (MCI) and Ministry of Trade and Industry (MTI).

With the support from NRF, IDMPO puts in place funding initiatives that interlock the R&D efforts to support Institutes of Higher Learning (IHLs), startups and Industry:

- i.ROCK (IDM Research Oriented Centres of Knowledge): Building world-class R&D capacity in local IHLs by partnering with the best research institutes in the world.
- i.JAM (IDM Jumpstart And Mentor): Fuelling grassroots innovation and entrepreneurship.
- IDM in Education: Harnessin IDM, Transforming Learners.

Through the investments of IDMPO, Singapore has been playing an important and increasingly visible role in the evolution of this next generation of interactive digital media.

- *Marine & Offshore*

The Marine & Offshore initiative seeks to establish Singapore as an integrated global marine and offshore hub. Key R&D programmes run by A*STAR include the Offshore Technology Research Programme, Maritime Port Authority (MPA) — Institute of High Performance Computing (IHPC) Maritime Research Programme, and Joint MPA-IDA Infocomm@Seaport Programme. A*STAR and EDB, working closely with NRF, drive the development of R&D capabilities and infrastructure, and the training of talent in support of the industry.

- *Satellite & Space*

This initiative aims to grow Singapore's R&D capabilities, forge internal and external collaborations, and develop talent for the space industry. Research focuses on satellite technology, satellite remote sensing applications, satellite communication services, satellite components, satellite integration, and satellite-based services. The Office for Space Technology and Industry is the designated office to develop the industry. It is driven by EDB, with support from MTI, MINDEF, MFA, MOE, NRF, A*STAR and satellite-based companies.

Stimulating Public Interest in Research, Innovation and Enterprise

In addition to the three focus areas above, there is a need to stimulate public interest in RIE to grow a strong local core of researchers that will bring Singapore forward in the next fifty years of science and technology. The National Science Experiment and NRF's two flagship events, Global Young Scientists Summit and Techventure, aim to engage the public, researcher and innovators to join our RIE efforts.

- *National Science Experiment*

The National Science Experiment organised by NRF and the Ministry of Education in Singapore is an islandwide outdoor science experiment to excite and interest young Singaporeans in science and technology. The first phase of the three-year experiment was conducted in 2015 and it involved over 43,000 students from primary and secondary schools and junior colleges carrying a specially-designed sensing device to collect mobility data as well as environmental data such as temperature, humidity, atmospheric pressure, light intensity and sound pressure levels. The data was transferred wirelessly to an online portal where students could view their own results as well as the aggregated data of fellow participants.

From this experiment, students learnt about big data and how it can be applied to manage real world issues. The data also offered meaningful insights for urban planners in terms of environment maps, Wi-Fi coverage and the participants' travel patterns.

- *Global Young Scientists Summit*

The inaugural Global Young Scientists Summit@one-north (GYSS) was held in 2013. Themed "Advancing Science, Creating Technologies for a Better World", GYSS is a gathering of young scientists and researchers (PhD students and post-docs) from all over the world, with eminent international science and technology leaders in Singapore. Since the inaugural year, close to three hundred bright young researchers gather in Singapore every year to be inspired and mentored by eminent scientists and technologists. Speakers invited to the GYSS are globally recognised scientific and technology leaders, and who are recipients of the Nobel Prize, Fields Medal, Millennium Technology Prize, Turing Award and IEEE Medal of Honour.

- *Techventure*

Singapore has since built a reputation as a startup nation, and Techventure is a key event that showcases Singapore's entrepreneurial and technology innovation eco-system. The event promotes Singapore startups and research projects to the global investment community and generates international mindshare of Singapore as an entrepreneurial hub where high quality startups and investors congregate to build a vibrant ecosystem.

Outcomes: Yielding Returns From Our R&D Efforts

Highly Ranked Research-Intensive Universities in Singapore

Today, Singapore has a strong scientific base. Singapore's research quality has improved over time, now ranking well above the world average. The number of PhDs being trained locally continued to increase from 7,522 in 2011 to 7,850 in 2015. The stock of Research Scientists and Engineers (RSEs) in the workforce has also experienced sustained growth.

Singapore's universities have steadily risen up in global rankings and improved their research influence internationally. In 2015, the annual World University Rankings placed the National University of Singapore (NUS) and the Nanyang Technological University (NTU) in the 12th and 13th positions respectively,[2] up from 22nd and 39th the previous year. From 2006 to 2015, NTU's field-weighted citation impact (FWCI) increased by 42%, while NUS' increased by 19%. The FWCI of NUS and NTU in 2014 were higher than other top Asian universities such as the University of Hong Kong, the University of Tokyo, and the Peking University.[3]

The growth of Singapore's universities as top research institutions is due to our focus on excellence in research and education, and our strong research

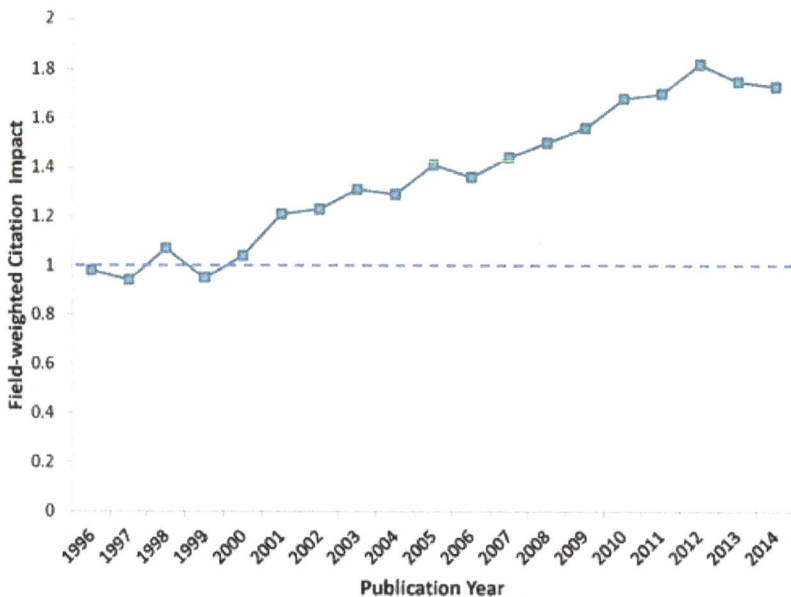

Relative Quality of Research in Singapore.[1]

infrastructure. This has enabled our universities to build up a strong faculty of world-class researchers.

Spurring World-Class Research

Our R&D investments have reaped outstanding results in the areas of social, health and economic benefits to Singapore. Some key outcomes are:

- *Growing a Thriving Water Ecosystem*

Facing challenges of water scarcity, Singapore invested in water research to help the country achieve water sustainability and security. Supported by strong R&D capabilities, Singapore now has a thriving water industry cluster comprising 180 water companies and 26 private research centres establishing business activities in Singapore. Key global companies include Memstar's membrane manufacturing expansion and headquarters, Mann+Hummel's Asia Pacific HQ, and Meidensha, a Japanese ceramic membrane specialist. Local companies are also established players in the global water markets. Hyflux, for instance, has built one of the world's largest membrane-based seawater desalination plant in Algeria. Within the thriving business environment, 1,460 jobs were created for the water industry. In overcoming our own challenges, we have transformed a national vulnerability into a strategic advantage.

Singapore has also grown to be an ideal test-bed for environmental and water technologies. One example is De.mem Pte Ltd, a Singapore-based company that designs, builds and operates decentralized water treatment facilities for the Southeast Asian market. The company leverages membrane integrity sensor technologies developed by the Nanyang Technological University's Nanyang Environment and Water Research Institute for early detection of issues affecting a water plant's performance. The company successfully launched its first overseas water treatment plant in Vietnam in 2013, and received new investments to build more membrane-based water treatment plants there.

Singapore will continue to draw our water supply from four water sources. These are: water from local catchments, imported water, NEWater, and desalinated water. However, desalination and production of NEWater require energy. Our aim is to continue investing in technology and R&D to reduce the energy consumption and to improve the long-term sustainability of Singapore's water resources.

- *Revolutionising Research in "Two-Dimensional" Materials*

In 2010, NUS set up the Graphene Research Centre (GRC), which boasts state-of-the-art fabrication facilities and a faculty comprising several world-renowned researchers. The Centre was created for the conception, characterization, theoretical modelling, and development of transformative technologies based on two-dimensional crystals, such as graphene. It aims to be a world leader in innovative and emergent materials science, with strong ties to the industry and academia. The Centre was immediately one of the best equipped and advanced graphene research centres in the world.

In 2014, NRF awarded NUS with funding to support the expansion of the GRC into a new Centre for Advanced 2D Materials (CA2DM). This move builds on the success of the GRC and expands research into other 2D materials beyond graphene, such as phosphorene and molybdenum disulfide. CA2DM started with about 50 researchers from multiple disciplines such as biomedicine and engineering.

Within a brief span of five years, CA2DM and its predecessor, the Graphene Research Center, reached a significant milestone in academic research. The Centre achieved an H-index of 50, which signified that 50 of its papers have at least 50 citations by peer publications, a reflection of the centre's work being acknowledged by global peers. Papers in two specific areas of research stand out amongst the 50 in being classified by the Web of Science as both Hot Papers and Highly Cited Papers. The two areas are black phosphorus and Weyl semimetals.

- *Advancing Biomedical Sciences*

Gastric cancer is a leading cause of global cancer mortality accounting for 700,000 deaths worldwide annually. It is particularly common in East Asia countries such as China, Japan and Korea, including Singapore where it is the fifth most common cancer in males who have a 1 in 50 lifetime risk of developing the disease. Gastric cancer is curable if detected early. Singapore needs to do research on gastric cancer ourselves because it is less prevalent in the West, and thus not much research on it is done there.

The Singapore Gastric Cancer Consortium (SGCC), supported by the Translational & Clinical Research (TCR) Flagship Programme in Gastric Cancer under NRF's Biomedical Sciences (BMS) TCR Strategic Research Programme (SRP), developed the world's first clinically proven real time *in vivo* molecular diagnostic system that enables early detection and intervention of the disease. Together with a customised online-software control system to analyse the biomolecular information for a quicker and more accurate diagnosis, the system can provide objective diagnosis of cancer almost instantly. This is unlike conventional endoscopic techniques and pathologist analysis which require several days. This system has been used on more than 500 cancer patients in Singapore, including stomach, colon and cervical cancers and has 40 peer-reviewed publications and four US and four UK patents.

Nurturing a Robust Startup Culture, Raising the Innovation Capacity of Our Companies

We have developed a vibrant startup ecosystem in Singapore and are seeing encouraging outcomes from our investments in research, innovation and enterprise. Today, Singapore ranks 10th in the world (and first in Asia) for best startup nations in The Global Startup Ecosystem Report 2015.

NRF has helped to catalyse the ecosystem with seed funding and accelerator programmes, innovation grants and venture capital schemes such as the Interactive Digital Media (IDM) Jump-start and Mentor (i.JAM) programme, Technology Incubation Scheme, and Early Stage Venture Fund. About 50 startups have since exited from our funding schemes. In the information and communications technology sector, we have had successful startup exits and the first unicorns born in Singapore, Garena and Razer Edge.

Garena, an online game-cum-gaming platform, is the first Singapore technology startup valued at US$1 billion in World Startup Report 2014. Razer Edge, a high-definition, full-PC, multi-touch display gaming console, attained eight awards at the 2013 International Consumer Electronics Show. It closed a round of funding with Intel Capital in 2014 and is now valued at US$1 billion.

Looking Ahead: Transforming Singapore Through Research, Innovation and Enterprise

The Singapore Government is committed to provide a conducive environment where research leads to innovations that are translated into technologies that are impactful, useful to industry, relevant to society, and address national needs. This intent will be realised with these key initiatives:

- *RIE2020 Plan*

In the next five years (2016 to 2020), under the sixth science and technology plan for Singapore — the Research, Innovation and Enterprise (RIE) 2020 Plan — the government has committed S$19 billion to invest in research, innovation and enterprise, to take Singapore to the next stage of development.

Given our current stage of development and the strong base of research capabilities that have been built up over the years, a key focus in RIE2020 will be to ensure that there is even stronger economic outcome from our R&D investments. NRF will do more to connect researchers with industry to create value by turning technologies into products, services and solutions. We will also help Singapore and Singapore-based enterprises to build up their innovation and technology-absorptive capacity, to support their growth in the changing economy and also meet our national needs. To encourage innovation and enterprise, we will continue with work to connect startups with VCs and accelerators, and in the process help our companies grow and generate more good jobs.

To maximise impact, funding will be prioritised in four strategic technology domains where Singapore has competitive advantages and/or important national needs. These are:

- Advanced Manufacturing and Engineering.
- Health and Biomedical Sciences.
- Urban Solutions and Sustainability.
- Services and Digital Economy.

Activities in the four strategic technology domains will be supported by three cross-cutting programmes to ensure excellent science, a strong pipeline of skilled manpower, and value creation. These are:

- Academic Research.
- Manpower.
- Innovation and Enterprise (I&E).

In the startup space, NRF will continue to strengthen connection between startups with VCs and accelerators, and also facilitate strategic partnerships between startups and larger enterprises through corporate venture schemes. In the long run, we aim to create a self-sustaining ecosystem where innovators, companies, investors and government collaborate to benefit from each other and capture value as a whole for society. Importantly, resources such as manpower and knowledge must be cycled back into the ecosystem to create sustainability.

- *SG-Innovate*

To help innovative startups commercialise, SG-Innovate was launched in April 2016. SG-Innovate will provide a focal point to connect entrepreneurs with

Technology Domains

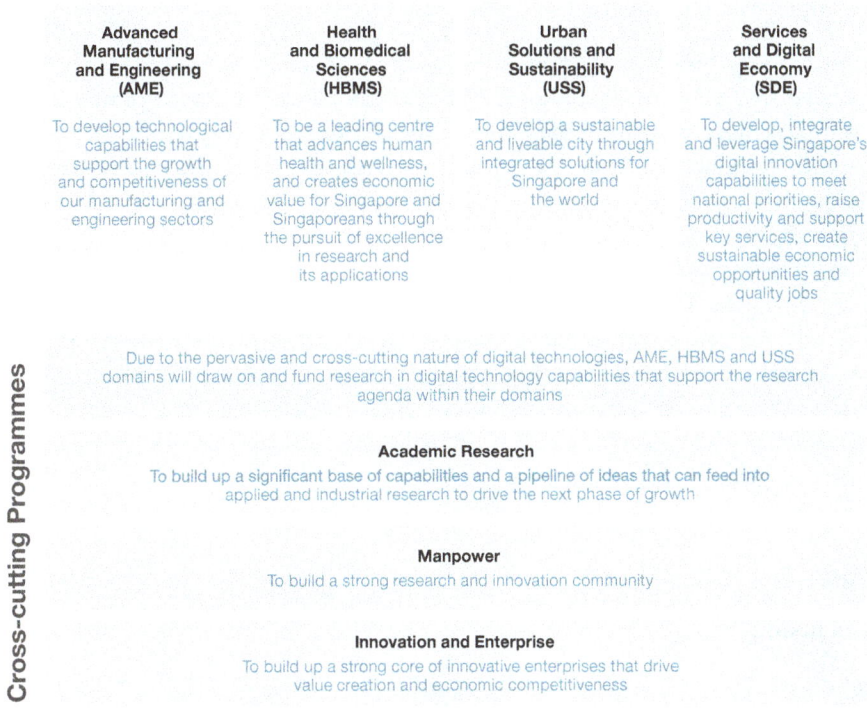

Advanced Manufacturing and Engineering (AME)	Health and Biomedical Sciences (HBMS)	Urban Solutions and Sustainability (USS)	Services and Digital Economy (SDE)
To develop technological capabilities that support the growth and competitiveness of our manufacturing and engineering sectors	To be a leading centre that advances human health and wellness, and creates economic value for Singapore and Singaporeans through the pursuit of excellence in research and its applications	To develop a sustainable and liveable city through integrated solutions for Singapore and the world	To develop, integrate and leverage Singapore's digital innovation capabilities to meet national priorities, raise productivity and support key services, create sustainable economic opportunities and quality jobs

Due to the pervasive and cross-cutting nature of digital technologies, AME, HBMS and USS domains will draw on and fund research in digital technology capabilities that support the research agenda within their domains

Academic Research
To build up a significant base of capabilities and a pipeline of ideas that can feed into applied and industrial research to drive the next phase of growth

Manpower
To build a strong research and innovation community

Innovation and Enterprise
To build up a strong core of innovative enterprises that drive value creation and economic competitiveness

Cross-cutting Programmes

RIE2020 Framework.

venture capitalists and industry mentors. Creating our own success stories is key, hence, training and anchoring a local core of experienced innovators who continually lead the process of innovation and translation of technologies into products, businesses and value is important.

- *Smart Nation*

NRF will also value-add to Singapore's transformation to become a Smart Nation. For example, NRF is developing Virtual Singapore, a dynamic 3D city model and collaborative digital data platform with 3D maps of Singapore, to be opened for use by the public, private, people and research sectors. With it, users can derive insights, develop solutions and run simulations using realistic large-scale scenarios of a Virtual Singapore. Virtual Singapore will support programmes and activities that transform Singapore into a Smart Nation.

About Virtual Singapore

Virtual Singapore is a dynamic three-dimensional (3D) city model and collaborative data platform, including the 3D maps of Singapore. When completed, Virtual Singapore will be the authoritative 3D digital platform intended for use by the public, private, people and research sectors. It will enable users from different sectors to develop sophisticated tools and applications for test-bedding concepts and services, planning and decision-making, and research on technologies to solve emerging and complex challenges for Singapore.

This project is championed by NRF, the Singapore Land Authority (SLA) and the Infocomm Development Authority (IDA). NRF will be leading the project development, whilst SLA will support with its 3D topographical mapping data and become the operator and owner when Virtual Singapore is completed. IDA will provide expertise in information and communications technology and its management as required in the project.

Conclusion

R&D has been a key strategy for Singapore since we started our drive towards becoming a knowledge-based economy and society. This has helped enable continued GDP growth in recent years through anchoring of new MNC investments, and catalysing the growth of large local enterprises (LLEs) and small medium enterprises (SMEs), so as to create good jobs for Singaporeans, and position our economy for the future.

Singapore' continued commitment to research, innovation and enterprise is augmented by our efforts to translate research to use, build up the innovation capacity of our companies to drive economic growth, and leverage science and technology to address our national challenges. Through long-term planning and effective implementation, our investments in research, innovation and enterprise will secure our future. These will contribute significantly to our economy, and create more good jobs and opportunities for Singaporeans; improve healthcare for our population, especially our seniors; and transform our urban landscape for greater liveability and sustainability.

NRF will continue its commitment to transform Singapore into a vibrant R&D hub — one that contributes towards a knowledge-intensive, innovative and entrepreneurial economy and society; and makes Singapore a magnet for excellence in science and technology.

References

[1] Field-weighted citation impact (FWCI) tracks how the number of citations received by Singapore's publications compares with the global average (represented by a FWCI of 1.00). For example, Singapore's FWCI of 1.82 in 2012 means that Singapore's publications received 82% more citations than the world average. Data source from Elsevier SciVal.

[2] The ranking is by London-based education consultancy Quacquarelli Symonds and is based on normalised-weighted research citations.

[3] NTU's and NUS' FWCI were 1.9 and 1.7 respectively in 2015, while the University of Hong Kong, the University of Tokyo and the Peking University were scored at 1.7, 1.3, and 1.4 respectively.

Chapter 5

The Faculty of Science, National University of Singapore — A Brief History

B T G Tan

Raffles College

The Faculty of Science is one of the oldest faculties in the National University of Singapore (NUS), and dates back to the formation of Raffles College in 1928/1929. The College had just two Departments, Arts and Science, when it was officially opened on 22 July 1929 by Sir Hugh Clifford, Governor of the Straits Settlements.[1]

The Science Department offered courses in three subjects — Mathematics, Physics and Chemistry, and students graduated with a Diploma in Science. At the outbreak of the Second World War from 1942–1945, the College was closed and its buildings used as a medical facility. During the Japanese Occupation, it was used as the headquarters of the Japanese army in Singapore.

The University of Malaya

After the war, Raffles College and King Edward VII College of Medicine were merged to form the University of Malaya on 8 October 1949.[2] This marked the establishment of the Faculty of Science, whose first Dean was Norman S Alexander, a physicist from New Zealand who later became involved in the establishment of many Commonwealth Universities.[3] At the establishment of the Faculty, the three founding disciplines of Raffles College, Mathematics, Physics and Chemistry, became founding departments of the Faculty of Science.

There had been a Department of Biology in the Medical College, but in 1949 with the founding of the University of Malaya, it was converted into the Department of Parasitology.[4] Two new biological departments — Botany and Zoology — were formed in 1949 to bring the number of founding departments of the Faculty of Science to five. The Department of Pharmacy which had been founded in 1935 as a department in the Faculty of Medicine and had then become an independent School joined the Faculty of Science in 1974.

Norman Alexander was succeeded in 1951 by a chemist, Robert A Robinson who served as Dean till 1955. Robinson was succeeded by John C Cooke who served as Dean for just one year, and was then succeeded in 1956 by a biologist, R D Purchon. Another biologist, H B Gilliland, became Dean in 1960 and served in this position until 1965, the year of Singapore's independence. Political developments since 1949 had resulted in the University of Malaya splitting into two divisions in 1959, one in Singapore and the other in Kuala Lumpur. The Singapore division subsequently became the University of Singapore in 1962.

The University of Singapore

The first Singaporean Dean of Science in the University of Singapore was Hon Yung Sen, a physicist who had studied at Raffles College before the war. He was succeeded in 1968 by Kiang Ai Kim, a noted chemist who was also a product of Raffles College. Another chemist, Ang Kok Peng, took over as Dean in 1971 and served till 1973, when Gloria Lim, a botanist, became the first female Dean of Science of the University of Singapore. Ang Kok Peng returned to the Deanship in 1977 and in 1980, when the National University of Singapore (NUS) was formed by a merger of the University of Singapore with Nanyang University, Gloria Lim returned as Dean of Science.

In 1979, I was appointed Vice-Dean of the Faculty serving under Gloria Lim. In those days, the Dean's Office ran on what would be regarded today as a skeleton staff. Apart from the Dean, there was only one Vice-Dean (myself), one secretary/executive officer and one office boy. Later on, we added a couple more secretaries, but there was no one at the administrative officer level until very much later.

Nanyang University

Nanyang University (Nantah) was established as a Chinese language University on 15 March 1956. Nantah's College of Science had academic staff members who were well qualified and often worked with their colleagues in the University of Singapore to advance their academic disciplines. Its Physics Department was long headed by its founding Head, distinguished spectroscopist Choong Shin-Piaw, who was also Dean of Science at Nantah.[5] Later, before the merger with the University of Singapore, the Department was headed by high energy physicist Hsu Loke Soo.

Graph theorist Teh Hoon Heng was the long-standing and well-respected Head of the Mathematics Department at the time of merger, and was also Dean of Science. Nantah's Chemistry Department's founding Head was Prof YS Wang who retired in 1963. The headship then rotated between Koh Lip Lin, Lim Swee Yong, TD Mao and Gan Leong Ming, with Kiang Ai Kim as its Head just before the merger. Anne Johnson was the highly respected Head of its Biology Department. The College of Science was also the birthplace of Nantah's Computer Science Department whose first Head was Hsu Loke Soo, and which eventually became the Department of Information Systems and Computer Science (DISCS) in NUS.

In physics, for example, the faculty members in this discipline from both universities had come together in 1972 to form The Institute of Physics, Singapore, a national academic and professional society for all physicists in Singapore. Hence when the decision was made to bring the two universities together, first in a joint campus in 1979, and then in a full merger in 1980, the scientists from both sides were able to effect the amalgamations of their respective departments relatively smoothly.

The National University of Singapore

At the time of the merger of the two universities to form NUS, the Dean of Science at Nantah was Koh Lip Lin, a chemist who was also a Member of Parliament. The two Universities formed a Joint Campus in 1979, and at the merger in 1980 Gloria Lim stepped down as Dean and Koh Lip Lin became the first Dean of Science of NUS. He asked me to stay on as Vice-Dean and assist him in the complex job of merging the respective science departments of the two universities.

This was without doubt a complex task which was potentially fraught with many difficulties and challenges. The fact that the merger of the two Science Faculties did not run into any major difficulties and was accomplished with relative ease must be credited mainly to Koh Lip Lin's leadership qualities and the calm and dignified way in which he always conducted himself. Already a highly respected figure in Nanyang University, his unruffled demeanour quickly won him the same high level of respect from my University of Singapore colleagues.

Koh Lip Lin's tenure as Dean of Science from 1980 to 1985 established a firm foundation for the newly merged Faculty of Science. The fact that the scientists from both pre-merger universities had already worked together, particularly in establishing their respective national academic and professional scientific bodies, greatly helped in welding them together into the cohesive merged departments. The enlarged departments generally became stronger in both research and teaching, and laid the groundwork for the subsequent rise of NUS as a research university.

Added to Koh Lip Lin's already onerous duties as Dean was his additional burden as a Member of Parliament. He attended to his political duties with the same dedication and seriousness as he devoted to his position as Dean. As his Vice-Dean, I learnt much from his dignified and even temperament, which he maintained even during times of stress and aggravation. He was a person of the highest integrity for whom I and many others always had the highest regard, and those of us who knew him well and worked closely with him regret his recent early passing.

In 1985, I succeeded Koh Lip Lin as Dean of Science. My most important initial duty was to select two Vice-Deans to assist me in the task of developing the Faculty of Science. Tony Tan Keng Yam had been appointed as Vice-Chancellor of the new NUS, and this began a period in which NUS began its rapid development from a mainly teaching university to become one of the world's leading research universities. Having Tony Tan as Vice-Chancellor of the merged university for the first year of its existence was a huge vote of confidence from the Government and augured well for the future of NUS.

Lim Pin succeeded Tony Tan in 1981 and initiated a new era for NUS in which it was to rapidly make its way upwards amongst the ranks of the world's leading universities. Hence it was important that the Faculty of Science should be well equipped to respond to these developments and play its part in the rise of NUS. As the sole Vice-Dean prior to my appointment as Dean, I had had to play the combined roles of Vice-Dean, administrative officer and even systems programmer (when I had to write computer programs to read and analyse data from the departments). Hence I strongly felt that more than one Vice-Dean would be needed for the challenges ahead.

Expansion of the Science Deanery

For my two Vice-Deans, I was fortunate to be able to persuade Chong Chi Tat of the Mathematics Department and Goh Chong Jin of the Botany Department to agree to serve as Vice-Deans. Both of them had particular individual qualities which I admired, and they both performed outstandingly as Vice-Deans and contributed much to the building up of the Faculty. Indeed, throughout my tenure as Dean, I was fortunate in having Vice-Deans who acquitted themselves with distinction. Many of them went on to higher office after serving in the Faculty of Science Deanery, a notable example being Chong Chi Tat who became Deputy Vice-Chancellor and the first Provost of NUS.

The next Vice-Dean who was appointed was Lee Soo Ying from the Chemistry Department who succeeded me as Dean and went on to senior positions in the NUS administration and at NTU. Next came Tan Teck Koon of the Botany Department who is at the time of writing still the Dean of Students, and Lai Choy Heng of the Physics Department who followed Lee Soo Ying as Dean of Science and subsequently became Vice-Provost and Executive Vice-President of Yale-NUS College.

Later on, we established the position of Sub-Dean in the Science Deanery, and the two Sub-Deans we appointed not only performed excellently but also deservedly went on to higher office in the NUS administration. Tan Eng Chye of the Mathematics Department succeeded Lai Choy Heng as Dean of Science and was subsequently appointed Provost of NUS, a position he still holds at the time of writing. Andrew Wee of the Physics Department followed Tan Eng Chye as Dean of Science and is currently NUS's Vice-President (Global and University Relations).

Computer Science and Computational Science

The merger with Nanyang University brought a new academic discipline into the NUS Faculty of Science. The Department of Computer Science had been formed in Nanyang University several years before the merger, and this Department came over into NUS as a fully-fledged department alongside the six existing departments (Mathematics, Physics, Chemistry, Botany, Zoology and Pharmacy). This soon evolved into the NUS Department of Information and Computer Science or DISCS as it became known.

At a time when Information Technology or IT was becoming a major part of the industrial and commercial infrastructure, computer science quickly became a popular major, and DISCS soon expanded into a large Department. The demand for computer and IT-related subjects seemed to be insatiable, and we had to cater for the IT needs of students in the other departments too, by offering courses such as Computer Programming and Applications or CPA which was run by the Physics Department (causing a certain amount of friction between DISCS and the Physics Department).

The availability of powerful and relatively inexpensive computers to researchers led to a rapid development in the use of computational techniques to model and simulate natural phenomena such as physical and chemical processes. In effect, this became a third paradigm of scientific research and investigation alongside the traditional theoretical and experimental paradigms long pursued by researchers. This computational approach demanded that researchers possess an array of mathematical and computing tools and skills.

A new major was thus initiated in the Faculty of Science to prepare graduates who would be equipped with the tools and skills to attack scientific problems using this computational approach. This new major — Computational Science — was virtually a double major, requiring students who were already reading a science major such as physics or chemistry, to read a bunch of subjects in mathematical and physical modelling as well as in computational and simulation techniques.

Two Vice-Deans, Chong Chi Tat and Lai Choy Heng, and a brilliant young physicist, Lim Hock, took the lead in establishing this new Computational Science programme in 1991, and its success led to the formation of a new Computational Science Department in 1996. It is clear now that this was a move well ahead of its time, and the computational techniques which the new Department taught and promoted are now standard and routine research tools made even more effective with the enormous computational power available today as compared to the 1980s. Unfortunately, the momentum we created could ultimately not be maintained and the Department was shut down in 2005.

Materials Science

Parallel with the introduction of Computational Science as a new major in he Faculty, we spent much thought on how the important subject of materials science could be brought into the Faculty as a new subject and possibly as a new major. We initially looked into the introduction of a new Master's course in materials science which could be read by graduates in physics and chemistry. However, we could not justify this Master's course with the projected take up rate, so we went back to the drawing board and decided to introduce an undergraduate major in Materials Science.

This new programme was led by Ng Ser Choon of the Physics Department and Gan Leong Ming of the Chemistry Department, and was successfully launched in 1991. The popularity of this new major meant that it became a separate department in 1996, with Ng Ser Choon as its first Head. The new Department spearheaded materials science both as an academic discipline sought after by students who wanted to enter this increasingly important technological field, as well as a rapidly expanding research area. In 2005, the Department was transferred to the Engineering Faculty where it remains today.

The SRP, USRP and the Special Programme in Science

Lim Pin's tenure as Vice-Chancellor saw a rapid rise in the funding available for research as well as a steady increase in the student and staff strength of the University. With increased funding support, the Faculty of Science was not only able to greatly increase its number of peer-reviewed publications, but also the overall quality of these publications as measured against internationally accepted benchmarks. As for student numbers, it became a severe challenge to maintain the overall quality of students admitted to the Faculty with the large increase in student numbers.

In 1989, the Faculty started working with the Gifted Education Programme of the Ministry of Education on a new programme for students at the Junior College level.[6] This programme allowed talented junior college students to work on small research projects with mentors who were academic staff from the faculty, as well as with other researchers from the rest of the University. The programme became known as the Science Research Programme or SRP, and is still in operation today, having introduced many cohorts of junior college students to science research under the guidance of an experienced principal investigator.

The Faculty's experience with the SRP which had been such a success with junior college students provided a pointer as to what could be achieved in introducing research to undergraduate students in the Faculty. In 1992, the Faculty launched the Undergraduate Science Research Programme or USRP, which offered Science students a chance to pursue a research project under the guidance of an academic staff member. The USRP laid the foundations for a University-wide undergraduate research programme, UROP or Undergraduate Research Opportunities Programme, which is still in operation.[7]

In 1996, the Faculty launched the Special Programme in Science or SPS which was aimed at science undergraduates who showed a particular aptitude for research. Tan Eng Chye and Chan Onn of the Mathematics Department took the leadership roles in establishing this new programme which allowed students to work on multidisciplinary research projects. The SPS predated the current University Scholars' Programme (USP) which came about as an amalgamation of the SPS with the University's Talent Development Programme (TDP) in 2001. Indeed the SPS is still going strong as an integral part of the USP for Science students.[8]

The Zoological Reference Collection

Consisting of over a million specimens whose collection was initiated by Stamford Raffles, the Zoological Reference Collection (ZRC) was a part of the Raffles Museum in colonial days, and portions of the collection were displayed to the visiting public. When Raffles Museum became the National Museum, the collection was removed from the Museum, and it took many years before this precious collection could find a permanent home.

The Department of Zoology had tenaciously and carefully looked after the collection, and in 1988 the ZRC, which gave the collection a proper home, was officially opened by Tony Tan. In 1998, the ZRC together with the equally precious Herbarium collection of plant specimens built up by the Botany Department was incorporated into the new Raffles Museum for Biodiversity Research (RMBR).

TMSI and the School of Biological Sciences

The ZRC was of course not the only achievement of the Zoology Department. Over the years, the Department had built up a formidable research reputation in the field of marine biology, though it also had excellent research in other areas such as entomology and taxonomy. The Head of Zoology, Lam Toong Jin, himself internationally known in marine biology, must be largely credited with driving the development of the department's expertise in this area and raising its international profile. Building on the strength of Zoology's marine biology research, the Tropical Marine Science Initiative (TMSI) was formed in December 1996 with the Science Faculty as one of its major partners. TMSI became the Tropical Marine Science Institute in April 1998 as a university-level research institute.

The Botany Department had also gradually developed its research expertise in a number of areas, and particularly in plant tissue culture as well as in orchids, notably by its Head, Goh Chong Jin as well as by Hew Choy Sin.

In the academic year 1969/70, the two Departments had come together to jointly offer Biology A and Biology B which combined both plant and animal biology to give students a more integrated biological curriculum. In 1990, these two subjects were replaced by three new subjects — Cell and Molecular Biology, Developmental and Systems Biology and Integrative and Organismal Biology to recognise the growing importance of the revolution in biology wrought by the discovery of DNA.[9] In 1991, Cell and Molecular Biology was elevated to an Honours specialisation alongside Botany and Zoology.

The closer integration of the two departments continued in 1993 with the establishment of the BioScience Centre, to serve as a common-core research facility for both departments. The two departments finally merged in 1996 to form the School of Biological Sciences, the term "School" being carefully chosen to give the new entity a special status having been formed as a merger of two departments. Lam Toong Jin became the first Director of the School which brought together the research resources of both departments.

Clinical Pharmacy

The Pharmacy Department had been born within the Faculty of Medicine, which was most logical as Pharmacy is intrinsically a medically-orientated discipline. Initially, Pharmacy was regarded as just a subject within the Faculty of Science, which was not really appropriate. To recognise the demands of the profession, Pharmacy was reinstated as a full three-year course in 1994 after a thorough review of its curriculum. In 1995, Pharmacy was recognised as a distinct and separate discipline with students being admitted to the Pharmacy course directly for the first time.[10]

The growing importance of the pharmacist as a member of a clinical team in hospitals and medical centres made it imperative that clinical pharmacy training be instituted as an integral part of the curriculum. To accommodate these demands, the course was extended to four years in 1997, and by 2009, the department had established a Doctor of Pharmacy programme as well as a Clinical Pharmacy residency programme. To fully recognise that the Pharmacy profession stands as an equal alongside other medical professionals and clinicians, the transformation of the department into a School of Pharmacy (as it had been) would be highly appropriate.

The Diamond Jubilee Celebration

In 1989, the Faculty of Science and the Faculty of Arts and Social Sciences jointly celebrated their Diamond Jubilee. A series of events was held to commemorate the 60th anniversary of both faculties which were born with the establishment of Raffles College. Though classes at Raffles College actually started in 1928, the official opening of the College was in 1929, which was the reason that the Jubilee was celebrated in 1929. A commemorative publication retold the history of the Faculty of Science and its departments. One novel souvenir which was produced was a transparent plastic cube within which was encased a diamond shaped structure on whose faces were the crests of NUS and its predecessor institutions.

This unique object was in fact designed by a distinguished mathematician, none other than Louis Chen!

As part of the Diamond Jubilee celebrations, the Faculty launched its Centre for Industrial Collaboration (CIC) which was officially opened by Ahmad Mattar (Minster for the Environment and a graduate of the Physics Department) on 5 December 1988.[11] Two other research centres, the Surface Science Laboratory and the Laboratory for Image and Signal Processing (LISP), were also opened by the Minister on the same day. The CIC was in later years combined with the Engineering Faculty's Innovation Centre to form the NUS Industry Liaison Office which is part of NUS Enterprise today. The Diamond Jubilee celebrations culminated with a grand gala dinner on 9 December 1989, with the President of the Republic, Wee Kim Wee, as Guest of Honour.[12]

An important event in the Diamond Jubilee celebrations was a talk on "Beauty and Truth in Mathematics" by the distinguished mathematician, Michael Atiyah, who was visiting NUS as the Lee Kuan Yew Distinguished Visitor.[13] Indeed, the Faculty was honoured over the years by a glittering array of scientists and mathematicians, many of whom were either Nobel Laureates or Fields Medallists. Some of those I particularly remember as I was honoured to chair their talks were Herbert Simon, Arno Penzias, Tony Hoare, Ilya Prigogine, Leo Esaki, Alec Broers and Jerome Friedman.

Computational Finance

With the apparent success of the Computational Science major, we shifted our attention to how computational techniques were being used in the field of finance and banking. Over the years, it was clear that many scientists in the US who were originally trained in areas such as high energy physics had moved over to work in finance or banking. With their mathematical and computational skills, they became highly sought after for the development of complex new financial products, particularly in the derivative markets.

In 1996. Computational Finance was introduced by the Faculty of Science as a new subject,[14] and its popularity allowed the Faculty to elevate it to the status of a major in 1997. This major was a joint effort by three Departments: Mathematics, DISCS and Computational Science, and was the very first move by NUS into this (then) novel field. The current undergraduate and graduate courses in Quantitative Finance offered by the Mathematics Department can be considered direct descendents of the Computational Finance course.

CRISP and SSLS

A number of NUS university-level research institutes can directly trace their origins to the Faculty of Science. Two of them, the Centre for Remote Imaging, Sensing and Processing (CRISP) and the Singapore Synchrotron Light Source (SSLS), were conceived in the Physics Department and still have strong links with Physics. CRISP was itself preceded by the Laboratory for Image and Signal Processing (LISP) which was a part of the Physics Department established in 1987, with Lim Hock as its first Director. CRISP, which started operating in 1996, and which was built on the success of LISP, was driven (as was LISP) by Lui Pao Chuen and was also initially helmed by Lim Hock.

The SSLS had its origins in the Physics Department in 1995 with significant inputs from the Materials Science Department in the person of Antony Bourdillon. Before funding was obtained from the National Science and Technology Board (NSTB) to acquire the synchrotron which is the centrepiece of SSLS, we had to go through an arduous process. But SSLS was finally commissioned and up and running by September 2000, and it now serves academia and industry, particularly in materials science research and development. This would not have been possible with the support of NSTB, which also funded CRISP.

The Growth of the Faculty

When the Faculty of Science was at Bukit Timah, the Dean's Office was generally located at the general office of the Department whose Head happened to be the Dean at that time. The first Dean's Office designed specifically for this purpose when we moved to Kent Ridge was Block S10, a small two-storey building located in the heart of the Faculty, amidst the Faculty's other buildings. With the expansion of the Deanery after 1985, it was obvious that this space was inadequate and the Deanery moved to the new Block S16, a nine-storey building incorporating the Dean's office, common teaching facilities, expansion space for DISCS and a new lecture theatre LT13 which was better suited for public talks than the standard lecture theatre

By 1997, the Faculty of Science had grown considerably in staff and student numbers since 1985. In September 1985, the undergraduate student population numbered 2,697 which included the students in DISCS.[15] By October 1997, there were 1,641 undergraduates in Computer Science (still within the Faculty but with separate admission figures) and 2,956 undergraduates in the rest of the Faculty,[16] making almost 4,600 undergraduates, an increase of 70%

over 12 years. This rapid rise was matched by the increase in academic staff and postgraduate numbers. In 1995, the Science Faculty was one of the pioneering faculties to introduce the new modular system which profoundly changed the undergraduate degree structure.

The expansion in staff and students was matched by a significant increase in the research output of the Faculty, both in quantity and quality. This was made possible by a phenomenal rise in the amount of research funding available, which was due to the efforts of the Vice-Chancellors, Tony Tan and Lim Pin. The academic staff of the Faculty must also be credited with their response to the increased funding which was borne out by the increasing numbers of papers in top international journals, especially in the face of their heavy teaching loads with the rapid rise in student numbers. By the 1990s, NUS had evolved into a credible research university, as recognised by a news article in the journal *Science* featuring the Faculty's research progress.[17]

Towards the Millennium

In July 1997, Lee Soo Ying of the Chemistry Department who had been Vice-Dean took over as Dean of Science and energetically continued the Faculty's expansion of its research and teaching programmes. An important development for the Faculty was the creation of a new Department, the Department of Statistics and Applied Probability in 1998.[18] The Department of Mathematics had grown considerably in staff strength over the last couple of decades, and the statisticians and probabilists had long felt that the distinctiveness of their disciplines warranted a separate department. The new department fulfilled that objective and also absorbed the statisticians in the Department of Economics.

1998 also saw the departure of the computer scientists from the Faculty. During my tenure as Dean, I had initiated a committee to look into the setting up of a School of Computer Science, and this progressed into the plan for DISCS to eventually become an entity independent of the Faculty of Science. This entity became known as the new School of Computing and is recognised as a Faculty within NUS.[19] The links between the computer scientists and the scientists in the Faculty of Science which were established during the days of DISCS still remain strong, as they should be.

The Faculty also introduced an entirely new degree, the Bachelor of Applied Science, in 1998 to give its students more adequate and appropriate preparation for a career in industry. A number of new majors were also introduced, notably Biology and Biotechnology (to replace the existing biological majors), Food Science and Technology, Applied Physics, Applied Chemistry, and Mathematics with Management Science. The Biomedical Science major was introduced as a broad–based major in 1998 bringing together the involvement of 10 departments in the Science and Medical Faculties.

70th Anniversary of the Faculty and the IMS

In 1999, the Faculty of Science celebrated its 70th Anniversary. Highlights of the celebrations were a public lecture by George Porter, the distinguished chemist and Nobel Laureate, and a gala dinner.[20] To mark the Anniversary, the Faculty established the Distinguished Science Alumni Awards. This Award was complemented by the Outstanding Science Alumni Awards established by the Faculty in 2005 to mark the centennial of NUS.

Lee Soo Ying was succeeded as Dean in July 2000 by Lai Choy Heng of the Physics Department who had also been a Vice-Dean. One of the first developments in Lai Choy Heng's tenure as Dean was the establishment of the Institute for Mathematical Sciences (IMS) as a university-level research institute in July 2000.[21] The IMS served the important function of focusing attention on mathematical research issues of both fundamental and applied importance. The first Director of IMS was Louis Chen, a distinguished probabilist and a much valued member of the Mathematics Department.

The Biomedical Sciences programme which the Science Faculty had established jointly with the Medical Faculty underwent further developments, with the setting up of an Office of Life Sciences in NUS. The Biomedical Science major was integrated into the new Life Sciences curriculum which, like the Biomedical Science major, was to be jointly taught by the two Faculties. Other important developments during 2001 in Biological Sciences included the setting up of the Functional Genomics Laboratories[22] and the opening of the Raffles Museum of Biodiversity Research's public gallery by Minister for Education Teo Chee Hean.

Centre for Financial Engineering

In 2002, the Centre for Financial Engineering (CFE) was transferred to the Faculty to be hosted as a faculty-level research centre by the Mathematics Department. This was a clear recognition of the pioneering steps which the Department had made in the field of financial mathematics and computational finance (which continues today as its quantitative finance course). The CFE's importance was recognised when it was transformed in 2006 into the Risk Management Institute, a university-level research institute.

Quantum Information and CIBA

In Physics, two important developments took place. One was the opening of the Research Centre for Nuclear Microscopy in May 2001 by the Chairman of NSTB, Philip Yeo, which is today better known as the Centre for Ion Beam Applications or CIBA, now one of the world's leading centres in nuclear microscopy. The other was the establishment of a Temasek Professorship Programme in Quantum Information Technology in 2002.[23] The first such Professor appointed was Artur Ekert, internationally known for his work on quantum information. Indeed, this directly led to the setting up of the first Research Centre of Excellence in NUS, the Centre for Quantum Technology or CQT, in 2007 with Lai Choy Heng and Oh Choo Hiap of the Physics Department providing much of the impetus for the project.

In 2003, Tan Eng Chye of the Mathematics Department succeeded Lai Choy Heng, who went on to become a Vice-Provost, as Dean of Science. The same year, the new Structural Biology Research Corridor of the Department of Biological Sciences was launched by Acting Minister for Education, Tharman Shanmugaratnam.[24] This new facility highlighted the Department's shift of research emphasis into genomics, proteomics and structural biology, and brought together the existing Functional Genomics Laboratory, the Structural Biology Laboratory and the Protein and Proteomics Centre (itself transformed in 2000 from the BioScience Centre).

Computational Biology and NUSNNI

The following year brought the successful implementation of the new Life Sciences curriculum which involved both the Science and Medical Faculties. 2004 also saw the launch of a new Computational Biology Cluster by the Faculty, which was a joint effort by the Department of Biological Sciences and three other departments.[25] Later that year, the Faculty celebrated its 75th Anniversary with a gala dinner on 9 October 2004 at the Suntec City Ballroom.

In 2004, NUS launched NUSNNI — the NUS Nanoscience and Nano-technology Initiative — with the Science Faculty as one of its principal part-ners, largely driven by Andrew Wee of the Physics Department. This Initiative subsequently became a university-level research institute, and has now incor-porated SSLS into its structure, with SSLS Director Mark Breese as one of the two Co-Directors of NUSNNI.

In 2005, the Materials Science Department was transferred to the Engineering Faculty. The importance of Materials Science as a key discipline of strategic significance to Singapore's technological progress had been amply demonstrated by the Department's progress within the Faculty of Science.

The potential for synergy between the Science and Engineering Faculties was further enhanced the following year in 2006 with the establishment of the new Engineering Science major, a joint project between the two faculties which clearly underscored the importance of the science disciplines (and physics in particular) to the practice of engineering.

In 2006, Tan Eng Chye relinquished the Science Deanship to become Provost of NUS, and was succeeded by Andrew Wee. That same year, the Faculty launched the Centre for Computational Science and Engineering to further develop the path in Computational Science which had been laid down by the Computational Science Department. In 2007, the Faculty with its partner faculties of Law and Medicine launched a new double degree programme in Law and Life Sciences.

CQT and Science Communication

2007 also saw the Faculty's earlier initiative in Quantum Information came to fruition when the first Research Centre of Excellence (RCE) in NUS, the Centre for Quantum Technologies (CQT), was launched as an integral part of the Faculty.[26] CQT has since firmly established itself as one of the top centres in the field internationally.

An innovative and ground-breaking programme was jointly established by the Faculty and the Australian National University in 2009.[27] This new Master's degree in Science Communication was targeted at science graduates who wanted to become effective communicators in science. They included science teachers as well as others in education, policy making and the media who had an interest in making science better understood by bringing it to a wider public.

Another degree launched in 2009 was the Master's in Quantitative Finance by the Mathematics Department which could be traced back to the Faculty's initial moves to establish Computational Finance. The Doctor of Pharmacy programme was launched by the Pharmacy Department, also in 2009, as a capstone to its long-standing efforts to bring Clinical Pharmacy into its curriculum structure as an integral part of its training for the Pharmacy profession. The year 2009 also marked the Faculty's 80th Anniversary, celebrated with a grand dinner on 24th October.

Mechanobiology and Environmental Studies

A second RCE was announced by the Faculty in 2009 as well — in the field of Mechanobiology — to follow the first one in the Faculty, the CQT.[28]

The Mechanobiology Institute was officially opened on 6 October 2010, to focus on the quantitative and systematic understanding of basic biological functional processes in living cells. The Mechanobiology Institute was also a key partner of the Centre for Bioimaging Sciences launched on 6 December 2010 together with the Science and Engineering Faculties and the Duke-NUS Medical School.

The Bachelor of Environmental Studies, a new four-year degree, was launched jointly in 2011 (driven primarily by Leo Tan) by the Science and Arts & Social Sciences Faculties as a truly interdisciplinary programme, with participation by several other faculties and academic units of NUS.[29] Another significant development in 2011 was the introduction of the Integrated Science Curriculum for the Special Programme in Science, with modules integrating Mathematics, Physics, Chemistry and Biology as well as research-oriented modules.

Two significant physics-related developments took place in 2012, the first being the establishment of the new Graphene Research Centre, reflecting the rapid developments in the materials science area since the discovery of graphene. The other was the Physics Department's unveiling of its new Teaching Observatory, a facility housing the largest aperture astronomical telescope in Singapore for the learning of advanced astronomy.

Andrew Wee was succeeded as Dean by Shen Zuowei of the Mathematics Department in 2012, going on to become NUS Vice-President for University and Global Relations. An agreement for a concurrent degree programme was jointly launched in 2013 by the Faculty and the Department of Biomedical Sciences, King's College London, leveraging on resources in both institutions including that of the Mechanobiology Institute.[30,31]

The Centre for Advanced 2D Materials and the Lee Kong Chian Natural History Museum

2014 saw the establishment of the Centre for Advanced 2D Materials with a $50 million grant from the National Research Foundation.[32] This new university-level research centre built on the success of the Graphene Research Centre set up by the Physics Department and was a significant new research facility for the Faculty. 2014 also marked the 85th Anniversary of the Faculty which was marked by a grand dinner on 8 November 2014 graced by the President of the Republic, Tony Tan.

2013 saw the closing of the public gallery of the Raffles Museum of Biodiversity Research to prepare for the move of its collections to a new Natural History Museum. On 18 April 2015, the Lee Kong Chian Natural

History Museum was officially opened by the President of the Republic, Tony Tan.[33] The new Museum was a fitting and magnificent permanent home for the Zoological Reference Collection.

Indeed the Science Faculty itself had come a long way since its early days in Raffles College, and by 2015 had expanded its undergraduate student numbers to more than 5,200. The Faculty is well placed to contribute even more to the nation as it deepens and broadens its research and teaching capabilities to keep pace with the unceasing advancement of science.

References

[1] National Heritage Board. [Online]. http://www.nhb.gov.sg/places/sites-and-monuments/national-monuments/former-raffles-college-now-nus-campus-at-bukit-timah

[2] Wikipedia. [Online]. https://en.wikipedia.org/wiki/National_University_of_Singapore

[3] University of Malaya. University of Malaya Calendar, 1951/52.

[4] R.S. Desowitz, "Review of Research on Parasitology in Singapore," *Singapore Medical Journal*, vol. 4, no. 1, pp. 30–33, March 1963.

[5] Nantah Friends. Alumni and Friends of Nanyang University. [Online]. http://www.geocities.ws/nantahfriends/nantah/choong.html

[6] "Seminar kicks off research programme," *The Straits Times*, March 14, 1989.

[7] National University of Singapore, *National University of Singapore Prospectus 1996/1997.*, 1996.

[8] Special Programme in Science (SPS). [Online]. http://sps.nus.edu.sg/alpha/

[9] "New biology subjects at NUS," *The Straits Times*, June 21, 1990.

[10] National Universty of Singapore, *NUS: To Meet the Challenges of A Developed Nation: 90th Anniversary, 1905–1995.*, 1995.

[11] "NUS Science Faculty hopes new centre will get more to take up research," *The Straits Times*, December 6 1989.

[12] "President praises key role of two faculties," *The Straits Times*, December 6, 1989.

[13] "Two NUS faculties to mark 60th year joy in a big way," *The Straits Times*, November 29, 1989.

[14] "New subjects at NUS science faculty," *The Business Times*, July 5, 1996.

[15] National University of Singapore, *National University of Singapore Annual Report 1985/1986.*: National University of Singapore, 1986.

[16] National University of Singapore, *National University of Singapore Anual Report 1997/1998.*: National University of Singapore, 1998.

[17] June Kinoshita, "Topnotch, Targetted Researh Propels National University," *Science*, vol. 268, no. 5211, pp. 633–634, May 5, 1995.

18 *Campus News*, no. 133, p. 4, 1998.

19 NUS Computing. [Online]. htpps://www.comp.nus.edu.sg/about

20 *Campus News*, no. 143, p. 5, 2000.

21 "National University of Singapore. Department of Mathematics Annual Report, 2000/2001," 2001.

22 *Knowledge Enterprise*, p. 11, July 2001.

23 "National University of Singapore. Science Faculty Annual Report 2002," 2002.

24 Ministry of Education. [Online]. https://www.moe.gov.sg/media/speeches/2003/sp20031111a.htm

25 National University of Singapore, "National University of Singapore. Faculty of Science Annual Report 2004."

26 Centre for Quantum Technologies. [Online]. www.quantumlah.org/main/aboutus.php

27 Faculty of Science. Joint Master of Science in Science Communication. [Online]. http://www.science.nus.edu.sg/education/graduate/pg-joint-msc-science-comm

28 National University of Singapore, "National University of Singapore. Annual Report 2008/2009."

29 National University of Singapore, "National University of Singapore. Annual Report 2010/2011."

30 National University of Singapore, "National University of Singapore. Annual Report 2009/2010."

31 National University of Singapore, "National University of Singapore. Faculty of Science Annual Report 2012/2013."

32 National Research Foundation, "Press Release 3 November 2014," 2014.

33 National University of Singapore. NUS News. [Online]. http://news.nus.edu.sg/press-releases/8874-nus-launches-singapore-s-first-and-only-natural-history-museum

Chapter 6

Department of Mathematics at Nanyang University (1956–1980)

Lee Peng Yee

Research in Mathematics

This is a small part of the oral history of Nanyang University or briefly Nanyang. It tells the story of Nanyang graduates in mathematics holding university positions in Singapore, Malaysia, North America and other far away countries. A significant number of the graduates joined the educational service in Singapore, as school principals, teachers, and officers in the Ministry of Education. In 1979, a Southeast Asian regional centre for mathematics was formally established at Nanyang. In this section, we start the story with research in mathematics. It was and still is our belief that research should always be an integral part of a university.

Nanyang accepted the first batch of students in January 1956.[2] In 1980, it merged with the University of Singapore, and was renamed the National University of Singapore (NUS). The logo of NUS carries the original logo of Nanyang, which consists of three circles and one star. However a list of the alumni of Nanyang was kept with Nanyang Technological University (NTU), which currently occupies the former Jurong campus of Nanyang. The alumni of Nanyang organised their first global reunion in Toronto in 1992. The word, "global", refers to participants coming from all over the world. Initially, it was held once a year in different countries, then later once every two years. So far, there seems to be no sign of it dwindling in size.

The author was a first-batch graduate of Nanyang, serving as a graduate assistant after graduation, involved in the examination of the final-year students in 1969, and eventually becoming a staff member of Nanyang for 10 years from 1971 to 1980. If the author were asked to name one event or one person of importance during his undergraduate days, he would say Professor Chin or 靳宗岳. Professor Chin came from the Guizhou province in China, and was a graduate of the University of Chicago. He may not be a great classroom lecturer. Nevertheless, he was an enlightened teacher.

He used a newly-published textbook. He showed us a way to go in mathematics. He introduced topology in class because it was an important area in modern mathematics at the time. He encouraged us to write mathematical essays in English. It was a prelude to research. Due to the essay the author wrote under his supervision, the author was admitted to a British university for graduate studies.

In the '60s, Nanyang was little known overseas. At home, the author worked as a university graduate and was paid the salary of a high school leaver. Ten years after the establishment of Nanyang, the graduates who went away started coming home. One thing they did was to build up the research capacity for the Department.

First of all, a necessary condition for research was a good library. In the language of the current digital world, it means access to information. As it was, the library had limited funds. Fortunately there were other sources. We bought books from a book seller at a reduced price. We had donations from the French and German embassies. We published a half-yearly journal, *Nanta Mathematica*, and exchanged it with 65 institutions around the world. It was possible because at the time many journals were published by an institution or a professional society, and not commercially. When Nanyang merged with the University of Singapore, all library books in Nanyang were re-catalogued before being put in the NUS library except the mathematics books. The mathematics books were displayed as a single collection in NUS for a period of time. Nanyang library had a relatively complete collection of newly published mathematics books at the undergraduate level and also at the graduate level in certain areas.

Secondly, again for research, it was important to focus on a few selected research areas. Our choices were graph theory, operational research, and real analysis. Both graph theory and operational research were new research areas, fast growing, and had applications in industry. Here is a story of the graph theory. We had a friend who worked in Bell Labs for over 30 years. Every time he solved a problem in graph theory and saved the company tons of money, he was given a round-the-world air ticket. On the way he visited us. Graph theory, being a new research area, required less pre-requisites as compared with real analysis. Consequently, the research group in graph theory was able to promote it locally in Singapore and also regionally, in particular, in the Philippines.

Operational research was a product of the war effort during World War II. After the war in the '50s, it was applied to the fields of social sciences, management, and so on. The research group in operational research at Nanyang worked jointly with the government departments, including the port authority, in the applications of operational research to some practical problems. Real analysis took a longer time to take off. We shall not elaborate here.

Thirdly, we must not work alone. We need to build up an international network of friends. At the time, Singapore was not such an attractive destination for visitors. If they were our friends, they would come. We name a few below. Frank Harary (US) and Claude Berge (France) were frequent visitors of Nanyang. Both were among the best-known researchers in graph theory and produced standard textbooks for the subject. In fact, many people had been here. Robin J. Wilson (UK), son of a former British Prime Minister and working in graph theory, was also here. Alan Mercer (UK), a professor of operational research at the University of Lancaster, visited Nanyang under the sponsorship of the British Council. At the time, there were only two universities in England having a department of operational research. Lancaster was one of them. We see more names of mathematicians in what follows.

There were immediate benefits from the above activities to the staff development and teaching. Four young faculty members after joining the Department in the late '60s obtained their PhD degrees locally at Nanyang. Teaching algebra was adjusted to include more combinatorics. Note that graph theory is a special branch of combinatorics. Operational research was introduced to the undergraduates in the early stages of their studies. In summary, to build up the research capacity, we must have a good library, focus on selected research areas, and develop a network of friends. In short, we must create a rich environment for teaching and research, and we did.

The Department and its Graduates

Nanyang University started as a regional university. It was initially sponsored by people coming from the region, and it took in students also coming from the region. The Department of Mathematics graduated 11 students in the first batch (January 1956–December 1959). The Department was smaller than other science departments. As time went on, it grew. By the seventh batch, it had 29 graduates. Starting from the seventh batch, more mathematics graduates went overseas for further studies. In this section, we shall report on the curriculum of the Department, its activities and finally its graduates.

As far as the university structure was concerned, Nanyang adopted the Chinese system. Recall that Tsinghua University in Beijing was in fact patterned after an American university. So Nanyang actually adopted the American system. Americans have a totally different educational philosophy from the British. In a British university at the time, only a small portion, say less than 15%, of the third-year students were selected for the fourth year or the honours year. In an American university, every third-year student could advance to the fourth year. In the British system, students specialised much earlier. In the American system, all science students studied the same subjects in their first year. In a way, the British system is more elitist, whereas the American system is more "education for all". The author was not aware of this difference at the time. Probably, many others were equally not aware of it. The grading system was also different. In recent years, the universities in Singapore have adopted the American grading system.

In the '50s, mathematics meant pure mathematics. Pure mathematics meant geometry, algebra, and analysis. In Nanyang, geometry meant affine and descriptive geometry. As in other universities, geometry was eventually dropped and new subjects were introduced. Nanyang was the first university in Singapore introducing courses on operational research and numerical analysis in 1971. It was also the first university in Singapore encouraging the use of computers, and it established a computer centre and later a research institute in mathematics and computer science. Another innovation in the Department was the introduction of a thesis for the fourth-year students. Furthermore, the system was flexible enough for other innovation. For example, a course on life contingencies in actuarial mathematics was offered in 1975. Two students in the class pursued the actuarial profession afterwards.

It was a policy of the Department to engage in school activities. The Department organized competitions, exhibitions, and lecture tours. The exhibitions were open to the public. The tours included tours to peninsular Malaysia. All were popular among schools. It published *Mathematical Garden*, a magazine for school teachers and students. The Department was also involved in the in-service training of teachers. It was the time when *Syllabus C* was introduced into schools, and *Syllabus C* contained new content not previously taught in schools. So workshops were organised for teachers to learn new content in *Syllabus C*. The first locally produced secondary school textbooks for *Syllabus C* were done by the faculty members of the Department. These activities were extremely helpful in recruiting students who were interested in mathematics to the Department.

A significant number of mathematics graduates of Nanyang joined the educational service in Singapore and elsewhere. Among the first-batch graduates, one was a senior officer in the Income Tax Department, and the rest were either lecturers at the university or teachers in colleges or schools. For those in schools, most of them were principals. Similarly, a high proportion of the early graduates, especially the first six batches, taught mathematics in schools. Some went overseas for higher degrees. After their PhDs, they lectured at the universities in different countries. There were those who came back to Singapore and Malaysia. There were also those who stayed on in North America and other places. They spread across the American continent from Vancouver in the west to Halifax in the east of Canada, and many cities in the United States. One went to Venezuela, and another to Central Africa.

All the happenings in the '70s of the Department of Mathematics at Nanyang were recorded faithfully by a visiting professor, Haruo Murakami, in a research report for the Japan Foundation.[6] In particular, he mentioned the team effort of the Department and the leadership provided by a group of three who were the first-batch graduates of Nanyang.

Though Nanyang was restructured a few times over the years, the core subjects and teaching approach did not change. Certain traditions also did not change. The best lecturers were allocated to teach the first-year students. Some lecturers walked into the classroom with two chalks and no notes. It remained a common practice. Ragging of first-year students, a British tradition, was replaced by a welcoming party. The Department trained many generations of mathematics graduates. A high number of them went overseas for higher degrees. An even higher number of them joined the educational service in Singapore, in particular, teaching in the classrooms. For those who taught in schools, they had made great contribution to what mathematics education in Singapore is today. This had not been acknowledged.

Our graduates had no problem getting into the universities overseas for further studies. Some ended up holding university positions afterwards. The undergraduates also had no problem getting a transfer to the universities overseas. This gave us confidence that we were doing the right thing for the Department, more precisely, good teaching, active research, and staff development. A greater contribution of the Department was to train a big group of competent and dedicated mathematics teachers for the Singapore schools.

Outreach Programmes

In the early '70s, we in Singapore did not have a critical mass to engage in research, to attract good speakers to our conferences, and to obtain financial support from international institutions. There was one way we could do it if we were working together with our neighbouring countries as a unit. Hence the first move was to go regional. The decision was strategic. Consequently, Nanyang benefited in the process. As it happened, Nanyang played a major role in the development of mathematics and mathematics education in the region. Here the region refers to the Southeast Asian region.

There was nothing new about regional activities. As far as we know, there were two regional conferences in mathematics held in Vietnam in the '60s. Nanyang sent one representative to one of the conferences. Also, there were other regular mathematics conferences in the '60s. The universities in peninsular Malaysia and Singapore took turns to host them. The organised effort came about only after the formation of the Southeast Asian Mathematical Society (SEAMS). The first SEAMS meeting was held in 1972 at Nanyang.[3]

In the late '60s and early '70s, there was a shortage of lecturers holding PhD degrees in the universities in the region. When they did, they were promoted quickly to administrative positions. The staff development was a major issue. Sending lecturers overseas for further studies did not solve the problem. Those who went away might not come back. So countries in the region started training and upgrading staff members locally. It was definitely a viable programme in Singapore, including Nanyang, and also in the region.

The next issue was to keep staff members academically alive. They needed constant stimulation and fellowships. Regional meetings and academic exchanges were stimulants. Fellowships were definitely incentives for staff members to upgrade themselves. At Nanyang, we invited visiting mathematicians. We sent our own staff to other universities. We initiated conferences in the region. These were bonuses for lecturers in the region and for our own staff. It helped make an academic job more attractive.

It was under such circumstances that SEAMS served as a catalyst at the time and provided an umbrella for regional cooperation. It is interesting to note that regional activities preceded national activities. At the time, not every country had a national society of mathematics. Gradually, the national societies were formed. Eventually they took over and initiated many activities formerly done by SEAMS. Finally in the later years, the contact was made with East Asian countries in the '80s and internationally in the '90s. The Institute at Nanyang, named after Lee Kong Chian, left its footprint in the process. It was possible due to the fact that the region was growing and in need of regional collaboration and international input. It was also due to the local and international support of the individuals, agencies, and the administration at Nanyang. To name one person, Lu Yaw played a crucial role in managing the Institute, in particular, the financial aspect of the Institute. He was a senior officer of the Ministry of Education, Singapore, seconded to Nanyang, and served as the deputy Vice-Chancellor in the '70s.

We could not have possibly solicited all the international support that we received without going regional. In the process, we benefited the most. In fact, the Institute turned out to be a *de facto* centre for the region. A natural step to move forward was to formalise it. We shall elaborate on this in the next section.

A Regional Centre

The Regional Coordinating Centre for Pure and Applied Mathematics was inaugurated at the Franco-SEA Conference, Nanyang University, on 28 May 1979.[4] The story should begin with the establishment of the Lee Kong Chian Institute of Mathematics in 1969.[5]

Dr Tony Tan Keng Yam (current President of the Republic of Singapore) (far left), His Excellency Mr J Gasseau, Ambassador of France (second from right) and Dr Lee Peng Yee (far right).

The Institute was established with a generous grant from the Lee Foundation. Locally it supported research activities. Other than the library and training courses, fellowships were available for our own staff members to go overseas and also for staff members from other universities, mainly in the Southeast Asian region, to visit Nanyang for one or two months. Furthermore, regular seminars were held with invited speakers from overseas. In the eight years from 1972 to 1979, 26 regional and national meetings were held in Southeast Asia, over one third in Singapore. Some conferences were organised by the Institute jointly with other organisations, for example, the Operational Research Society of Singapore. We worked under the condition that there were no common holidays among the universities in the region at the time. These activities helped upgrade teaching and research in mathematics. Regionally, the Institute shared the library facilities with the visitors from the region, in particular from the Philippines and Indonesia. The number of mathematical periodicals quadrupled in a period of 10 years. In 1979, the mathematics library received 200 journals in pure and applied mathematics. The Institute also shared overseas visitors with other universities in the region. The visitors would often extend their visits beyond Singapore. Nanyang had a guest house on top of a hill within the Jurong campus. A log book was kept. It recorded a long list of international guests to Nanyang. On record, between January 1975 and August 1976, the Institute had received 45 overseas visitors from 15 countries. That was to say, at least two visitors every month.

In the year 1969 when the Institute was established, a computer centre was set up within the Institute. Later on it was separated to become a service centre for the whole university. In 1975, a computer science department, the first in Singapore, was established. After that the Institute changed its name to include computer science. As mentioned above, the Institute published a half-yearly research journal called *Nanta Mathematica*. For a long time it was the only research journal in mathematics published in the region. It served as a means of communication with mathematicians overseas. The Institute had other publications, including the quarterly *Newsletter of SEAMS* and the *Southeast Asian Bulletin of Mathematics*, which were published on behalf of SEAMS.

Again, as mentioned above, four young staff members of Nanyang were upgraded locally at the Institute. In fact, the service was extended to the Philippines and Indonesia. Three universities in Manila formed a consortium in 1974 running their own doctoral programme. Also in 1974, a lecturer from Nanyang offered graduate courses in Manila. Afterwards, almost yearly for several years, one or two lecturers from Nanyang or universities in Malaysia conducted graduate courses in the summer there. Their students would come to Nanyang, making use of the library facilities and the supervision provided by the staff members of Nanyang. This arrangement had proved to be mutually beneficial. Half of those early PhD candidates in mathematics had been to Nanyang. A similar arrangement was made for Indonesia and Thailand. For example, a lecturer from the University of Sumatera Utara studied operational research at Nanyang so that he could initiate an undergraduate course on operational research in his own university. Graduate courses were organised, with the participation of the Institute, in Chiangmai and Bangkok for the lecturers and students from the Southeast Asian region.

At the time of writing (2015), there are over 30 academic descendants who hold PhD degrees in the Philippines as a result of the extension programme.

Also, there are four generations of PhD students in Indonesia, taking the lecturers from Nanyang as the first generation.

Concerning funding, some of the graduate students from the Philippines came to Nanyang under the fellowships given by the Institute. Others were sponsored by their own institutions in the Philippines. There were also fellowships available for Indonesians. The fellowships were donated by a Singapore company having business in Indonesia. There was also funding from the French, Japanese, and UNESCO.

Traditionally our contact has been mainly with the English-speaking countries. However the Institute was able to foster closer links with French and Japanese mathematicians on a regional basis. The first exchange between a staff member from Nanyang and a French mathematician was realised in 1973. Since then, there had been many exchanges involving not only mathematicians but also computer scientists. The French embassy supported the regional activities in a number of ways including providing the initial financial assistance to publish the SEAMS Bulletin. The highlight was of course the first Franco-Southeast Asian Mathematical Conference in 1979. Jacques-Louis Lions (France) was the person instrumental in bringing about the conference and in connecting us to the international agencies. He was Secretary of the International Mathematical Union (IMU) from 1975–78 and 1979–82.

The first contact with Yukiyoshi Kawada (Japan) took place in Tokyo in 1976. He was Secretary of the International Commission on Mathematical Instruction (ICMI) during 1975–78. He initiated a series of conferences on mathematical education in the region. The first was held in Manila in 1978. He also linked Nanyang with the funding agency in Japan. Hence, visiting fellowships sponsored by Japan followed thereafter. Both France and Japan sent the best speakers to our conferences and workshops.

At a council meeting of SEAMS in Manila in 1975 the idea of having a regional centre was first brought up. It was decided a year later in another council meeting in Bandung that the Society should set up such a centre. The decision was finally made at a meeting in Bangkok in 1978. It was proposed that the centre be located at Nanyang. The Centre had the support of IMU, UNESCO and the mathematical community of Southeast Asia. UNESCO provided the necessary funding for the participants from the region to attend the regional conferences. The Centre had been recommended as a model for regional cooperation.

The Franco-SEA Conference was held at Nanyang in conjunction with the inauguration of the Centre. It was held from 14 May to 1 June 1979, a total of 18 days. For the first two weeks, there were two workshops, one on graph theory and another on analysis, preparing the participants for the conference that followed. This was to make sure that the participants could benefit as fully as possible from the conference. The speakers included Jean Dieudonné, Laurent Schwartz (France) and Kiyoshi Itô (Japan). Dieudonné was a leading figure in the Bourbaki group. Schwartz was a Fields Medalist (1950). Itô received the Gauss Prize in 2006. There were also mathematicians from mainland China, Taiwan, Vietnam, and Papua New Guinea. The last day was a special day on mathematical education, held at the University of Singapore. The conference set the pattern for other regional conferences that followed.

Experience showed that with the region as a unit, we could achieve much more than we could individually. It took time to build up a network. It took an even longer time to reap the fruits of the network. The regional centre was a legacy. In 1980, Nanyang moved out of the Jurong campus to join NUS at the Bukit Timah campus and finally at the Kent Ridge campus.[7,8] Everything concerning the Centre and all the activities of the Institute came to an abrupt halt. The building of the Institute that was constructed half way was demolished. The research journal ceased publication. The last issue was Volume 13 Number 1. We know we have disappointed greatly our colleagues in Southeast Asia. Over 16 years of tactical evolution, Nanyang merged with NUS and turned British. It was the worst of times, it was the best of times.[1]

References

[1] Charles Dickens, A Tale of Two Cities, 1859.

[2] Nanyang University Tenth Anniversary Souvenir 1956–1966, edited by the *ad hoc* editorial committee, Nanyang University 1965.

[3] Lee P.Y., Development of Mathematics in Southeast Asia: The Experience of the Southeast Asian Mathematical Society, *The Chronicle* Nov/Dec 1978 No. 4, pp. 2, 7.

[4] Regional Maths Centre set up at Nantah, *The Chronicle* May/June 1979 No. 3, pp. 1, 4–5, 8.

[5] Lee P.Y., Ten Years of Lee Kong Chian Institute of Mathematics and Computer Science, *The Chronicle* Jul/Aug 1979 No. 4, pp. 2–3.

[6] Haruo Murakami 村上温夫, Research Report: Nanyang University 1977–1978 (in Japanese), 日本研究课资料第 33号, The Japan Foundation.

[7] Peter H.L. Lim, Chronicle of Singapore 1959–2009, Editions Didier Millet, 2009.

[8] Ong Chu Meng 王如明, The Chronology of Nanyang University 南洋大学文献, Global Publishing, 2015.

Notes. A brief history of Nanyang University can be found in Ref. 7. See also Refs. 2, 6 and 8. For more details concerning the Centre, see Refs. 3–5. The quote of the last sentence above was taken from Ref. 1 and stated in a reverse order.

Chapter 7

Fifty Years of Mathematics in Singapore: A Personal Perspective[a]

Chong Chi Tat[b]

In tandem with the development of Singapore as a nation, mathematics in Singapore underwent a dramatic transformation in the first 50 years of the country's history. What follows is a brief account, based on both my own recollection and inputs from colleagues, of the events that shaped the current state of mathematics in Singapore.

When Singapore became independent on 9 August 1965, it had two mathematics departments, one at the University of Singapore and the other at Nanyang University. The history of the Department of Mathematics at the University of Singapore traces back to 1929 when Raffles College was established.[5] The major mathematical figure at the time was Alexander Oppenheim who was a student of G H Hardy at Oxford University and received his PhD from the University of Chicago under L E Dickson. Oppenheim joined Raffles College in 1931 and was Head of the Department for the next 28 years until 1959, when he was appointed Vice-Chancellor of the University of Malaya which evolved from Raffles College in 1949. In 1965, Oppenheim retired from his position and left Singapore. He was later honoured by the government of Malaya with the award of Tan Sri and was knighted by the British monarchy. Apart from his contributions to the development of mathematics in the early days of Singapore, Oppenheim also published extensively in number theory and is best known for formulating the conjecture named after him on the representation of numbers by quadratic forms of several variables. A definitive solution of the conjecture was given by the Fields Medalist Grigori Margulis in 1987.

The Department of Mathematics at Nanyang University was established in 1955 when the university was founded. In the '60s and '70s, driven by a pioneering spirit, a group of the Department's earlier graduates who returned

(*left to right*) J-L Lions, L Schwartz and J Dieudonné.[c]

from the UK and Canada with doctoral degrees succeeded in securing funding from the Lee Foundation to set up the Lee Kong Chian Centre for Mathematical Research. During this period, the Department played a significant role in promoting mathematical activities in Singapore and Southeast Asia. These included supporting overseas visitors to Singapore, publishing the journal *Nanta Mathematica*, taking the lead in forming the Southeast Asian Mathematical Society, establishing exchange and training programmes in mathematics with universities in the region, as well as organising the First Franco-Southeast Asian Mathematical Conference in 1979 (jointly sponsored by the French-Singapore Scientific Cooperation Programme). The conference was a major event as it featured 10 leading French mathematicians among the list of invited speakers. These included Jean Dieudonné, Jacques-Louis Lions and Fields Medalist Laurent Schwartz.

The departments of mathematics at the two universities differed in research interests. Due to its colonial origin, the Department at the University of Singapore was headed by expatriate faculty from the very beginning, and this continued until 1980 when K K Sen retired after a brief one-year headship to return to India. The research areas during the period 1950–1970 were wide and diverse: number theory (Oppenheim), classical algebraic geometry (Dan

---

[c] Photo courtesy of Leong Yu Kiang.

Pedoe), combinatorics (Eric Milner (especially combinatorial set theory) and Richard Guy (who also worked in game theory and number theory)), combinatorial group theory (Malcolm Wicks), analysis (U C Guha), topology (E J Brody), statistics (P H Diananda (who also worked in analysis)), applied mathematics (Peter Lancaster and Rex Westbrook[d]), astrophysics (K K Sen and S J Wilson), history of Chinese mathematics (Lam Lay Yong). Many were publishing high-quality scholarly papers. In fact, the first paper in the *Annals of Mathematics* from Singapore (by Brody on lens spaces) appeared in 1960. On the other hand, it was not until 1965 that the Department saw its first Singaporean student return from abroad with a PhD (Peng Tsu Ann, PhD in group theory from Queen Mary College). Peng led the Department as Head from 1982 to 1996 and oversaw the Department's rapid growth during that period. By the early 1970s, the Department had among its junior faculty several alumni who returned from overseas studies. They included Leong Yu Kiang (PhD in group theory, Australian National University), Ng How Ngee (PhD in representation theory, University of Illinois at Urbana-Champaign), Louis Chen (PhD in statistics, Stanford University) and Cheng Kai Nah who took the unusual step for a Singaporean in those days of pursuing a doctorate in continental Europe (PhD in group theory, Johannes Gutenberg-Universität Mainz).

At Nanyang University, research was very much focused on graph theory (Teh Hoon Heng and his students). While the faculty's research interests included areas such as lattice theory (Chen Chuan Chong and Koh Khee Meng), analysis (Lee Peng Yee) and operations research (Chew Kim Lin), graph theory was arguably the most prominent among them. In 1985, the journal *Graphs and Combinatorics* was published by Springer-Verlag with Teh as its founding managing editor.

When Singapore declared independence, the University of Singapore had awarded its first PhD (Lancaster, 1964, under the supervision of Dan Pedoe) and the Singapore Mathematical Society (formerly the Malayan Mathematical Society) was 13 years old. The total number of mathematicians engaging in teaching and research in the country was not more than 20. However, by the time I joined the Department of Mathematics at the University of Singapore in April 1974 as a lecturer, it already had a faculty strength of 17. Research was for personal interest and not a requirement for career advancement. Nevertheless,

---

[d] Richard Guy, Peter Lancaster, Eric Milner and Rex Westbrook later moved successively to the University of Calgary between the 1950s and '60s, and were sometimes jokingly referred to as "the Singapore mafia".

the pursuit of scholarship for its own sake was very much alive. I remember that together with some junior colleagues, we ran a weekly seminar to read *Cours d' arithmetique* (*A Course in Arithmetic*) by Jean-Pierre Serre, and a seminar on John Milnor's *Topology from the Differentiable Viewpoint*, even though none of us worked in these related fields.

The Singapore Mathematical Society viewed promoting public interest in mathematics to be a key mission. To this end, it regularly held public lectures delivered by local mathematicians or, on rare occasions, visiting mathematicians. In addition, the Society also played an active role in identifying budding mathematical talents. From the early days, it organised an annual inter-school mathematical competition. Not long after I joined the Department, I was recruited by several junior colleagues (primarily Louis Chen, Leong Yu Kiang, Ng How Ngee) to get involved in the organisation of the annual competition. This included liaising with schools, setting competition problems, marking the solutions, getting the prizes for winners, arranging for the engraving of names of winners on the plaque, as well as planning and organising prize giving ceremonies. Separately, there were two publications under the auspices of the Society: *Mathematical Medley* (founded in 1973) aimed at the general public, school teachers and students, and the *Bulletin of the Singapore Mathematical Society* (formerly *Bulletin of the Malayan Mathematical Society*, founded in 1955) which was devoted to mathematics at the research level. The latter has since been discontinued as attention and interest of local mathematicians shifted to publishing their research papers solely in international journals.

Public research funding for mathematics, and basic science in general, was not available before 1980. Despite this, mathematical research in Singapore was not impeded. In a compendium on research publications between 1949 and 1980 at the University of Singapore printed in the *Straits Times*, no less than 320 papers were published during this period, of which 174 appeared in international refereed journals. On the other hand, attending a conference outside Singapore was very rare and required the use of personal funds, support from a private foundation or an international funding agency. My very first participation in an international conference was at the *1978 International Congress of Mathematicians* (ICM) in Helsinki and was funded by the International Mathematical Union. The first time I attended an international meeting in my area of research was six years after joining the Department, on my way to sabbatical leave in the US, making a stopover in Europe for the conference. It was also the first time that I met members of the research community with whom I had corresponded for a number of years. (As an aside, to give a sense of the limited facilities available at the time, my office in the

Chern Shiing Shen, Singapore, June 1980.[e]

Department during the first six years had no telephone. To make or receive phone calls, I had to use the one on the department secretary's desk on another floor of the building. A buzzer in my office would sound if the secretary received a phone call for me.)

Despite the difficulties, with some effort it was possible to secure funding from private sources, such as the Lee Foundation, to organise small-scale meetings in Singapore. This enabled, for example, the visit of Chern Shiing-Shen, one of the 20th century's great geometers, to Singapore in June 1980. When I met him at the Helsinki Congress and extended an invitation to him to visit Singapore, an English translation of his autobiographical article 学算四十年 (*Forty years of mathematical studies*) had appeared in the *Mathematical Medley*[3] (my colleague Leong Yu Kiang and I were writing a series of articles called "Notes on Mathematicians" for the *Medley*. The translated article was one of them). Chern accepted the invitation and spent a week in Singapore. More than 400 people attended Chern's public lecture. The auditorium was packed and latecomers had only standing room. This was an unprecedented success for a public mathematical lecture in Singapore (10 years later, the Fields Medalist Michael Atiyah gave a public lecture to an audience of almost 700, a number which has not been surpassed to date for such lectures).

---

[e] Photo courtesy of Leong Yu Kiang.

Singapore Group Theory Conference, June 1987.[f]

Research funding for the basic sciences began in 1980 upon the merger of Nanyang University and the University of Singapore to form the National University of Singapore (NUS). The 1980s was a period of rapid growth in mathematical activities. With the merger, the Lee Kong Chian Centre for Mathematical Research was then a centre in the new department in NUS. It provided financial support for mathematical conferences organized by the Department and the Society. An international conference was held almost every year during this period, each devoted to a specific field. These included group theory, analysis and topology. Each conference brought in leading figures in the field.

Particularly worth noting is the Singapore Group Theory Conference held in 1987[2] where the invited speakers included Walter Feit, John Thompson, Graham Higman, Michio Suzuki and Jean-Pierre Serre. Between 1980 and 1990, the Centre supported many short-term overseas visitors, most prominent of whom were Fields Medalists René Thom, Michael Atiyah, Jean-Pierre Serre and John Thompson. Serre, in particular, first visited Singapore in February 1985.[4] He made several return visits, most recently in 2009.

The second 15 years (1980–1995) of Singapore's independence also saw the Department grow in faculty strength. At one point, it exceeded 106.

---

[f] Photo courtesy of the Department of Mathematics, National University of Singapore.

Back row (left to right): Tan Sie Keng, Leong Yu Kiang, Brian Hartley, Karl Gruenberg, Michio Suzuki, Michel Broué, Derek Robinson, Jean-Pierre Serre, John Cannon, Malcolm Wicks. *Front row* (left to right): Peng Tsu Ann, Cheng Kai Nah, Walter Feit, Bernhard Neumann, Graham Higman, S.I. Adian, John Thompson, A.I. Kostrikin, Noboru Ito.
Singapore Group Theory Conference, June 1987.[8]

Most major fields of research were represented: logic, combinatorics, algebra (especially group theory), number theory, representation theory, analysis, differential geometry, algebraic and differential topology, differential equations, mathematical physics, computational mathematics, operations research, probability and statistics.

This was also a period during which talented NUS graduates pursued PhD studies at some of the best departments in the US, funded either by the University or by the host institutions. However, not all of them stayed on in academia. In 2015, 10 were on the faculty at the NUS: Tay Yong Chiang (PhD in computer science, Harvard University), Tan Eng Chye (now Provost of NUS) and Lee Soo Teck (both PhD in representation theory, Yale University),

---

[8] Photo courtesy of Peng Tsu Ann.

Jean-Pierre Serre, Singapore, 1985.[h]

Chan Heng Huat (PhD in number theory, University of Illinois at Urbana-Champaign), Victor Tan (PhD in number theory, UCLA), Toh Kim Chuan (PhD in operations research, Cornell University), Loke Hung Yean (PhD in representation theory, Harvard University), Loh Wei Liem and Chan Hock Peng (both PhD in statistics, Stanford University), and Teo Chung Piaw (PhD in operations research, MIT). The Department also made a concerted effort to recruit promising young mathematicians, some of whom are today acknowledged leaders in their fields. In recent years, the search for global talents, at both the junior and senior levels, has intensified with notable success.

In 1998, the University set up the Department of Statistics and Applied Probability, and the group in statistics moved to the new department. This move resulted in the reduction of faculty size in the Department of Mathematics, and the number of tenure track faculty since then hovers between 55 and 65. Despite this, the quality of mathematical research continued in the upward trend. Evaluation of research performance shifted from the number of papers published in international refereed journals to the quality and significance of work done.

Mathematical research in Singapore was marked with a new milestone in 2000 with the establishment of the Institute for Mathematical Sciences (IMS) at the National University of Singapore. The idea of a mathematical institute in Singapore dates back to the 1980s following the enormous success of the MSRI — Mathematical Sciences Research Institute — in Berkeley. (I recall at the party given by Chern Shiing-Shen at his residence (with a spectacular view of

---

[h] Photo courtesy of Leong Yu Kiang.

Institute for Mathematical Sciences.[1]

the San Francisco Bay) during the 1986 ICM in Berkeley, Chern showed Peng Tsu Ann and me his "Director's chair", a gift from the Institute at which he served as its Founding Director from 1981 to 1984. The idea of a mathematical institute in Singapore modelled after the MSRI later became a regular topic of conversation between Peng and several of us in the Department). Proposals for such an institute were submitted to the University during the interim years. With the country's decision to move into knowledge economy in the new century, the Ministry of Education gave approval to provide initial funding for such an institute to be set up within NUS. The IMS organises programmes and workshops with cooperation from the Department of Mathematics. Louis Chen was appointed the Founding Director and served until 2013. Roger Howe of Yale University was the Chair of the IMS Scientific Advisory Board during the period 2001–2010 and was succeeded by Siu Yum Tong of Harvard University. Occupying two double-storey colonial houses in Prince George's Park of the hilly Kent Ridge campus, IMS hosts more than 700 mathematical scientists every year from Singapore and around the world, who give lectures or participate in research collaboration at the Institute. Together the visitors generate a high level of vigour, vibrancy and scientific activities not previously seen in Singapore.

---

[1] Photo courtesy of Louis Chen.

In 2005, Nanyang Technological University established the School of Physical and Mathematical Sciences with Ling San as its Head. The Division of Mathematical Sciences now numbers more than 20 on its tenure track faculty and covers research areas in coding theory, logic, algorithms and theoretical computer science, computational mathematics, number theory, probability and topology. Within a short span of 10 years, NTU has established a research-intensive division in mathematics. A number of its research groups, for example the group in coding theory, are highly regarded internationally.

Along with increased research funding, the Ministry of Education also provided more scholarships for graduate education. This led to a significant growth of the doctoral programme in the mathematical sciences. By 2005, the number of PhD students in the Department had exceeded 60. That number went past 100 by 2015, including all the students enrolled in doctoral mathematical programmes in Singapore. Many graduates have successfully secured postdoctoral positions at leading research universities in Europe, US and Canada (for example, Oxford, Vienna, Münster, Stanford, Purdue, Toronto).

The Department of Mathematics at NUS moved to its newly-renovated building in two phases, beginning in late 2009 and completing more than a year later. The new home was designed with the objective of creating an environment conducive to teaching, learning and research. For the first time since expanding its doctoral programme, faculty members and graduate students have offices in the same building which also houses seminar rooms and a lounge — a working environment that is a far cry from that which I experienced many years before.

Mathematics in Singapore entered a new phase in 2010 when Shen Zuowei of NUS was invited to give a 45-minute lecture at the ICM in Hyderabad, in the Numerical Analysis and Scientific Computing section. While an ICM invitation may not be considered a major event in a country with a long mathematical tradition, it nevertheless marked "the coming of age" for mathematics in Singapore, as the invitation was based on work done at NUS.

The international recognition continued four years later at the ICM 2014 in Seoul, where four members from NUS (Bao Weizhu, Gan Wee Teck, Shen Weixiao and Yu Shih-Hsien) were invited to give 45-minute lectures (in the sections of Mathematics in Science and Technology, Number Theory, Dynamical Systems and Ordinary Differential Equations, and Partial Differential Equations, respectively). In the development of mathematics in the country, this was a high point and an historic achievement.

Indeed, the progress was dramatic overall. By 2015, members of the two departments of mathematics were collectively on no less than 60 editorial boards of major mathematical journals. Publishing in the most elite pure

Mathematics Lounge, NUS.[f]

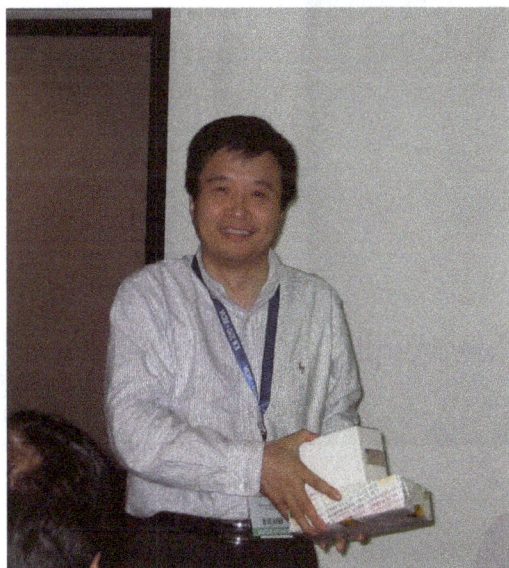

Shen Zuowei, ICM 2010.[j]

_____

[j] Photo courtesy of Sherman Riemenschneider.

Gan Wee Teck, ICM 2014.[f]

Bao Weizhu, ICM 2014.[k]

---

[k] Photo courtesy of Bao Weizhu.

Shen Weixiao and Yu Shih-Hsien.[f]

mathematics journals such as the *Annals of Mathematics*, the *Journal of the American Mathematical Society*, and *Inventiones Mathematicae* was no longer considered an unattainable goal, as demonstrated by several colleagues. Invitations to speak at major conferences in one's research area were no longer a novelty. Several members have held senior positions in international scientific societies. For example, Louis Chen served as President of the Institute of Mathematical Statistics as well as President of the Bernoulli Society, while Chong Chi Tat was a member of the Executive Committee of the Council of the Association for Symbolic Logic. A few have their names affixed to mathematical terms, for example the Chen-Stein method in probability, the Ron-Shen duality principle in time-frequency analysis, and the Gan-Gross-Prasad conjecture in representation theory and number theory.

Along with advances in pure mathematics, research in applied and computational mathematics at NUS went through a major transformation during the last 25 years. In many respects this was in line with the rapid changes one witnessed in the role mathematics — particularly applied and computational mathematics — played in modern science and technology. From the early 1990s, the Department sought to strengthen its group in these areas. Today it is known internationally for its research in imaging science, scientific computing, operations research and quantitative finance. Members of the group serve on the editorial boards of top journals in the fields, and Shen Zuowei was one of the 27 invited speakers at the *International Congress of Industrial and Applied Mathematics* (ICIAM) in 2015. The ICIAM, held once every four years like the ICM, is the most prestigious international meeting in applied mathematics.

The first 50 years of mathematics in Singapore since independence ended on a high note. Looking back, this came as the combined effort of many individuals with the shared vision that Singapore would one day make a presence on the world mathematical stage. The journey was arduous but not without its rewards. The next 50 years will be even more challenging: in a world of steep competition for talent and excellence, maintaining the current position will be hard and making progress will be harder still. There is no assurance that the path taken will still lead to trees that bear fruits. And a structure that was painstakingly erected may still crumble over time. How history will record the first 100 years of mathematics in post-independence Singapore will depend on how the journey will be taken, and how the department's course is charted in the coming decades.

Acknowledgement

The author is indebted to Louis Chen, Leong Yu Kiang and Peng Tsu Ann for sharing their memories of the department during the period 1965–1980. Thanks are also due to them as well as to Shen Zuowei, Sun Yeneng and Zhu Chengbo for their many helpful comments, corrections and suggestions. This article is very much the result of a group effort and their invaluable contributions are deeply appreciated.

References

[1] Louis Chen Hsiao Yun, Mathematical Reflections, *Science NUS: Internal Monthly Newsletter — Faculty of Science*, November 1999.

[2] K. N. Cheng and Y. K. Leong, *Group Theory: Proceedings of the Group Theory Conference held at the National University of Singapore, June 8—19, 1987*, Walter de Gruyter, 1989.

[3] Chern Shiing Shen, Forty Years of Mathematical Studies (translated by C. T. Chong), *Mathematical Medley* 3 (1975), 22–44.

[4] C. T. Chong and Y. K. Leong, An interview with Jean-Pierre Serre, *Mathematical Medley* 13 (1985), 11–19 (Reprinted in *The Mathematical Intelligencer* 8 (1986), 8–13).

[5] Peng Tsu Ann, *A brief history of the Department of Mathematics*, http://ww1.math.nus.edu.sg/misc/History%20of%20Department.pdf.

Chapter 8

Physics at NUS

Tang Seung Mun, Feng Yuan Ping, Oh Choo Hiap, Lai Choy Heng, Ong Chong Kim, Sow Chorng Haur and Andrew T S Wee

The Pre-War Period

It is probably not known to many that the teaching of Physics at university level in Singapore began before Raffles College was established. Physics was a compulsory subject for the medical students in King Edward VII College of Medicine and it was taught by the Government Chemist and other offices of the Government medical service. In 1927, Dr E Madgwick was appointed Reader in Physics at King Edward VII College and was later appointed Professor of Physics at Raffles College.

The first science students of Raffles College were enrolled in 1928, and in 1929 the Manasseh Meyer Science Building at Bukit Timah campus was opened, half of it being made available to the Physics Department. Professor Madgwick ran a one-man show for about three years until 1931 when Mr C G Webb joined the department as Demonstrator in Physics. In August 1935, Professor Madgwick resigned and the work of the department was carried on by Mr Webb with the assistance of two demonstrators. Dr N S Alexander from New Zealand arrived in Singapore in late 1936 to occupy the Chair in Physics. Together with Mr Webb and Mr Hon Yung Sen, a local graduate who was appointed as Demonstrator in Physics in 1937, they were the only teaching staff of the Physics Department until the Japanese invasion in 1941.

In those days, the department provided a three-year course for the College diploma, covering mechanics, heat, light, sound, electricity and magnetism. The courses given in each year consisted of lectures and a series of laboratory experiments. In addition, there was a course for medical and dental students, which was similar to but simpler than the first-year course for the science students. The number of medical and dental students taking physics was usually much larger than that of science students. In the academic year 1935/36, for example, 44 medical and dental students read physics while only 11 science students took first-year physics, five took second-year physics and two took final-year physics.

The Years Immediately after the War

The physics laboratories, almost empty and without services, were re-opened in October 1946. At that time, Professor Alexander was away from Singapore and Mr Webb took charge of the restoration of the laboratories and attempted to secure the delivery of books and apparatus from the UK. After a year of improvisation, the Physics Department was back in working order, though on a minimal basis.

Professor Alexander returned in December 1947, but Mr Webb went on leave in early 1948. Mr Hon was away on study leave in UK. The department was short of staff and this retarded to a considerable extent the work of rehabilitation, of which a good deal remained to be done at that time. During that year a number of third-year students gave valuable assistance as part-time demonstrators in the practical classes for first-year medical and dental students.

The Early Years of the University of Malaya

In the seven-year span between 1949 and 1956 after the University of Malaya was formed from the amalgamation of Raffles College and King Edward VII College of Medicine, as many as 12 teaching staff were recruited. However, all except two of them stayed for only one or two terms of contract. During this period, both Professor Alexander and Mr Webb had resigned, the former in 1951 and the latter in 1956. Mr Hon Yung Sen became the Acting Head. He took charge of the department for three years until 1959 when Dr K M Gatha arrived to assume the headship.

Since the end of the War up to 1956, the Physics Department occupied a section of the Manasseh Meyer Block and a temporary hut built by the Japanese during the war. Space was very limited and this placed a constraint on student intake. The problem was not solved until late 1957 when the new Physics Building was completed. The total number of students attending Physics courses each year fluctuated around 85 throughout these years. More than half of these students were medical, dental and engineering students. The Honours course was introduced in 1949. The Department produced about 15–20 Pass degree physics graduates and two to five Honours degree graduates each year. Many new physics subjects were introduced, but mainly for the Honours students. Among them were quantum mechanics, nuclear physics, X-ray crystallography, relativity, electrodynamics and ionosphere physics.

The Sixties to the Eighties at the University of Singapore

In 1960, the governments of the then Federation of Malaya and Singapore indicated their desire to change the status of the divisions into that of a national university. Legislation was passed in 1961 establishing the former Kuala Lumpur division as the University of Malaya while the Singapore division was renamed the University of Singapore on 1 January 1962.

From 1959 to the mid-sixties, the teaching staff number in the Physics Department increased slowly from seven to 10. It further went up to 16 in 1973 as many returned scholars joined the Department. This number remained about the same for the rest of the decade. Professor Gatha passed away in 1967 and Mr Hon again acted as Head of the Department until 1969 when Professor A Rajaratnam was appointed the Department Head.

The number of Pass degree physics graduates increased steadily from 20 in 1959 to a peak of 79 in 1967. After this peak, it dropped to a low of 22 in 1973 and varied between 30 and 40 for the next few years. The number of Honours degree graduates also followed the same trend, rising from three in 1959 to 31 in 1968, and then declining to 13 in 1974 (see Fig. 1). The drop in the physics student enrolment after 1968 was due to the establishment of the Engineering Faculty and the fact that very few students from Malaysia gained entry to the

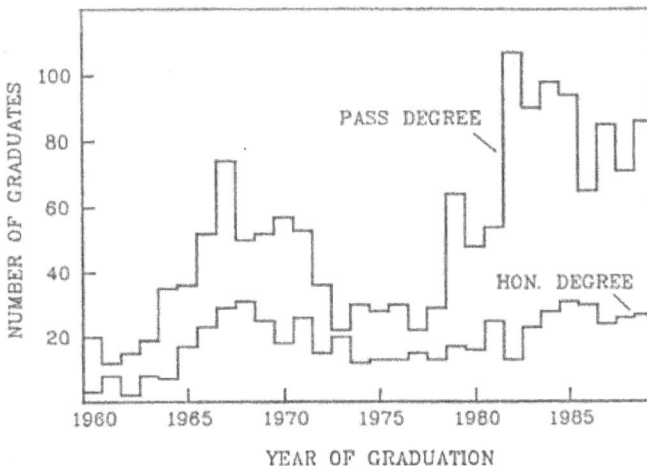

Fig. 1.   Number of Physics graduates over the period 1960–1989.

University. Interestingly, since the 1990s to 2015, the number of Honours degree physics students has remained approximately constant at around 30–40, suggesting that despite new degrees and curricula structures being implemented, the number of "die-hard" physics students has remained constant over the years.

The physics course content and structure changed a great deal in these 20 years. In 1964, an additional course was introduced for third-year students and henceforth, there were two physics courses — Theoretical Physics and Experimental Physics. Students intending to read Honours in physics had to take both courses in their third year. In 1973, the courses were revised and re-structured and the syllabi expanded in all years. Theoretical Physics and Experimental Physics were renamed Physics and Applied Physics respectively. The revision of the physics courses was done to provide students with training relevant to local industries. Subjects such as workshop technology, linear programming, computer programming, applied electronics and materials science were included in the applied physics course.

In 1966, the extension to the Physics Building in the Bukit Timah campus was completed. The building was formally declared open by Prime Minister Mr Lee Kuan Yew on 1 July 1966.

In the area of research, the productivity was generally low in the sixties and seventies. Emphasis was placed on teaching and producing graduates who were in demand by the local industries.

The Merger to form the National University of Singapore

The number of physics academic staff members increased from 15 to 23 with the formation of the Joint Campus and the subsequent merger of the University of Singapore and Nanyang University in 1980 to form the National University of Singapore. Thereafter, it further rose to 29 before the end of 1983. In 1982, Professor Lim Yung Kuo was appointed to head the Department. His term of office lasted six years until he retired on 31 October 1988.

Together with the other departments of the Science Faculty, the Physics Department moved to the Kent Ridge campus in June 1981 before the buildings were fully completed. The laboratories became operational in August 1981 and all equipment purchased under the IBRD (International Bank for Reconstruction and Development) loan were installed within the next few months.

The physics student number doubled immediately after the merger but declined slightly in the next few years. The main reason for this was that the

Department of Information Systems and Computer Science had increased their student intake during this period, with the Physics Department having typically about 250 first-year students, 120 second-year students, 90 third-year students, 30 Honours students and 20 higher degree students. From 1988, the Department took over the running of the Computer Programming and Applications course from the Department of Information Systems and Computer Science. The number of students taking this course approached 700 that year.

A new subject General Physics was introduced in 1988. This was intended as an optional course for first-year science students who did not plan to major in physics and for second-year students who had not read Physics in their first year. The physics courses were re-structured and revised again before the 1989/1990 session began. The new course structure enabled students to concentrate on fewer topics at a time and allowed them greater flexibility in their choice of elective courses.

The research facilities in the Department improved tremendously during the 1980s. This was largely due to the new University policy of increased importance to research. Research grants were generously awarded to academic staff and research scholarships were readily available to students with good Honours degrees for graduate studies. Consequently, the number of research publications by staff members increased significantly.

By 1989, the Department had grown from a one-man operation to a full-fledged department with 34 academic staff and 50 technical and office staff. Since its establishment in 1928 until 1989, the Department had produced about 1,600 Pass degree physics graduates and over 500 graduates with Honours and higher degrees. Photographs of some of our past staff and students are shown in Fig. 2. The Diamond Jubilee in 1989 marked the end of the first 60 years and also the beginning of the era of increased support for research and development (R&D) in Singapore.

The Nineties and Beyond

The National Science and Technology Board (NSTB) launched the first National Technology Plan (NTP) in September 1991. Focusing on economically driven R&D, the government allocated S$2 billion for a five-year plan which served as the blueprint for R&D development in nine sectors: information technology; microelectronics; electronic systems; manufacturing technology; materials technology; energy, water, environment and resources; food and agrotechnology; biotechnology; and medical sciences.

Table 1.   Heads of Department (1929–2015).

1928–1935	Madgwick, E	1959–1967	Gatha, K M
1935–1936	*Webb, C G	1967–1969	*Hon Yung Sen
1936–1951	Alexander, N S	1969–1982	Rajaratnam, A
1951–1956	Web, C G	1982–1988	Lim Yung Kuo
1956–1959	*Hon Yung Sen		

(* Acting)

(*Continued*)

Not available

1928-1935
Madgwick E

1935-1936
C G Webb (Acting)

1936-1951
Norman Stanley
Alexandra

1951-1956
C G Webb

1956-1959
Hon Yung Sen (Acting)

1959-1967
Gatha K M

1967-1969
Hon Yung Sen (Acting)

1969-1982
Rajaratnam A

1982-1988
Lim Yung Kuo

Table 1. (*Continued*)

1988–1995	Tang Seung Mun	2000–2006	Oh Choo Hiap
1995–1997	Tan Kuang Lee	2006–2007	Andrew Wee
1997–1998	Tan Tiong Gie	2007–2014	Feng Yuan Ping
1998–2000	Lai Choy Heng	2014–	Sow Chorng Haur

1988-1995
Tang Seung Mun

1995-1997
Tan Kuang Lee

1997-1998
Bernard Tan Tiong Gie

1998-2000
Lai Choy Heng

2000-2006
Oh Choo Hiap

2006-2007
Andrew Wee

2007-2014
Feng Yuan Ping

2014
Sow Chorng Haur

Fig. 2.   Photographs of some of our visitors, staffs and students between 1961 and 1986.

The Physics Department underwent major expansion in its research facilities from this period. Major facilities and centres set up include the van de Graff accelerator (Box Story 1), the Centre for Ion Beam Applications (Box Story 2), the Surface Science Laboratory (Box Story 3), the Centre for Superconducting and Magnetic Materials (Box Story 4); the Centre for Quantum Technologies (Box Story 5), and most recently the Centre for Advanced 2D Materials (Box Story 6).

The Department further expanded to about 60 academic staff and 45 technical and office staff by the end of 2015. With the increased investment in research funding and strategic hiring of high quality academic staff over the past two decades, the quality and quantity of physics research has rapidly increased. This is reflected in various publications, citations and ranking indicators. For example, the QS World University Rankings by Subject 2015 for "Physics & Astronomy" saw NUS ranked 23rd in the world.

The Heads of Departments since the establishment of the NUS Physics Department are shown in Table 1.

Box Story 1: XRF and PIXE Research (1982–2002)
Tang Seung Mun

The XRF (X-Ray Fluorescence) and PIXE (Proton-Induced X-ray Emission) research facilities were developed from the teaching equipment acquired to establish the nuclear teaching laboratory when the Physics Department moved from the Bukit Timah campus to the Kent Ridge Campus in 1981. With funding to set up the nuclear teaching laboratory, the following major instruments were purchased: a 2.5 MeV van de Graaff accelerator, several types of radiation detectors (Si(Li), Ge(Li), NaI(Tl), surface barrier) with associated electronics, a multi-channel analyzer, a number of radioisotope excitation sources (Co-57, Fe-59, Cd-109 and Am-241). With these instruments, a number of nuclear experiments for teaching purposes were set up for the second-year, third-year and the honours students. For second-year students, there was the experiment to detect charged particles using a surface barrier detector. Two experiments were designed for third-year students. They were the Compton scattering experiment and the Mossbauer spectroscopy experiment. An experiment on X-ray fluorescence spectroscopy and another on coincidence spectroscopy were available for honours students. The van de Graaff accelerator was mainly used for honours projects on neutron activation analysis with fast neutrons generated from bombarding a tritium target with a deuteron beam.

Shortly after the setting up of the radioisotope source excited XRF facility, Prof C T Yap realised that this facility was very suitable to authenticate Chinese porcelains made in different periods in China because the kaolin clay and the glaze of porcelains contained elemental signatures which depended on the time and the place where they were made. With the XRF facility, he carried out an extensive study of his own vast collections of Chinese porcelains made in the past 300 years and published many research papers in the 1980s. The XRF facility was also used by other faculty members of the Physics Department to develop XRF techniques for precision elemental analysis.

In the early 1980s, the use of the van de Graaff accelerator for research was confined to the study of conversion coefficients in decays of radioisotopes produced using fast neutrons generated by the (d,n) reaction. In 1983, at the request of the International Atomic Energy Agency (IAEA), the Physics Department hosted an IAEA regional workshop on Particle-Induced X-ray Emission (PIXE). The van de Graaff accelerator was re-configured for PIXE applications with equipment (a PIXE chamber, a Si(Li) detector, a multichannel analyser and some signal counting electronics) given by IAEA to conduct the workshop. Since then, the PIXE facility was further developed and used by Prof S M Tang and members of his research team for numerous research projects on elemental/stoichiometric

(Continued)

Box Story 1 *(Continued)*

analysis of gemstones, coastal waters, marine sediments, aerosol particles and Y-Ba-Cu-O superconductors.

In 1992, a quadrupole focusing magnet system was acquired with the teaching laboratory upgrading fund to upgrade the broad-beam PIXE system

(Continued)

Box Story 2: The Centre for Ion Beam Applications (2001–Present)
Frank Watt and Thomas Osipowicz

The applied nuclear physics group set up by Prof Tang Seung Mun grew in stature with the addition of the new state-of-the-art nuclear microscope, where the ability to focus million volt protons and helium nuclei down to sub-micron spot sizes, led to a wide range of new and additional microscopy application areas, from advanced material research to biomedicine. The purchase of a 3.5-million volt high stability dynamitron type accelerator (HVEE Singletron©) to replace the ageing Van de Graff accelerator improved the performance of the nuclear microscope. In 2001, the facility was named the Research Centre for Nuclear Microscopy (Figs. 3 & 4).

Fig. 3. Mr Philip Yeo, Chairman of the National Science and Technology Board, opening the Research Centre for Nuclear Microscopy with the Centre Director, Prof Frank Watt, in 2001.

(Continued)

Box Story 1 (*Continued*)

to a nuclear microscope capable of delivering 100-pA of proton beams with a sub-micron spot size. This nuclear microscope was used for student projects and a wide range of research works for a decade until it was replaced by a new nuclear microscope purchased under a research grant by Prof Frank Watt in 2003.

Box Story 2 (*Continued*)

Fig. 4.   In 2001, NUS President Prof Shih Choon Fong was presented with a momento, that of a Singapore dollar coin where the NUS logo was written on to the dollar sign, using the new technique of proton beam writing, pioneed in the Research Centre for Nuclear Microscopy.

In early 2000, it was realised that high energy ions (such as MeV protons and alpha particles), as well as being useful as a novel form of microscopy, had a much broader range of application areas. The Centre then expanded to include high depth resolution surface studies, proton beam writing (a new 3D micromachining technique pioneered in the centre which has application areas in microphotonics, microfluidics, lab-on-a-chip technology and silicon nanodevices), and more recently proton beams for cancer therapy — medical physics).

In mid-2000, in order to reflect this rapid expansion in application areas, the Research Centre for Nuclear Microscopy was then renamed the Centre for Ion Beam Applications. At present, the Centre has five distinct beam lines facilities, many of which are state of the art or unique, and has 30 members made up of six academic staff, post-doctoral researchers, research assistants, technicians and graduate students. The Centre can boast world record performances in terms of focusing MeV ions to nanodimensions, and a diverse and multidisciplinary applications programme.

Box Story 3: The Surface Science Laboratory (1986–Present)
Andrew T S Wee

The Surface Science Laboratory (SSL) was established in 1986 at the Department of Physics, National University of Singapore.[1] The first ultrahigh vacuum (UHV) system delivered in 1987, possibly the first in Singapore, was a Vacuum Generators (VG) ESCALAB Mk 2/SIMSLAB (Fig. 5), acquired with a S$1.5 million Science

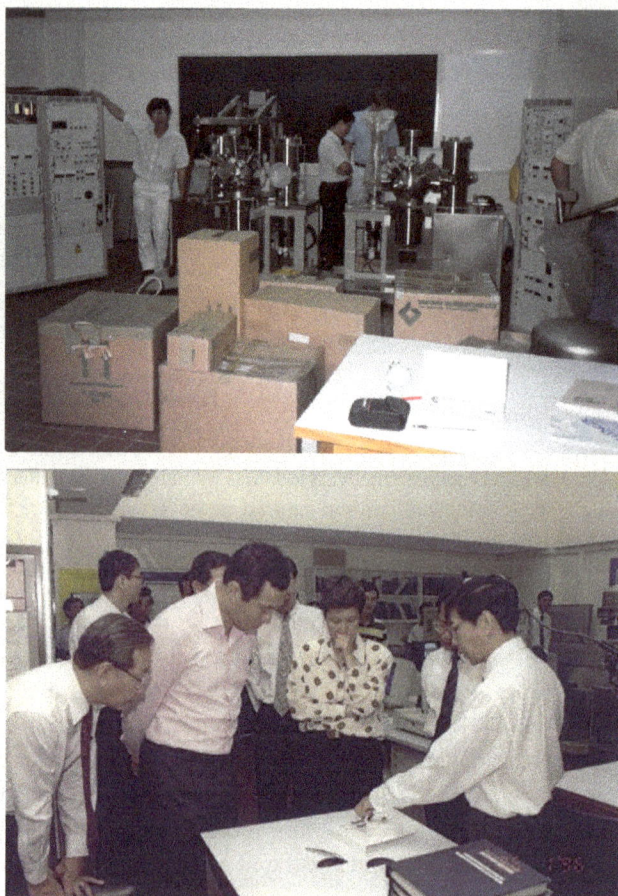

Fig. 5.   (top) Delivery and installation of the VG ESCALAB Mk 2/SIMSLAB at the NUS Surface Science Laboratory; Professor Tan Kuang Lee is standing on the left. (bottom) Prof Tan hosting a 1996 visit by PM (then DPM) Lee Hsien Loong.

---

[1] http://www.physics.nus.edu.sg/~surface/

(Continued)

Box Story 3 (*Continued*)

Council grant. The then Director, Professor Tan Kuang Lee, and his chemical engineering collaborators Professors Kang En-Tang and Neoh Koon Gee did prolific and pioneering research in photoelectron spectroscopy of polymers on this instrument, and their achievements were recognised when they won the 1996 National Science Awards.

Prof Andrew Wee joined NUS in 1990 and succeeded as lab director after Prof Tan Kuang Lee retired. He and other surface science colleagues helped grow the Surface Science Laboratory from a single UHV system to several state-of-the-art UHV systems today. As one of the earliest major research laboratories in NUS, the Surface Science Laboratory is recognised today as one of the world's leading surface science groups producing numerous international journal publications and patents, and participating in numerous international collaborations. The lab also actively engages local industry, and has organised surface and interface analysis workshops to raise Singapore companies' competency in state-of-the-art materials characterisation techniques.

The laboratory currently houses five ultrahigh vacuum (UHV) systems (as of mid-2015), including the ESCALAB, Omicron low-temperature scanning tunneling microscope (LT-STM), Omicron variable-temperature (VT)-STM, ultraviolet/X-ray photoelectron spectroscopy (UPS/XPS) system, and Cameca IMS 6f Secondary Ion Mass Spectrometry (SIMS) system, along with an array of thin film growth systems, and other characterisation equipment. In 2000, a S\$3 million NSTB[2] grant was received to set up the Surface-Interface-Nanostructure-Science (SINS) beamline and end-station (Fig. 6) at the Singapore Synchrotron Light Source (SSLS).[3] The synchrotron source produces highly monochromatic and tunable X-rays, serving as a unique probe to study surfaces and interfaces.

From 2005, the Surface Science Laboratory published numerous papers on graphene, 2D monolayers, and hybrid organic-inorganic heterostructures. The LT-STM system has been particularly instrumental in allowing the visualisation of atomically resolved images of graphene and adsorbed molecules. Figure 7 shows a photograph of the LT-STM with growth chamber attached, as well as a typical STM image of epitaxial graphene, which was highlighted in *Nature News* on 25 March 2009.[4] As of 2015, the lab engaged in STM studies of 2D materials

---

[2] National Science and Technology Board (NSTB), the predecessor of the Agency for Science, Technology and Research (A*STAR).
[3] http://ssls.nus.edu.sg/
[4] Nature News 25 March 2009; *Nature* 458, 390–391 (2009).

(*Continued*)

Box Story 3 (*Continued*)

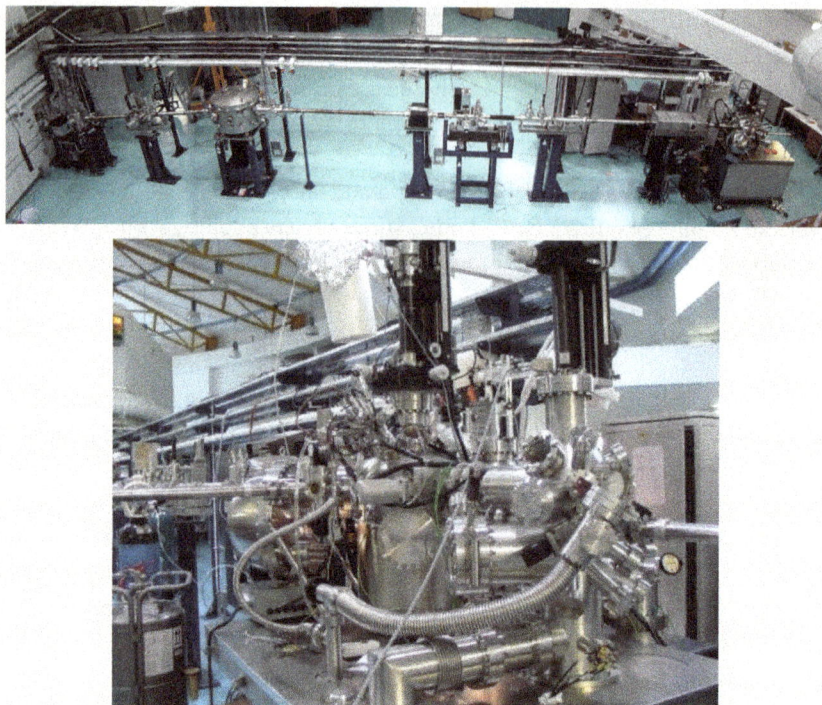

Fig. 6.   The Surface-Interface-Nanostructure-Science (SINS) beamline (top), and end-station (bottom) at the Singapore Synchrotron Light Source (SSLS).

such as transition metal dichalcogenide (TMD) monolayers and phosphorene. Unlike graphene, TMDs such as MoS_2 and WSe_2, are semiconductors with tunable direct bandgaps dependent on the number of atomic layers, and have potential electronic and optoelectronic applications.

Indeed, surfaces are critically important in modern technology, such as in the fields of heterogeneous catalysis, optical and electronic devices, protective

(*Continued*)

Box Story 3 (*Continued*)

coatings, adhesion, sensors, energy storage and generation and so on. We expect an exciting future for the Surface Science Laboratory in the years to come.

"God made the bulk; surfaces were invented by the devil." Wolfgang Pauli[5]

Fig. 7. (left) Low temperature scanning tunnelling microscope (STM) with growth chamber attached; (right) STM image of epitaxial graphene with model schematic superposed (*Nature News*, 25 March 2009; *Nature* 458, 390–391 (2009)[6]).

---

[5] As quoted in *Growth, Dissolution, and Pattern Formation in Geosystems* (1999) by Bjørn Jamtveit and Paul Meakin, p. 291.

[6] http://www.nature.com/news/2009/090325/full/458390a.html

Box Story 4: Centre for Superconducting and Magnetic Materials
(1998–Present)
Ong Chong Kim

The Centre for Superconducting and Magnetic Materials (CSMM) at the Department of Physics was established in May 1998. Professor Ong Chong Kim was the founding and current director. The centre is internationally known for its research in superconductive physics and devices. Other fields of research include materials science, optics, ferroelectrics, and ferromagnetics.

In February 1999, soon after its founding, CSMM played a key role as one of the four founding centres in the inception of the Institute of Engineering Science (IES). IES was a university-level research institute for talented researchers who desired to focus on longer term high profile and high quality multi-disciplinary research.

The past, present and future areas of research at CSMM include:

- Developing functional metamaterials for microwave electronics, especially high T_c superconducting, magnetic and ferroelectric nanostructures.
- Using thin films to make miniature high performance devices.
- Exploring new design ideas and innovative applications.
- Developing "proof of concept" products and prototypes.
- Designing and building sophisticated materials fabrication and characterisation facilities in-house to advance research.
- Researching fundamental materials science.

Since its inception, the CSMM has been actively involved in organising conferences in NUS for materials researchers and industry to share their latest findings, explore directions for research and collaboration, and to transfer new findings to industries. The inaugural workshop organised by CSMM was held in Singapore on December 1999 under the title "JSPS-NUS workshop on Superconducting Thin Film and Devices". The workshop later evolved into the East Asia Symposium on Superconductive Electronics series, which is currently organised once every two years rotating among Singapore, Japan, Korea, Taiwan and India. Figure 8 shows a photo of the participants at the inaugural conference in 1999.

As the economic and technological importance of magnetism and electromagnetic materials grew in Singapore, CSMM took the lead by organising symposiums on magnetism and superconductivity under ICMAT in Singapore, China and India. CSMM helped to organise a workshop on high temperature (Tc) Superconducting Materials and Applications in Hanoi in 2004. CSMM was also a key contributor to the "International Conference on Space Charge Physics in

(Continued)

Fig. 8. The inaugural East Asia Symposium on Superconductive Electronics held in December 1999 in Singapore.

Dielectric" organised by the French Vacuum Society and the IEEE-DEIS technical committee.

Starting in September 1999, CSMM and the Data Storage Institute began hosting an annual workshop on magnetic materials and devices, which has provided a forum for academics, members of research institutes and industrial researchers to meet.

As of 2015, the CSMM occupies a floor area of $220\,m^2$ at the Department of Physics and consists of well-equipped research laboratories that have been staffed by generations of talented and committed researchers. CSMM has been well funded by the Defence Science Organisation (DSO) and other government agencies, and has made important contributions in the study of materials physics and the development of hybrid multi-layer thin film devices. Most of the state-of-the-art facilities are designed and constructed in-house.

One of CSMM's strengths has been its ability to integrate materials research and in its preparation and characterisation of film and nanostructures. Its approach in the measurement of permittivity and permeability has been well received in the scientific community. Figure 9 shows the shorted microstrip fixture used to characterise dynamic properties of magnetic thin films. It can measure up to 12 GHz at elevated temperatures of up to 150°C. An external magnetic field can be applied at any angle to the samples through Helmholtz coils. The device has now become an international standard for characterising

(*Continued*)

Fig. 9. The shorted microstrip fixture for characterising magnetic thin film which became an international standard.

the dynamic properties of ferromagnetic thin film. The fixture later extended into the sensitive detection and quantification of Spin Rectification Effect (SRE). It is useful for spin current measurement and spin wave characterisation.

CSMM is also highly regarded in the scientific community for its capabilities in device design and implementation. It uses two parallel approaches to improve the performance of devices by studying the underlying science of the materials and by improving the device design through technology development and theoretical simulation.

CSMM is constantly breaking new grounds in R&D. The core of its success lies in a conducive research atmosphere, in-house design expertise, close collaboration with research institutions in Singapore such as Temasek Labs, DSO, IMRE, and DSI and the ability to serve the R&D needs of the industry. This has yielded excellent results. For example, CSMM demonstrated that crystalline yttrium-stabilised zirconium oxide as a high k dielectric can be used as an alternative to SiO_2 in MOS devices. This solution was recommended by the industry journal, *Semiconductor International*, in May 2001 and it resulted in high k dielectrics being adopted for use in commercial products in the semiconductor industry.

One of CSMM's achievements in the field of metamaterials was the fabrication and implementation of an omnidirectional retroreflector, a device once

(*Continued*)

Box Story 4 (*Continued*)

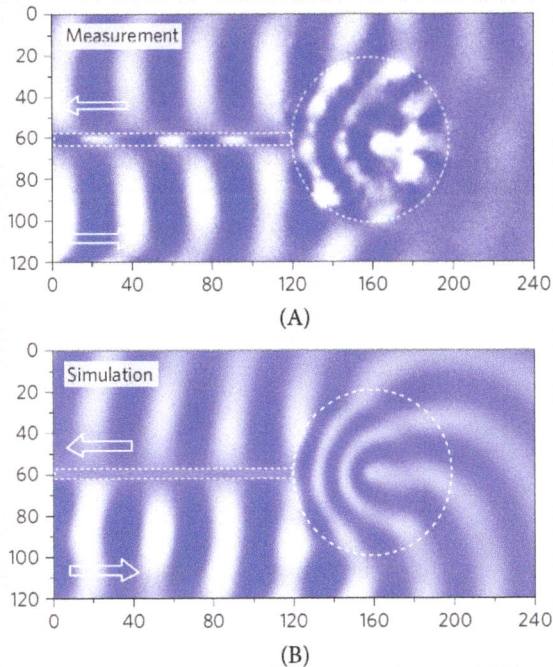

(A)

(B)

Fig. 10. The omnidirectional retroreflector, a device once deemed scientifically impossible.

deemed impossible to construct, that can reflect microwaves perfectly back to their source from all angles (Fig. 10). This has opened up a world of practical applications, especially for radar tracking.

The paper was published in *Nature Materials* (volume 8, pages 639–642, 2009) entitled "An Omnidirectional Retroreflector Based on the Transmutation of Dielectric Singularities". The authors were Ma Y G, Ong C K, Tye T and Leohardt U. The paper was introduced and described by Dennis M R of the University of Bristol in the same issue of *Nature Materials* as, "A Cat's Eye for all Directions". The discovery was also reported in the *Straits Times* on 4 July 2009, page 8. It was also promoted by Discovery Channel News, on MSNBC's website.[7]

In 2004, experts from CSMM wrote a 680-page textbook on microwave measurement entitled "Microwave Electronics: Measurement and Materials

---

[7] http://www.msnbc.msn.com/id/31968012/ns/technology_and_science-innovat

(*Continued*)

Box Story 4 (*Continued*)

Characterisation" published by John Wiley and Sons Ltd. This book provides an in-depth coverage of both established and emerging techniques in materials characterisation based on microwave electronics. It represents the most comprehensive treatment of the topics up to that time. The materials characterised range from dielectrics, semiconductors, conductors, magnetic materials to artificial materials, while the electromagnetic properties characterised include permittivity, permeability, chirality, mobility and surface impedance.

CSMM received a US patent in 2005 for the "Electrically Tunable Microwave Devices with Patterned Ferroelectric Thin Film". This invention enabled the fabrication of patterned ferroelectric thin films without degrading either the

(*Continued*)

Box Story 5: Development of the Centre for Quantum Technologies (2007–Present)
Oh Choo Hiap and Lai Choy Heng

In tandem with NUS progressing into a world-class research university, the Physics Department firmed up a research road map in the year 2000.

The purpose of the roadmap was to identify, select and develop key areas and objectives as the short- and long-term research goals in the department. As with all good plans, the roadmap must be reviewed and modified regularly to suit changing needs and goals in Singapore and in the physics world. Five broad yet closely related areas were selected, incorporating existing research and proposing new areas:

• Physics of Nanostructures.
• Condensed Matter and Materials Physics.
• Quantum Computation and Information.
• Nonlinear Dynamics and Complex Systems.
• Biophysics.

The Quantum Computation and Information and Biophysics areas were new. These new directions turned out to be fruitful. One was the basis of the research centre of excellence, the Centre for Quantum Technologies (CQT), and

(*Continued*)

Box Story 4 (*Continued*)

substrate or conductive circuit layers, thereby improving the performance of the microwave devices. The device also has the advantage of being readily incorporated into most current fabrication processes for ferroelectric thin films. This invention has widespread applications in modern devices such as phase shifters (used in phased array antenna for radar tracking systems, multi-point communication systems and satellite broadcasting), tunable filters (for cellular base stations and relay satellites), and tunable matching networks (for amplifiers, filters and balances).

To date 33 PhD, 34 MSc and 57 final year BSc students have completed theses based on their research at CSMM. In recognition of his career achievements and work at CSMM, Prof Ong Chong Kim received the NUS Outstanding Researcher Award in 2010.

Box Story 5 (*Continued*)

the other formed part of the effort in the formation of another research centre of excellence, the Mechanobiology Institute.

The Centre for Quantum Technologies plays a key role in positioning Singapore to become one of the leading nations in cutting-edge quantum technology. This important Centre had its origin in 1998 in a quantum information science team led by Lai Choy Heng and Oh Choo Hiap. With the support of NUS and then NSTB/A*STAR, the team grew in strength to become the Quantum Information Technology (QIT) Group in the physics department. With the effort and help from Lai Choy Heng (then Dean of Science), Chow Shui Nee (Director of Research), and Lim Khiang Wee (NSTB, A*STAR), a proposal for the Temasek Professorship (TP) was approved by Lee Tong Heng (Director of Research) and submitted to NSTB in Aug 2001. Artur Ekert (Oxford University, UK) was appointed as TP on 1 February 2002, with the TP Project in Quantum Information Technology (QIT) and a grant amount $4,999,750 for three years with Oh Choo Hiap as the Project Manager. A state-of-the-art experimental research laboratory in quantum information technology was established (Fig. 11).

Experiments studying the generation of entangled photons and free space quantum cryptography were accomplished. The QIT Lab attracted many visitors including Nobel Laureate Professor Tony Leggett in April 2005 (Fig. 12).

(*Continued*)

Box Story 5 (*Continued*)

Fig. 11.   Artur Ekert and other members of CQT .

With the A*STAR appointment of Artur Ekert as the Temasek Professor, the existing local expertise, and newly appointed researchers as NUS faculty members, the Quantum Information Technology group at NUS quickly became internationally recognised as one of the most dynamic and promising groups in the field. The group gained world recognition for its work with several breakthrough results published in top physics journals. It had established extensive collaborations around the world, including the University of Cambridge and Oxford in UK, the Moscow State University, the National Institute of Standards and Technology (USA), the University of Science and Technology (China) and Tsing-Hua University (China). The popular science magazine *New Scientist* in 2004 featured the team's research and cited the story as one of the 10 top science stories of the year.

In 2006, the team comprising Oh Choo Hiap, Berge Englert, Dagomir Kaszlikowski and Kwek Leong Chuan (NIE/NTU), received the National

(*Continued*)

Box Story 5 (*Continued*)

Fig. 12.   The visit by Nobel Laureate Professor Tony Leggett (first from left) in April 2005.

Science Award for their outstanding contributions to theoretical research on quantum coherence and its applications. Their work have potential for further experimental and industrial exploitation of coherent quantum phenomena (secure communication included) in Singapore.

In 2007, the NUS proposal for a Research Centre of Excellence (RCE) in quantum information science and technology was approved and the Centre for Quantum Technologies (CQT) was established officially on 7 December 2007. The NRF and MOE awarded CQT S$195 million in core funding over 10 years. The aim is to develop a world-class research programme focused on understanding, controlling and exploring potential applications of quantum phenomena. Research activities span both theory and experiment. CQT has since brought together quantum physicists and computer scientists to explore the quantum nature of reality and quantum possibilities in technology.

Box Story 6: Graphene Research Centre & Centre for Advanced 2D
Materials (2014–Present)
Feng Yuan Ping

Graphene-related research in the Physics Department NUS, began before graphene became a hot research topic. In 2000, Chen Ping *et al* published a paper in the journal *Carbon*,[8] in which graphene was extensively referenced. Following the successful fabrication of single layer carbon by mechanical exfoliation by Andre Geim and Konstantin Novoselov in 2004, more researchers in the department started working on this amazing two-dimensional (2D) material. Research groups of Chen Wei, Andrew Wee, Feng Yuan Ping, etc joined the gold rush and started publishing their results from 2007. In late 2007, the Physics Department recruited Barbaros Özyilmaz, a leading young experimentalist in graphene research, through the NRF Research Fellowship scheme. At the same time, NRF awarded one of six projects in its pioneering Competitive Research Programme (CRP) to a team led by Loh Kian Ping of Chemistry Department to conduct research on graphene. Physics faculty members Andrew Wee, Feng Yuan Ping, and R. Mahendiran were among the co-principal-investigators of the project. With Barbaros joining the department and significant funding from NRF, graphene research in the department as well as in NUS gained momentum. In November 2008, the *Asian Conference on Nanoscience and Nanotechnology* (AsiaNANO2008) was held in Singapore, and a symposium on graphene was organised by Barbaros Özyilmaz and Loh Kian Ping. Many of the leading researchers in the young field of graphene research gathered in Biopolis, Singapore. They reported their latest results and shared their views on graphene research. Subsequently, a number of international workshops and conferences on graphene were organised in Singapore, which attracted the world attention to Singapore's efforts in graphene research.

---

[8] P. Chen, X. Wu, J. Lin, H. Li, K. L. Tan, "Comparative studies on the structure and electronic properties of carbon nanotubes prepared by the catalytic pyrolysis of CH_4 and disproportionation of CO", Carbon 38 (2000) 139–143.

(Continued)

Box Story 6 (*Continued*)

In 2010, NUS invested more than S$40 million to set up a Graphene Research Centre (GRC), the first in Asia. Antonio Castro-Neto joined the department and became the Director of GRC. The centre was hosted in the Faculty of Science and operated through the Physics Department. A number of new physicists were recruited, including NRF Fellows Goki Eda (2011), Shaffique Adam (2012), Slaven Garaj (2012), Quek Su Ying (2013), Lin Hsin (2013). In addition, Vitor Pereira and Jens Martin joined the department in 2010 and 2012, respectively. They conducted their research in the GRC and became core members of the centre. This further strengthened NUS' leadership as a world centre for graphene research. Over the years, the principal investigators of GRC have attracted significant additional research funding. Nobel Laureate Andre Geim has been a visiting professor at the centre and the department. Both Andre Geim and Konstantin Novoselov serve in GRC's Scientific Advisory panel. NUS is ranked among the top institutions for graphene research in terms of research output.

By 2014, research activities in the centre expanded to other 2D materials such as 2D transition metal dichalcogenides. To better reflect the research activities, the GRC was renamed the Centre for Advanced 2D Materials (CA2DM) and upgraded to a university level research centre in 2014. NRF provided additional funding of S$50 million to support research activities and commercialisation of research results. As of December 2015, CA2DM consists of four major research groups: Graphene, Other 2D Materials, Devices, and Theory, led by Barbaros Özyilmaz, Loh Kian Ping, Lim Chwee Teck and Feng Yuan Ping, respectively. 55 PIs from various departments in NUS as well as other local universities and research institutes, including 10 NRF research fellows, are conducting cutting-edge research on 2D materials. Physics faculty members, Antonio Castro-Neto, Barbaros Özyilmaz, Andrew Wee, Quek Su Ying, Hsin Lin, Shaffique Adam, Slaven Garaj, Chen Wei, Feng Yuan Ping, Ji Wei, Chua Lay-Lay, Peter Ho, Sow Chorng Haur, Utkur Misaidov, Zhang Chun are among the core members. In addition, an Industrial Liaison Office was set up in CA2DM, led by Tricia Chong, to facilitate industrial collaboration and commercialisation of research results. 2D materials has become a major focus area in NUS, and CA2DM is recognised as an internationally leading centre for 2D materials research.

Chapter 9

Department of Chemistry: A Historical Review*

Mok Kum Fun, Sim Keng Yeow, Andy Hor,
Richard M W Wong and Eugene Khor

Another milestone in the evolvement of the Department of Chemistry has been reached when an institution now known as the National University of Singapore celebrates its 60th anniversary (diamond jubilee) in December 1989. This review covers the development and progress of the department since 1928 and is divided into three broad periods: 1929 to 1965, 1965 to 1980 and 1980 to the present.

Period 1929–1965

Officially the Department of Chemistry started in 1929 with the opening of Raffles College for the teaching of Arts and Science, although chemistry was taught to pre-medical students at the King Edward VII College of Medicine since 1905 and the first science students of Raffles College were enrolled in 1928.

Dr G E Brooke delivered the first lecture on chemistry in the College and the teaching of chemistry took place initially in the Government Analyst's offices at Sepoy Lines.

Dr C W McOwan, who arrived in 1929, was the first Professor of Chemistry and he was joined in 1930 by Mr C T J Owen as lecturer. Professor McOwan resigned in 1940 and Dr P Purdie became Professor of Chemistry in 1940. During World War II, Professor Purdie and Mr Owen lost their lives in 1943 and 1945 respectively. In commemoration of Purdie and Owen and three other chemists from the Government Analyst Department who succumbed during the war, the local section of the Royal Institute of Chemistry (RIC) instituted in 1952 a silver Memorial Medal in Chemistry to be awarded annually to an outstanding graduate in Chemistry. This award was superseded by the present Singapore National Institute of Chemistry (SNIC) Gold Medal when the RIC was dissolved in the early 1970s.

---

*Reprinted with permission from *60 Years of the Faculty of Science, National University of Singapore 1929–1989*, published in 1989 on the occasion of the Diamond Jubilee celebrations of the Faculty of Science, NUS.

When Raffles College was reopened in 1946, Dr M Jamieson and Dr A Jackson of the Government Department of Chemistry (the forerunner of the present Department of Scientific Services) acted as Professor and lecturer respectively. They were assisted by Mr Kiang Ai Kim. Mr Toh Chin Chye and Mr Chan Chieu Kiat were part-time demonstrators. Dr Jamieson and Dr Jackson, together with Mr A W Burt and Mr A I Biggs, also of the Government Chemistry Department, continued to give lectures part-time until about 1952. At the first convocation of the University of Malaya in 1950 Dr Jamieson was awarded the degree of Honorary DSc for exemplary services he rendered.

Professor R A Robinson arrived in 1948 and he was instrumental in the rapid expansion and development of the Department in the next decade. Mr A K Kiang was promoted to a Lecturership, Mr Lim Chin Kuan joined as Assistant Lecturer and Mr Chan Chieu Kiat, Mr Chia Chwee Leong and Mrs H K Tong as temporary demonstrators in 1949. In 1950 Dr Leong Peng Cheong was appointed Senior Lecturer and Mr M C K Svasti as Assistant Lecturer. Staff strength steadily increased with the appointment of the following lecturers: Dr Rayson RL Huang (1951), Mr R W Green (1951), Dr E N A Sullivan (1954), Dr Ang Kok Peng (1955), Dr Francis Morsingh (1956) and Dr Lee Hiok Huang (1959). In 1959, Professor Robinson resigned and he was conferred the title of Emeritus Professor of Chemistry in 1960 in recognition of his service. He passed away in 1979 and a Robinson Memorial Lecture was endowed by the Faraday Division of the Royal Society of Chemistry (RSC) in 1980 from the proceeds of contributions by Professor Robinson's former students and colleagues.

Professor Kiang Ai Kim took over the Chair in 1960. Although Dr R L Huang was Acting Head of Department for a few months in the preceding period, Professor Kiang was the first local person in the history of the Department to be both Professor and Head of Department. The period between the late 1950s and early 1960s saw an unusual increase in staff movement. The University of Malaya in Kuala Lumpur was established in 1959 although teaching had commenced in 1957.

A few staff members left the Department to join the new university and with Singapore becoming self-governing most expatriate staff also left during this period. Vacancies in staff positions were increasingly filled by local returning scholars who had recently graduated with a PhD degree from British universities. These included Dr Huang Hsing Hua (the present Deputy Vice-Chancellor), Dr Tan Eng Liang (the first local Rhodes Scholar) and Dr Sim Keng Yeow (the present Acting Head of Department).

On the non-academic side, Mr Tan Yong Koon and Mr Tan Yong Kian were appointed as laboratory assistants in 1929 and 1930, respectively. They retired in the 1970s, each having served the Department for over forty years. Three laboratory attendants, Messrs. Awan b Haron, Pang Chin Yok and Paujan b Roji joined after the war. They have all since retired, the last in 1988. In 1951, Mr Abdul Rahman, a laboratory assistant, after attending a course in Poona, became the first qualified glass blower in the Department. He left for the new university in Kuala Lumpur but only after he had passed on the special skill to his son Mohamed Nahar, who then succeeded his father as the glass blower.

With inadequate facilities and equipment as well as heavy commitment in teaching, research was futile in the early days of Raffles College. The main emphasis was really on teaching, not only to science students but also to pre-medical and pre-dental students. Both Professors McOwan and Purdie made the attempt in research but their efforts were in vain. It was Professor Robinson who started the fruitful research tradition of the Department. He was able to solicit a considerable amount of research funds and to obtain much equipment for teaching and research. For a relatively small department the research output was very creditable. The main areas were in thermodynamics of electrolyte solutions (R A Robinson), free radicals (R L Huang) and phytochemistry (A K Kiang). During this period, Professor Robinson together with R H Stokes wrote the book 'Electrolyte Solutions' which to this day is a standard reference book on the subject.

At the formation of Raffles College the Department of Chemistry was housed in the left wing of the Manasseh Meyer Science Building at the Bukit Timah campus. This was adequate as both staff and student numbers were small. During the war, Raffles College was used by the Japanese armed forces. Although the war ended in 1945 it was only in 1947 when the chemistry laboratories were ready for use again. With rapid increase in the enrolment in the early 1950s the Department expanded into the upper left wing of the Manasseh Building and half of the F M S Building by 1953. It also occupied the lower left wing of the Manasseh Building in 1959 wherein was housed the newly established Microanalytical Laboratory.

Period 1965–1980

This period under review witnessed some notable changes in the history of the Department in terms of both staff and facilities. Although a series of lectures on industrial chemistry was introduced to the Honours students in 1962 and a new course in Applied Chemistry was started in 1963 as a Year III as well as an Honours Year subject, an Assistant Lecturer was appointed specifically to teach this subject only in the academic year 1965/66. The syllabus was drawn up with the consultation of Professor R T Fowler of the University of New South Wales and Professor P D Ritchie of the University of Strathclyde. Both of them were Visiting Professors to the Department at various times with Professor Ritchie being the first person to hold such a post in Term II of the 1964/65 session. It was due to Professor Ritchie's association with the University of Strathclyde that the Department saw a regular supply of Visiting Lecturers in Applied Chemistry from that institution. Course work in Applied Chemistry included lectures on industrial management and costing which were given by some members of the Department of Business Administration. Lectures on other specialised topics were given by part-time lecturers from industry.

Professor F H C Kelly, the first and only Professor of Applied Chemistry, was appointed in 1968 to take charge of the Applied Chemistry section of the Department. He resigned in 1974 and in April 1975 the Applied Chemistry section formally became the Department of Chemical Engineering within the Science Faculty. It was not until 1980 that this new Department was transferred to the Faculty of Engineering.

Apart from recruiting staff for the Applied Chemistry section, the Department also saw a steady increase in staff numbers for the traditional chemistry section. From mid-1960 to the beginning of the 1970s there was almost one new appointment every year. These new staff included Dr Yeo Ning Hong who joined in 1971. However, there was a period of stagnation between 1973 and 1979 when the Department made no new appointments. But there was a new development involving the Department towards the end of the 1970s. This was the idea of "Joint Campus" with the Nanyang University in Jurong. Initially some staff from the Department gave part-time lectures to chemistry students in Nanyang University and the Nanyang staff reciprocated by giving some courses to students in the University of Singapore. Later the two universities merged and chemistry staff from Nanyang University moved into the Bukit Timah campus.

Professor Kiang Ai Kim, who had held the Chair from 1960, retired in 1971 and was conferred the Emeritus Professorship. He was succeeded by Associate Professor Ang Kok Peng who had just returned to the Department after his stint as Singapore's ambassador to Japan since 1967. Shortly after assuming the headship of the Department he was appointed as Minister of State for Communications and later also as Minister of State for Health. During his absence from 1972 to 1975, Associate Professor Lee Hiok Huang, who rejoined the Department from the University of Malaya, became Head of Department.

With the introduction of Applied Chemistry a parallel set of non-academic personnel was required. In addition the total number of chemistry students was also increasing rapidly and reached 360 by 1965. Year I practical classes had to be repeated four times and Year II classes duplicated. It was therefore timely that a new building called the Science Tower was completed in May 1966 and allocated to the Department. This ten-storey building, reputed to be the tallest university structure in Asia at that time, was officially opened by Mr Lee Kuan Yew, the Prime Minister of Singapore on Friday, 1 July 1966. With a total floor area of about 6,500 sq m it had rooms and laboratories for 20 academic staff, all the teaching laboratories for inorganic, organic, physical and applied chemistry, student research laboratories, microanalytical laboratory, glass blower's room, chemical stores and a workshop. A Dexion hut which was left behind by the Engineering Faculty on its removal to Kuala Lumpur was converted into a chemical technology laboratory and workshop. Starting from the 1970/71 session the Department was also not required to run a service course for the pre-dental and pre-medical students. This relieved the heavy demand for teaching laboratories. However, with staff expansion during the initial period under review, rooms and research laboratories were at a premium again, shortly after moving into the new Science Tower. With the influx of eight to nine staff after the merger between Nanyang University and the University of Singapore, partition and renovation of staff offices were necessary. Once again, the Department was running out of space.

Analytical chemistry was introduced as a separate section in both the lectures and the practical classes in 1974 for the second and third year courses and this was also extended to the first year in 1983. Although the Department had been steadily acquiring new instruments during this period, sophisticated analytical instrumentation was limited and not very adequate for research. During this period the Department had only three ultraviolet visible spectrophotometers, three infrared spectrophotometers, three gas-liquid chromatographs, one polarograph, one atomic absorption spectrophotometer and a proton nuclear magnetic resonance spectrometer. This small pool was supplemented by a number of similar instruments from the Department of Chemistry of Nanyang University after the merger was instituted.

Period 1980–1989

This period witnessed the most dramatic change in the Department in terms of the large increase in student enrolment, staff recruitment and the provision of adequate teaching and research facilities.

With the complete merger between Nanyang University and the University of Singapore to form the National University of Singapore in 1980, the Bukit Timah campus was too small to accommodate the combined staff of 23 members and the expected increase in student population. However, planning and construction of a completely new campus had begun in the late 1970s and in June 1981 the Science Faculty moved to Kent Ridge. This began a period of rapid growth both for the University and the Department of Chemistry. New staff members, both academic and non-academic, now numbering about a hundred, were recruited to meet the increasing student population which today number more than 1000 enrolled in Chemistry (about 500 in Year I, 340 in Year II, 200 in Year III and 34 in the Honours course).

During this period under review a new post of Senior Tutor was also created. Two MSc graduates from the Department were appointed to such a post. They have since completed their PhD overseas and have returned to serve as lecturers in the Department. At the end of 1987, Professor Ang Kok Peng, who served as Head of Department for many years, retired after a long and distinguished career. Professor Huang Hsing Hua has been appointed as Deputy Vice-Chancellor since 1981. Presently, the academic staff strength of the Department stands at 38.

Parallel with the rapid expansion in students, staff and available teaching and research facilities, the Department has increasingly been focusing its attention towards avenues that would promote excellence both in teaching and research. The introduction of small group teaching through tutorials, in spite of much constraints in time-tabling, has become a common feature of our educational system. An Industrial Chemistry course was introduced in the final year to meet the increasing demand from industry for our graduates with wider applied chemistry background. The introduction of the Direct Honours programme was another innovation to cater for exceptional students who receive their Honours degree in three years. The Department is now examining the revision of its course structure to meet future demands for more research oriented graduates.

Staff members in the Department and their students are actively involved in research on various on-going projects in the following fields: conformational studies, electrochemistry, organic synthesis and structural studies, phytochemistry, coordination and organometallic chemistry, fluorine chemistry, environmental chemistry, polymers and materials chemistry, spectroscopy and theoretical chemistry. Increasingly significant is the number of multidisciplinary and interdisciplinary projects in which staff members from the Department and other departments participate. The NUS Polymer Science Group was specifically set up as one method of achieving this goal. Their collaborative efforts, especially through the use of sophisticated facilities at the Surface Science Laboratory of the Science Faculty, have resulted in a significant research output in the last few years. Our staff are also active as consultants to industry. Work on the more applied aspects of research has resulted in the filing of several patents by staff members. One staff member was one of the three recipients of the inaugural National Young Scientist & Engineer Awards in 1988.

The Department now occupies two buildings completely and two others partially. The facilities include six teaching laboratories, staff offices and research facilities, analytical services and instrument rooms, equipment and chemicals stores and a glass-blowing workshop. With adequate funding for teaching and research from the government the Department has acquired in the last few years a number of state-of-the-art instruments. These include two mass spectrometers with gas chromatograph interface capability (GCMS), a 90 MHz Fourier transform nuclear magnetic resonance spectrometer (FT-NMR), a Fourier-transform infrared spectrometer (FT-IR), an ultraviolet photoelectron spectrometer (UPS), a Laser Raman spectrometer and a thermogravimetric analyser/differential scanning calorimeter, two sequential inductively-coupled plasma atomic emission spectrophotometers (ICP-AES), a few atomic absorption spectrophotometers (AAS), a few high-performance liquid chromatographs (HPLC), an ion chromatograph, an amino-acid analyser and a number of gas-liquid chromatographs (GLC).

The technical staff has also increased correspondingly to provide essential technical support in teaching and research. The Microanalytical Laboratory, the Glass-blowing Unit and the various Spectral Services units of the Department have continued to provide services not only for the Department's teaching and research activities but also for other Departments in the University as well as for other organisations in Singapore and Malaysia.

As the Department enters its Diamond Jubilee year, the preceding passages chronicle its progress and achievements interspersed with fond memories and experiences. The Department is on the threshold of new and exciting challenges for the 1990s and is prepared. Due to the increase in enrolment during the last two years both for undergraduates and postgraduates laboratory space is again running short. The Department expects to move into the remaining three floors of Block S5 in the early 1990s when these laboratories are vacated by the other departments in the next phase of the Faculty's building expansion. The present one-year applied chemistry course will be expanded into a two-year course beginning from the second year. Active encouragement and a thriving research atmosphere are stimulating more honours students to pursue postgraduate work and the Department expects a steady increase of post-graduate enrolment in the coming years. More sophisticated research facilities will become available. Already on order is a single crystal X-ray diffractometer and a high field NMR spectrometer. Future directions of the Department will point to more emphasis on interdisciplinary and multidisciplinary research endeavour and the focusing of development of research excellence in selected areas. The Department will play its part in meeting goals and aspirations in nation building in Singapore.

Addendum: 1989–2016

Andy Hor, Richard M W Wong, Sim Keng Yeow and B T G Tan

Headship Changes

Sim Keng Yeow who had served as Head of the Chemistry Department with great distinction, steering the Department through a period of development and expansion, stepped down as Head in 1996, and was succeeded by Lee Soo Ying, then also Vice-Dean of Science. When Lee Soo Ying became Dean of Science in 1997, he was in turn succeeded by Hardy Chan who served till 1998, when Lai Yee Hing took over as Head.

In 1999 Andy Hor became Head of the Chemistry Department. He served as Head till 2009, except for a one-year sabbatical period from 2004–2005 when Lee Hian Kee was Head.

New Degree Programmes

The Chemistry Department was one of the sponsors of the new Materials Science programme which was launched by the Faculty of Science in 1991, and which became a separate Department in 1996. In 1996, the Department offered a new minor subject, Analytical Chemistry, as one of eight new minors launched by the Science Faculty.

In July 1997, the Chemistry Department inaugurated a new programme in Applied Chemistry.[1] This echoed an earlier introduction of the subject by the Department in 1963 (under the charge of K. Svasti and W. Hamilton), and when F.H.C. Kelly joined the Department in 1968 to take charge of this new subject and served till 1974. In 1975 the Applied Chemistry section of the Department became the Department of Chemical Engineering with P. Thurairajan as its Head. The new Department stayed in the Faculty of Science until 1980 when it was transferred to the Faculty of Engineering, where it remains today. The rebirth of Applied Chemistry within the Chemistry

Department was aligned with the Faculty's introduction of the new Bachelor of Applied Science degree, which was designed to prepare its graduates for a career in industry. In 1999, Applied Chemistry could be read as an Honours subject for the Bachelor of Applied Science (Hons) degree.

CMMAC Launched

In order to expand and advance its experimental and analytical capabilities, the Chemistry Department established a new chemical analysis centre in 1998, which was formally named as the Chemical, Molecular and Materials Analysis Centre or CMMAC in 2000.[2] This new centre was designed to be a central instrumentation facility which would provide high level professional support for teaching and research activities both in academia and industry.

The instruments in the CMMAC laboratories encompass a wide range of analytical techniques. Combined with the knowledge of the Department's academic staff, CMMAC is well positioned to provide solutions in chemical, molecular and materials analysis. CMMAC's objectives were to promote close research interactions among scientists, establish collaborative research programmes, provide training in chemical and spectroscopic instrumentation and conduct courses in relevant topics.

More New Degree Programmes

In 1999, the Chemistry Department, looking towards the new millennium, launched two major new programmes.[3] The first programme was a new major under the Bachelor of Science degree — Chemistry with Management. This major provided a path for those who had a deep interest in Chemistry, but who were also planning to move into corporate management at some stage of their careers.

The second new programme was a Graduate Diploma programme in Analytical Chemistry, which was indeed the first ever diploma programme launched by the Department, and which constituted a milestone in its efforts at providing continuing education opportunities for its graduates. The Diploma was targeted at graduates who were working professionals in the chemical industry and quality control sector, to upgrade their skills in analytical chemistry.

1999 was also the 70th Anniversary of the Department and the Faculty of Science, and in the Faculty's commemorative publication,[4] the Department could proudly declare that it was able to offer three separate Honours track degree programmes. Two of these programmes, the Bachelor of Science in Chemistry and the Bachelor of Applied Science in Applied Chemistry, were run entirely by the Department. The third programme, the Bachelor of Applied Science in Food Science and Technology, was a Faculty-wide programme, but was spearheaded by the Chemistry Department.

Indeed, the Food Science and Technology major, which had been introduced in 1998 by the Faculty with the Chemistry Department as its anchor Department, had proven to be immensely popular with students.[5] This was the first degree programme in food science in Singapore. It was introduced to meet the growing regional demand for food scientists and technologists, and in particular to equip them with the necessary skills and knowledge to handle the increasingly sophisticated production of modern-day foods. The launch of this programme subsequently led to the establishment of the Kikkoman Singapore R&D laboratory in Oct 2005 in the Department, opened by NUS President Shih Choon Fong and Minister for Education Tharman Shanmugaratnam. This was among the earlier corporate research laboratories in Public-Private Partnership in Singapore.

ChemConnections and the Ang Kok Peng Fund

In 2000, the Department launched its ChemConnections publication, which was established to provide a means of linking together the Department's past, present and future.[6] It was to be the Department's family album, shared by past, present and future students, as well as a platform on which the Department could publicise its activities, discuss its developments and develop its social contacts.

Later issues of ChemConnections highlighted the Department's 70th Anniversary get-together, and the launch of the Ang Kok Peng Memorial Fund. This fund, named for an important Head of the Department, had the objective of supporting educational programmes and activities for faculty, staff and student development. This was of course entirely in keeping with Ang Kok Peng's life work in chemistry which had significant impacts on chemistry education and research.

The Kiang Ai Kim Scholarship Fund

In 2000, the Chemistry Department set up the Kiang Ai Kim Scholarship Fund in honour of the 85th birthday of Kiang Ai Kim, the Department's first Singaporean Head of Department and a distinguished chemist specialising in the chemistry of natural products and organic heterocyclic compounds. The Scholarship was launched on 8 November 2000 by S R Nathan, President of the Republic and Chancellor of NUS. The objective of the Fund was to award scholarships and book prizes to outstanding chemistry students; recognize good and deserving chemistry undergraduates; and develop staff and student contributions in the chemical and pharmaceutical industries. A sum of S$608,871 was raised which was matched by the Government.

In 2001, the capabilities of the Food Science and Technology major were considerably expanded with the opening of the new Food Science and Technology Suite by Sidek Saniff, the Senior Minister of State for the Environment.[7] Located on level 6 of Block S3 within the Chemistry department, this new facility included a food processing area, a food safety and quality laboratory, a taste panel area, and a graduate research laboratory. It also included staff offices for the Food Science and Technology staff.

International Experience

The Chemistry Department with the support of the Faculty of Science launched the China Immersion Programme (CHIP) in 2005. It provided an opportunity for its students to conduct a month-long visit to the universities, industry, research institutes and corporate laboratories in China to gain deeper insights into their operations, developments and challenges. This was one of the earlier formal international activities organized for NUS students. The success of the CHIP programme promptly led to the launch of a similar programme in Europe, namely the European Immersion Programme (EURIP). Both programmes were conducted yearly over a decade.

Chemistry Olympiads

The Singapore National Institute of Chemistry (SNIC), in partnership with the Chemistry Department, first led the Singapore team of junior college students to participate in the International Chemistry Olympiad (IChO) in Finland in July 1988. This has developed into a major and annual international competition event for the schools. The Singapore Chemistry Olympiad, with the Chemistry Department playing a major organizational, advisory and coaching role, was promptly launched in 1989. After a decade of vigorous developments, the Singapore Junior Chemistry Olympiad (SJChO) was launched in 2010, which is now an annual event, with 2,000 secondary students participating yearly.

The Chemical Sciences Programme

In the academic year 2002/2003, the Chemistry Department and the Chemical and Biomolecular Engineering Department of the Faculty of Engineering jointly launched the Chemical Sciences programme.[8] This was done in consultation with both the Faculties of Science and Engineering, as well as with the Life Sciences undergraduate programme. The objective of this new programme was to provide students with a solid and broad foundation in life and chemical sciences which would enable them to pursue graduate research programmes in interdisciplinary areas such as medicinal chemistry and other life science related areas.

In January 2002, the Chemistry Department opened the Chemistry Graduates Centre, which provided a focal point for graduates of the Department, serving as the home of the Chemistry Graduates Club (CGC).

Partnership with TUM and the Medicinal Chemistry Programme

2003 saw the Chemistry Department establishing a partnership with the Technische Universität München or the Technical University of Munich (TUM).[9] The German Institute of Science and Technology (GIST) which brought together NUS and TUM was launched in February 2003. GIST was set up with a strong industry focus and with close links between its teaching and research strengths, and offered graduate programmes, executive training and research opportunities to meet the needs of globally active companies, and internship opportunities in companies in Germany and TUM. GIST was the biggest ever initiative established by a German university overseas, and the first and only from TUM. The Industrial Chemistry joint MSc programme offered by the Chemistry Department in partnership with TUM was hence the first launch project of GIST.

In April 2004, as part of the Experimental Therapeutics Initiative of the Office of Life Sciences, the Medicinal Chemistry programme was launched which served to play an integral role in drug discovery, to provide the important link between the identification of therapeutic targets and the delivery of experimental drugs for preclinical trials.[10] The Chemistry Department was one of the key departments from the Faculty of Science involved in this programme, alongside the Biological Sciences and Pharmacy Departments. Departments from the Faculties of Medicine and Engineering as well as industry players were also partners in this programme.

In 2009, Andy Hor, after a distinguished tenure as Head, completed his three terms and was succeeded by Xu Guo Qin, who served until 2012 when he was in turn succeeded by Loh Kian Peng, who was succeeded by Richard Wong in 2015.

The Singapore National Institute of Chemistry (SNIC) partnered with the Department to launch the first series of chemistry international symposium SICC (Singapore International Chemistry Conference) in the nineties. This was formalized in Dec 2001 with the 2nd SICC, and thereafter became a biennial event till to date. This became one of the earliest and most established international symposia in a basic science discipline in Singapore.

The success of the SICC conferences led to many other research conferences, symposia and workshops in Singapore. November 2010 saw another important event, the 1st International Collaborative and Cooperative Chemistry Symposium organised by the Chemistry Department. This was then evolved into a series of individual bilateral symposia with three of the most strategic partners of the Department, namely China, India, and Australia, which ran over a full decade. In 2011, the Chemistry Department held its first Environment, Energy and Catalysis Symposium jointly with its counterparts from the King Fahd University of Petroleum and Minerals (KFUMP), Saudi Arabia. This symposium was organized with the objective of enabling potential research collaboration between the two institutions. The research collaboration between KFUMP and NUS was formalized in 2015. Under the collaboration, six projects involving both departments received a funding of S$1.1 million over three years on environmental analytical chemistry, catalysis, nanostructured materials and computational electrochemistry.

Accreditation of Food Science and Technology Programme

In 2009, the Food Science and Technology programme celebrated its 10th anniversary after a decade of solid growth and contribution to the food industry. Four years later, in 2013, the Food Science and Technology programme received the international accreditation which it deserved, when the Bachelor of Applied Science degree and the Bachelor of Applied Science Honours degree in Food Science and Technology were accreditated by the International Union of Food Science and Technology or IUFoST.[11] In 2014/2015, the two degrees were strengthened with enhanced curricula. With the discontinuing of the Applied Science degrees in 2015, the Food Science and Technology major was reassigned to the Bachelor of Science Pass and Honours degree programmes.

Revamp of the Curriculum

In 2013, the curriculum of the Chemistry Department was revamped. This revamp introduced three new specialisations under the Chemistry major, which were Materials Chemistry, Medicinal Chemistry and Environment and Energy. This curriculum change was to take place from semester 1 of academic year 2013/2014.[12]

In 2014, the Chemistry Department introduced new topics which were highly relevant to the chemical industry and also updated its laboratory curriculum to include the teaching of sophisticated reactions and more relevant skills such as multistep synthetic reactions, as well as greater use of NMR reaction product analysis.[13] Laboratory classes were also revamped to include experiments in the emerging technology areas of environmental, energy and materials chemistry. New laboratory classes were being developed to train undergraduates in biochemical techniques.

The undergraduate chemistry programme was accredited by the Royal Society of Chemistry in 2015. This is an international benchmark of 'good practice' which incorporates the latest trends to produce a relevant and updated degree programme which addresses the needs of employers and students.

Inaugural Symposia

Many thematic and bilateral symposia were launched in the 2000's. Examples included the NUS-JSPS (Japan Society for the Promotion of Science) Symposium on Analytical Sciences in February–March 2002, the 1st Singapore International Symposium on Chemical Education in August 2002, the 1st Chemistry-IMRE Joint Symposium in September 2002, the NUS-HSA (Health Science Authority) Workshop in May 2006, and the NUS-SNU (Seoul National University) Joint Symposium: Frontiers in Fundamental & Applied Chemistry in February 2009 etc. The annual symposium dedicated to the presentation of the research projects conducted by final-year Honours students was also formalized in 2006.

In 2015, the Inaugural Singaporean Inorganic Chemistry Symposium (SICS2015) was organized by the Chemistry Department. The Symposium featured researchers from more than 13 inorganic based research groups from the Chemistry Department as well as from the NTU Division of Chemical and Biological Chemistry, with speakers covering a wide range of inorganic research ranging from medicinal chemistry to catalysis to fundamental main group chemistry.

Also in 2015, the Food Science and technology programme was one of the organisers of the inaugural Asia Pacific Symposium on Food Safety. The other organisers were the Southeast Asia Association for Food Protection, and the Asia Pacific Institute of Food Professionals.

The bi-annual Singapore International Chemistry Conference has been jointly organized by SNIC and NUS Chemistry since 1999. The series of conferences has attracted distinguished scientists all over the world and contributed growing international presence. The 9th SICC will be held in 2016.

In 2015, Richard Wong Ming Wah became Head of the Chemistry Department. A distinguished researcher in computational quantum chemistry, he assumed the Headship of a Department which has grown from its modest origins in 1929 to become one of the major players in the world of chemistry education and research (ranked first in Asia and seventh in the world by the QS World subject ranking),[14] and carries on its traditions to bring it to even greater heights of achievement.

In 2016, the Department of Chemistry has a total of 50 tenure-track faculty members and another 20 teaching track members. About 1,000 students were enrolled in various undergraduate and postgraduate programmes. The consistent high research performance of the staff was demonstrated by the fact that 10 chemistry members have won Outstanding Research Awards since the University Awards were introduced in 2002.

With the generous donations from Chemistry alumni members together with government matching fund, over S$500,000 has been raised for the Chemistry Alumni Fund. The Fund will be used to enhance student teaching and learning, support their studies, overseas experiential learning, research exposure, etc. A chemistry appreciation and fundraising dinner was held in May 2016.

References

[1] National University of Singapore, "National University of Singapore Faculty of Science Annual Report 1999".

[2] National University of Singapore. Department of Chemistry, Faculty of Science. [Online]. http://www.chemistry.nus.edu.sg/cmmac/about.htm

[3] National University of Singapore, "National University of Singapore Department of Chemistry Annual Report 1999/2000".

[4] National Universiy of Singapore, *Old Memories — New Challenges: A Celebration of 70 Years of Commitment to Science.*, 2000.

[5] National University of Singapore, *National University of Singapore Faculty of Science Annual Report 2000.*

[6] National University of Singapore, *ChemConnections*, vol. 1, no. 1, 2000.

[7] National University of Singapore. Faculty of Science Department of Chemsitry. [Online]. https://www.chemistry.nus.edu.sg/education/graduates/CurrentStudents/KAK/KAK.htm

8 National University of Singapore, *Science NUS: Internal Bi-monthly Newsletter, Faculty of Science*, July/August 2001.

9 National University of Singapore. The NUS Chemical Sciences Programme. [Online]. http://www.chemicalscience.nus.edu.sg

10 National University of Singapore, "National University of Singapore Annual Report 2003".

11 National University of Singapore, "National University of Singapore Faculty of Science Annual Report 2004".

12 National University of Singapore. Food Science and Technology Programme Home Page. [Online]. http://www.fst.nus.edu.sg

13 National University of Singapore, "National University of Singapore Faculty of Science Annual Report 2013/2014".

14 QS Quacquarelli Symonds Ltd. QS Top Universities. [Online]. http://www.topuniversities.com/university-rankings/university-subject-rankings/2016/chemistry#sorting=rank+region=+country=+faculty=+stars=false+search=

Chapter 10

Department of Botany: Achievements and Progress through the Years*

G Lim

The Botany Department was established in the year 1949, the same year that the University of Malaya was founded in Singapore. Its first Professor was R E Holttum (Fig. 1) who is now retired and living in Kew, England. I was among the first batch of students reading botany as one of 4 science subjects (the other subjects then available were Chemistry, Mathematics and Physics), in the first year, called the Preliminary year in those days. Why was it called by that name? The reason was that those of us who entered the University in October 1949 came straight in after passing our Senior Cambridge examinations (equivalent to present day 'O' Ievel).

The Department was small, both in staff and student numbers. Among the academic staff, besides Professor Holttum, there were only 3 others (Mr J Carrick, Mr I Enoch and Mr A Thompson) and Science students reading botany in the first year were only 10 in number (Fig. 2). I well remember the awe and curiosity with which we faced the laboratory classes examining botanical specimens and performing experiments in plant physiology — because most of us had not done any biology or even science in our school days. The Department also taught botany to 1st year dental and medical students. The academic staff were expatriates from England. We enjoyed their teaching, their concern and friendliness. The non-academic staff comprised only two laboratory technicians, one mandore for the department garden, and one department secretary.

Professor Holttum retired in October 1954, and Mr Thompson was Acting Head, until the arrival of Professor H B Gilliland who took over as Head from September 1955 to 1965 when he retired. The next Head was Professor A N Rao, who was Acting Head in 1965–67 and became the substantive Head

* Reprinted with permission from *60 Years of the Faculty of Science, National University of Singapore 1929–1989*, Published in 1989 on the Occasion of the Diamond Jubilee Celebration of the Faculty of Science, NUS.

Fig. 1.   Professor R E Holttum.

Fig. 2.   Students reading Botany in the Preliminary year 1949–50.

Fig. 3.   Honours class of 1968/69 & Department staff.

from May 1967 to November 1985. Although he has retired, Professor Rao still teaches part-time in the Department. The current Head of Department is Professor Gloria Lim, who took over the headship in December 1985.

Over the years, under the various Heads, the department has grown steadily with student numbers increasing and new staff being recruited. From an initial handful of students in the classes of earlier years we have currently hundreds of students. The increases however were often rather sudden at different times, and on each occasion the department had the resilience to cope with such dramatic increases.

The various courses taught have changed as time progresses, and this is as it should be, as botanical science advances, and new frontiers expand and an avalanche of scientific information in the various areas and aspects of the plant kingdom is published.

The first batch of Honours students comprised only 2 students, Wong Hee Aik and me. Hee Aik is presently Professor and Head of Biochemistry Department, National University of Singapore. Year by year as student numbers increased, so those reading Honours in Botany also increased. By the 1960's the Honours class varied in number from 3 to 7 (Fig. 3); in the 1970's these increased to 8 or 9 students each year (Fig. 4), and now in this decade of the 1980's the Honours classes have grown in numbers ranging from 11 to 16, while this year there are 13 of them (Fig. 5).

Fig. 4.   Honours class of 1969/70 in front row and Department staff.

Fig. 5.   Honours class of 1989/90.

Courses and Training in Botany

In the early years, Botany was read as one of the Science subjects in the Science Faculty. The courses taught comprised taxonomy of ferns and higher plants, plant physiology, algology, bryology, anatomy of higher plants, genetics and cytogenetics. By the late 1960's changes were in the air, with new thinking on biological sciences, and thus in 1969–1970, Biology A and Biology B came into being, with the Botany and Zoology Departments teaching the units in Biology A, while the units in Biology B were shared among 6 departments — Anatomy, Biochemistry, Botany, Microbiology, Physiology and Zoology — 4 of these are departments in the Medical Faculty. Further developments took place more recently around the early 1980's when the 4 Medical Faculty departments withdrew from teaching Biology B units in 1982–83 which then were and are now taught by only both the Botany and Zoology Departments.

Presently the courses offered include not only modified aspects of those taught to students in the 1950's–1970's, but also molecular biology and genetics, biochemistry, virology, tissue, cell and protoplast culture techniques as well as topical trends in modern plant sciences.

Honours students also undertake research projects and these are supervised by the lecturers. Research projects vary over a very wide range of specialised topics. Examples of some of them in recent years are:

- Tissue culture of trees and ornamental ferns
- The study of fungal pathogens
- Cytology and breeding of orchids
- Investigation of *Odontoglossum* ringspot and mosaic viruses in orchids
- Direct gene transfer to plant protoplasts

Field trips from the early days of the Department form an integral part of Botany courses for students, especially for the Honours class, which usually makes a field visit to the forests of Malaysia, for a week's duration (Figs. 6–8). This familiarises the students with plants and trees in their natural environment and enables them to understand better the interactions of plant ecosystems, and the rich floristic composition and variety of tropical plants.

The Department also undertakes the training of Asean and Colombo Plan awardees from time to time. Workshops, special training courses or individual training are provided to both foreign and local applicants. Some of the foreign participants were from the Asean countries, and Burma, who came here under the Asean, Colombo Plan or WHO scholarships.

Fig. 6.   Field trip to Taman Negara, Malaysia.

Fig. 7.   Field trip to Fraser's Hill. Honours class Aug 1966.

Seminars are regularly held in the Department, and this is by no means a recent innovation, as they were introduced as a regular department activity since the 1950's. Besides seminars, the Department has also from time to time organised conferences, regional meetings and symposia on various specialised topics of botanical interest, and thus enabled interactions and scientific discussions between staff, students and plant scientists from various countries.

Fig. 8.   Field trip to Maxwell Hills. Honours class 1967.

An active academic exchange programme has also been developed in recent years between the Department and Universities and Research Institutes in France under the French Academic Exchange Programme, and between the Department and Japanese Universities and Research Organisations under the JSPS (Japan Society for the Promotion of Science) and National University of Singapore Scientific Cooperation Programme, whereby our staff through short attachments, and the foreign scientists who visit the Department have enabled us to develop greater expertise in specialised techniques and exchange or learn botanical information especially in the new areas and rapidly advancing frontiers of plant science. Academic visits to other countries such as UK, USA, Australia and Europe have also been made by the staff from time to time, on British Council grants, Commonwealth Exchange Programme, Conferences, or Sabbaticals and these have benefitted the staff tremendously in their research and teaching.

Physical Changes

Not only have courses and syllabuses changed, became modified, curricula updated, and student and staff numbers increased, but the physical development of the Department has also altered beyond recognition. Just as a mighty oak from a small acorn grew, or to use an example nearer home, a mighty Dipterocarp timber tree from a small seed grew, so has the Department. The Bukit Timah campus housed the University from the 1950's to the 1970's. In the early days the Department comprised just a low one storey building (a hut actually), built of timber and bricks and sited in the Department garden. In the late 1950's to 1960's, the Nissen huts built during the 2nd world war on the campus were used as laboratories for the teaching of first year classes. Then as the Department expanded, It took over one side of the 2 storey blocks around the upper quadrangle of the Bukit Timah campus, for use as staff rooms, teaching laboratories and department office.

The Department moved to the Kent Ridge campus in May–July 1981. Presently at Kent Ridge, the teaching laboratories, research laboratories, staff rooms, honours and post-graduate laboratories are considerably more spacious and up-to-date in design, compared to the buildings at the old accommodation at Bukit Timah. However, despite this increase in physical space, space problems continue to get acute due to continuing increase in student numbers, staff, both academic and non-academic, as well as office staff, not to mention equipment, pieces of which are considerably sophisticated and some of which are large and need adequate space to house them.

Research

In the development of the Department, besides teaching, research from the early days of the department has played an important part. The research areas have expanded tremendously. In 1950's taxonomy, plant physiology, fern studies, anatomy and developmental botany were among the major research areas. Gradually, other specialisations such as mycology, plant pathology, tissue culture, ecology and cryptogamic botany were included in the research carried out by staff, honours and postgraduate students.

These research areas continued to expand and sub-specialisations came into existence, such that in this decade, with the intake of more staff, acquisition of new techniques and equipment tools, the Department has embarked on molecular biology, genetic engineering and manipulation, enzyme kinetics, environmental botany, biochemistry, protoplast and cell cultures among other modem areas of plant science in both teaching and research.

In the 1950's and 60's though staff numbers and postgraduate students were few and equipment was less sophisticated and inadequate, nevertheless the research done was of sufficiently high quality as seen in the earlier publications. This emphasis on quality in research has laid the foundation for some of the present areas of excellence in the Department, areas that include: (i) plant tissue, cell & protoplast cultures whereby mass and micropropagation, genetic engineering and selection for better quality, for drought and disease resistance of economically important plants such as orchids, vegetables, fruit trees and timber trees are undertaken (Figs. 9–11); (ii) plant pathology encompassing both fungal and viral diseases (Figs. 12–14), in which the causal pathogens of

Fig. 9.   Micropropagation of Guava.

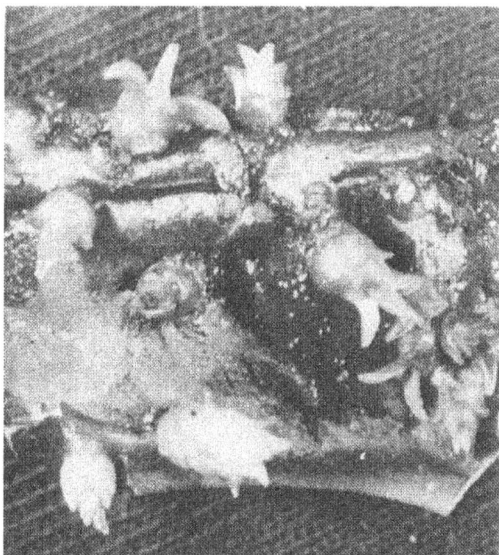

Fig. 10.   Direct school bud from leaves of Mangosteen.

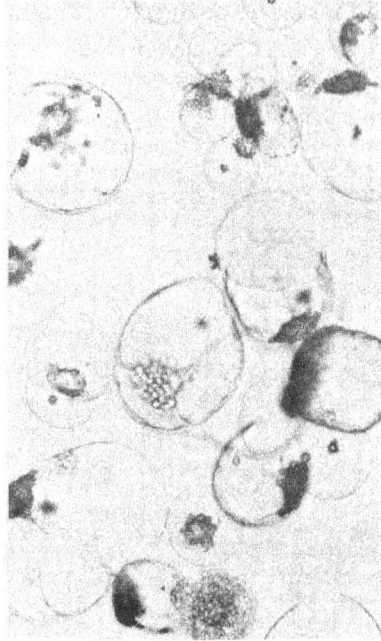

Fig. 11. Protoplasts of Orchid.

Fig. 12. Kang Kong leaves with white spot disease.

Fig. 13.   Bougainvillea leaves with common leaf spot disease.

Fig. 14.   Anthracnose disease of Papaya.

important local crops, ornamental and horticultural plants are determined, indexed and ways of prevention and control investigated and recommended and (iii) plant taxonomy, biosystematics and floristic studies where the local and regional flora such as angiosperms and ferns are compiled. Other areas of active research include:

- Molecular studies of viruses in orchids
- Development of algae resistant surface coating for external walls of buildings in the tropics
- Fungal growth on walls of buildings and stored food products
- Aquatic fungi
- Biological nitrogen fixation in legumes and non-legumes.
- Ecological adaptations of plants under stress
- The reproductive biology of mangrove plants
- The primary pathways of carbon metabolism in the light and in the dark characteristic of Cyanobacteria (blue-green algae)
- Biochemical systematics of tropical plants
- Carbon fixation in orchids and other tropical plants
- Single cell cultures of orchids
- Orchid cytogenetics

These research studies are vigorously pursued by both staff and postgraduate students reading MSc and PhD. Most of the research findings were and are published in international journals and books, and presented at local, regional or international conferences.

Some of the current facilities in the Department include a scanning electron microscope, spectro-photometers, gas chromatographs, HPLC, ultracentrifuges, computers, (one for each academic staff),computer facilities for Honours and postgraduate students, and many other instruments and equipment. All these are used both in teaching and research.

Future Directions

The department's future directions will be to continue to excel in its various areas of strength, and provide appropriate and relevant training to its students and postgraduates. With the increasing demand for University education among the population, the Department will expand and the post-graduate student numbers as well, may increase.

Teaching at undergraduate and postgraduate levels will continue to be of primary importance to the Department. Molecular plant biology and the techniques, involving DNA, RNA and genetic probing of plants, viruses and other lower organisms, such as algae and fungi will undoubtedly be areas to develop further and encouraged. Whole plant and whole organism function and physiology will continue to be a strong focus, while aspects that have useful applications especially for our country and topical pertinent issues will not be neglected but on the contrary emphasized. Thus the several areas of plant science will progress further and be strengthened vigorously.

The Department's areas of excellence and the high reputation of some of its academic staff have attracted and will continue to attract a number of overseas visitors to spend their sabbatical leave, or short term stay in the Department. Some of the staff also will continue to provide consultancy services to outside organisations, and industries due to their accumulated experiences and expertise.

Conclusions

The Department has thus over a period of 40 years developed in strength in many areas of plant science and achieved much, not only in producing graduates and postgraduates in Botany, but also in achieving excellence in a number of areas of tropical botany through its research and its research publications. Many of our graduates have done us proud in their different spheres of work as efficient administrators, bankers, scientists, researchers, teachers and principals of schools. We commend them, and wish them continuing success in their careers, and may they continue to uphold with pride the Botany Department and their alma mater, the University.

Acknowledgements

I thank Mrs J Wong, Deputy Registrar, National University of Singapore, for supplying some past records of the Department; Professor R E Holttum, Professor A N Rao and Associate Professor P N Avadhani for comments and suggestions; Dr C S Loh and Dr Hugh Tan for some of the photographs.

Chapter 11

Department of Zoology: Past, Present and Future*

T J Lam

Introduction

The Department of Zoology was established in the University of Malaya in 1950, a year after the University's founding. The first Raffles Professor of Zoology was R D Purchon who guided the development of the Department in the first decade. Then came a succession of Heads: D S Johnson (1960–1962), L Harrison (1962–1971), S H Chuang (1971–1977), C F Lim (1977–1978) and R E Sharma (1978–1981). I took over the Headship in February 1981.

In the meantime, the University had undergone several changes. It became University of Malaya (MU) in Singapore with the establishment of MU in Kuala Lumpur in 1959. The University was renamed University of Singapore in 1961. With the merger of Nanyang University and University of Singapore in 1980, the National University of Singapore came into being.

The nearly 40 years of history of the Department have seen important changes in facilities, teaching and research programmes, and staff and student profiles. These will be reviewed in the ensuing sections. The objective is to trace the development of the Department to its present state and to project into the future.

Facilities

The Department of Zoology had its humble beginning in huts in the Bukit Timah campus. Only in 1961, did it have a five-storey building of its own. With the shift of the University to the Kent Ridge Campus, the Department moved into its present building in 1980, but soon found the space inadequate. Additional space was provided in a nearby new block in 1985. A third phase of

*Reprinted with permission from *60 Years of the Faculty of Science, National University of Singapore 1929–1989*, Published in 1989 on the Occasion of the Diamond Jubilee Celebration of the Faculty of Science, NUS.

expansion is now being planned. It is anticipated that an adjacent new block to be shared with the Botany Department will be built in the near future.

Although space is a constraint for expansion, the Department enjoys good facilities. Major facilities include: 19 individual research laboratories; 2 common research laboratories (histology and radiotracer), 1 common research-cum-teaching laboratory; 2 large teaching laboratories (one shared with Botany); seawater recirculating systems; three large aquarium areas (wet laboratories); one boat; insectary; four cold rooms; animal culture room (46 mutant strains of *Drosophila melanogaster* and 11 mutant strains of mouse); electron microscope unit; confocal laser microscope and image processing unit; photographic unit (consisting of 2 dark-rooms and an equipment room); teaching museum-cum-computer room; tissue culture facilities; radioimmunoassay facilities; stereotaxis facilities; and facilities for molecular and transgenic work.

In addition, the Department acquired the Zoological Reference Collection (ZRC) in 1972 from the former Raffles Museum through the Science Centre Board. This is a priceless collection of animal specimens of all phyla collected since 1862 . It contains many type specimens. The value of ZRC to the study of regional Zoology cannot be overestimated.

After occupying temporary premises for a few years, the ZRC has since 1987 moved into its permanent home in the first three floors of the Science Library building. Excellent facilities are provided for storage and research. The ZRC has the potential of becoming an international centre for taxonomic/ systematics studies of Southeast Asian fauna.

A research facility which the Department had in the 50's and 60's and had since lost was the Marine Research Laboratory at Raffles Lighthouse, a small coral-fringed island about an hour's launch journey to the west of Singapore. Such a facility would be most useful for our marine biology teaching and research programmes.

Teaching Programmes

First Decade

The first decade was spent in the development of a sound teaching programme in Zoology relevant to tropical Malaya (Purchon, 1961). Emphasis was placed on using local fauna as examples in lectures and teaching materials in the practical classes. We owe much to the staff of this period (Table 1) for laying the foundation of our present Zoology programme in Biology A and for identifying local

Table 1.   Academic staff of the Zoology Department, NUS: past and present.

Staff	Period	Field
R. D. Purchon	1950–1960	Marine Biology
S. H. Chuang	1951–1977	Marine Biology
D. S. Johnson	1951–1972	Crustacean and Freshwater Biology
J. R. Hendrickson	1952–1959	Vertebrate Biology
A. G. Searle	1953–1958	Genetics
R. E. Sharma	1955–1981	Marine Biology
A. J. Berry	1956–1959	Coastal Ecology
S. S. Dhaliwal	1959–1960	Genetics
C. H. Fernando	1960–1964	Parasitology/Freshwater Biology
D. H. Murphy	1961–present	Entomology/Mangrove Ecology
J. L. Harrison	1962–1971	Mammals
K. Vigneswaran	1962–1964	Genetics
A. K. Tham	1963–1975	Fisheries
G. J. Pawsey	1964–1967	Genetics/Embryology
P. Ward	1964–1967	Ornithology
R. U. Gooding	1964–1968	Parasitic Copepods
C. F. Lim	1967–1978	Molluscs
P. H. Yuen	1967–1969	Parasitology
T. E. Chua	1967–1972	Fisheries Biology
M. C. Ting	1968–1971	Genetics
T. J. Lam	1969–present	Fish Physiology & Endocrinology/ Aquacultre
H. W. Khoo	1971–present	Fisheries Biology/Transgenic Fish
V. P. E. Phang	1971–present	Fish Genetics
W. K. Cheung	1973–1975	Insect Physiology
L. W. H. Tan	1974–1982	Marine Biology/Aquacultre
K. C. Lun	1974–1976	Genetics
L. M. Chou	1977–present	Marine Biology/Herpetology
S. H. Ho	1979–present	Insect Toxicology/Grain Storage Entomology
T. W. Chen*	1980–present	Animal Production/Live Fush Transport
V. M. Sin*	1980–present	Immunology/Toxicology/Aquacultre
K. F. Shim*	1980–present	Animal and Fish Nutrition/Aquacultre
L. H. Teo*	1980–present	Comparative Physiology/Invertebrate Aquacultre

(*Continued*)

Table 1.   (*Continued*)

Staff	Period	Field
K. L. Chan	1980–present	Vector Control/Entomology
T. Poole	1981–1982	Primate Behaviour
Y. K. Ip	1982–present	Comparative Biochemistry
J. B. Sigurdsson	1982–present	Marine Biology/Ornithology
J. J. Counsilman	1982–1987	Animal Behaviour
R. Hori	1982–1987	Developmental Biology
C. H. Tan	1983–present	Fish Reproduction and Growth
M. Nadchatram	1983–1988	Acarology
J. L. Ding	1984–present	Molecular & Cell Biology
A. D. Munro	1984–present	Fish Neuroendocrinology & Behaviour
D. J. W. Lane	1985–present	Marine Biology
D. Menne	1985–1987	Biophysics/Bat echolocation
T. M. Lim	1987–present	Developmental Neurobiology
V. W. T. Wong	1988–present	Molecular Biology
S. De Silva	1989–present	Fish Nutrition/Aquacultre/Fisheries Management
Kang Nee	1989–present	Bird Behaviour & Ecology

*Joined the Department from former Nanyang University where they had already served many years.

sources of, and describing, teaching materials still used in some of our practical classes today.

In the 50's and 60's, the Department also offered a teaching programme for pre-medical, pre-dental and pre-pharmacy students. The programme was terminated in 1970.

Second Decade

The second decade saw the development of Fisheries education in the Department (Lam and Khoo, 1988) in addition to the Zoology programme. A Fisheries Biology Unit was started in 1962 with Tham Ah Kow as its Director. The Unit was closed in 1973. During its 11 years of existence, it produced 44 Diploma of Fisheries, 14 Certificate in Fisheries Administration, 10 MSc and 2 PhD graduates (Lam and Khoo, 1988).

Fisheries training was also provided in part by the Regional Marine Biological Centre (RMBC), a research unit established in 1968 under an agreement between UNESCO and the Government of Singapore. The Centre maintained an international reference collection of marine biological materials, particularly zooplankton, an important food resource for Fisheries. It provided training and research facilities for students and scientists. The RMBC ceased operation in 1978. During its life-time, it had contributed greatly to Fisheries training and research in Southeast Asia.

Third Decade

The third decade saw the reorganisation of the Biology teaching programmes into Biology A and Biology B. The objective was to ensure that Biology students have at least one other science subject as their options in their first two years of study. This was deemed necessary to produce Science teachers capable of teaching more than one subject. The programmes were started in the 1969/1970 session. Originally, Biology A combined the programmes of the Botany and Zoology Departments, while Biology B combined the programmes of the Biochemistry, Physiology, Microbiology and Anatomy Departments with Botany and Zoology Departments participating in or contributing some courses. Thus the more traditional aspects of Zoology such as invertebrate and vertebrate Zoology, morphology, taxonomy/systematics, evolution, zoogeography, ecology and applied Zoology (e.g. Fisheries) are covered in Biology A, while cell biology, physiology, genetics, biochemistry, embryology and microbiology are dealt with in Biology B.

The Honours programmes in Zoology continued as before with some changes to accommodate the change in background of the students as a result of the introduction of Biology A and Biology B.

Fourth Decade

In the fourth decade, the Zoology curricula of Biology A and B were reviewed and revised in line with advances in zoological sciences and with the perceived needs of Singapore. Experimental and quantitative Zoology has been strengthened, and additional optional courses such as Developmental Biology and Animal Behaviour introduced in Biology B. Molecular Biology and Comparative Biochemistry have been included in the Biochemistry courses. These changes could be accommodated because of the withdrawal of the para-medical Departments (Biochemistry, Physiology, Microbiology and Anatomy) from Biology B since the 1982/83 session.

In addition, sectional options have been introduced in the third-year Zoology course (BA302) in Biology A. This is to enable students to receive greater depth of treatment of his chosen topics after a common background of core topics.

Another new feature is that final-year (third-year) students are required to carry out a group research project for one term. They are divided into groups of 5–6 students, each group carrying out a joint project under the supervision of a staff member. At the end, a 'congress' is organised where the student groups present their findings (oral as well as poster presentation) to the whole class, postgraduate students and all the academic staff. Each group also produces a joint written report.

The Honours programme has also been revised to reflect the strength of the Department and the changing emphasis of the Zoological sciences. Optional advanced or specialised courses have been introduced (Table 2) as well as an essay examination paper which requires students to write on an unseen topic for 3 hours. One feature has, however, remained unchanged throughout the history of the Department: every Honours student is required to carry out an independent research project and write a thesis on it.

In addition to the Honours programme, a Direct Honours programme has been introduced. This enables students who have obtained As for all their three

Table 2.   Short advanced courses in the Zoology Honours Programme, NUS.

Group A (Compulsory)	Group B (Optional)	Group C (Compulsory)
Biometrics	Aquacultre	Thesis Writing
Histology	Fisheries Management	Biological Illustration
Systematics	Animal Nutrition	Photography
	Animal Production	Seminar
	Animal Behaviour	Marine field course
	Marine Ecology	Research Project & thesis
	Marine Fouling	
	Fish Genetics	
	Fish Reproductive Physiology	
	Insect Physiology	
	Insect Toxicology	
	Insect Vector Control	
	Molecular Biology	
	Immunology	
	Bioenergetics	
	Developmental Neurobiology	
	Comparative Neurobiology	
	Comparative Biochemistry	

subjects in the first year to choose to study Zoology as the only subject in the next two years and obtain an Honours degree at the end of it. The Department has so far produced four Direct Honours graduates, three of whom have gone on to do PhD. One has completed a PhD from Cambridge and has joined the staff of the Department.

Research Programme

First Decade

Research in the first decade was focussed mainly on general biological studies of local and regional fauna (Purchon 1961). The animal groups which received the most attention were molluscs, corals, brachiopods, crustaceans, sipunculids, and vertebrates (tree shrew, amphibians and reptiles). In addition, studies were conducted on genetics of mouse, cat and rat, freshwater biology, and parasites.

Second Decade

Studies of molluscs, crustaceans, brachiopods, corals, freshwater biology, genetics and parasitology continued. Entomology, parasitic copepods and ornithology were now studied with the recruitment of new staff (Table 3).

With the establishment of the Fisheries Biology Unit and the Regional Marine Biological Centre, there was increasing research on Fisheries biology and ecology including studies of plankton.

Third Decade

Although the rich local fauna continued to receive considerable attention, there was noticeable increase in ecological, physiological, and mission-oriented research in this decade compared to the previous decades (Lam and Rao, 1980).

Table 3. Publication record of the Zoology Development, NUS.

Years	No. Publications
1950/51–1959/60	73
1960/61–1969/70	172
1970/71–1979/80	123
1980/81–1989/90	410

The aquatic organisms (particularly fish) remained the favourite subjects of study, although considerable amount of work was also done on insects. The mission-oriented research concerned mainly insect pest management and aquaculture.

Fourth Decade

With the merger of Nanyang University and University of Singapore, the Department's staff strength was increased by four. The four colleagues brought with them expertise in immunology/acupuncture, animal reproduction and production, animal nutrition, and insect physiology and control. Their research in the previous decade in the former Nanyang University was reviewed by A Johnson (1980).

With further expansion of staff (Table 1) and good research funding, the Department's research activities increased markedly. This is reflected in the marked increase in publications in this decade compared to the previous ones (Table 3).

The research activities now fall into several areas (Table 4). However, the major areas of strength have remained: (1) fish biology and aquaculture, (2) marine biology and management and (3) insect biology and vector/pest management. The following is a brief description of each of these areas:

(1) Fish/Shellfish Biology and Aquaculture. An active area is reproductive and larval physiology/endocrinology with a view to establishing or improving hatchery production of fry (juveniles) for aquaculture. The endocrine and environmental control of reproduction in fish and prawns is studied not only at the organismic level but also at the cellular and molecular levels. For example, we have a project on the molecular biology of vitellogenesis (yolk formation) in fish. We have already cloned the gene encoding vitellogenin synthesis by the-liver and are now studying the expression of the gene. We have also projects on purification of fish hormones and transgenic fish production. The latter involves introducing useful genes into fish eggs by microinjection or electroporation to produce fish with superior stocking characteristics such as fast growth.

Another active area is fish larval physiology and rearing. At present a constraint in fish production is the difficulty of rearing fish larvae to the fry stage because of high mortality. We have found that thyroid hormones promote fish larval growth, development and survival. We are also studying larval nutrition through the use of microencapsulated feed.

Table 4.   Research areas of the Zoology Department, NUS.

Fish Biology & Aquacultre	Marine Biology	Entomology & Pest/ Vector Management	Other Areas
• Physiology/ Endocrinology (reproductive, larval osmoregulatory & respiratory) • Molecular Biology • Nutrition • Genetics • Developmental Biology • Environmental Physiology & Biochemistry • Behaviour • Toxicology & Immunology • Prawn Reproductive Endocrinology • Broodstock Maturation Nutrition • Microencapsulated Feed & Larval Rearing • Transport of Live Fish Hormon Purification Gene cloning & Transgenic Fish & Prawn	• Coral Reefs • Mangroves • Marine Fouling • Marine Bioactive Compounds	• Mosquito Biology & Control • Stored-Products Pest Biology & Control • Taxonomy & Systematics	• Tissue Inflammation, Heavy-Metal Toxicology & Acupuncture • Comprative Biochemistry • Bird Biology • Taxonomy & Systematics of Crustaceans and Other Animal Groups

In addition, growth of fry is being studied in relation to nutrition, endocrinology and genetics. Another aquaculture-related project is on improving the packaging of live ornamental fish for transport.

Other aspects of fish biology studies include broodstock nutrition, genetics of colour variants, osmoregulation, toxicology, behaviour, immunology, development neurobiology and respiratory/stress physiology.

(2) Marine Biology and Management. The ecology of coastal areas including coral reefs and mangroves is being studied. The objective is to achieve a more rational approach to coastal zone management.

A survey of marine fouling organisms in Singapore waters is also being conducted. These are organisms which cause fouling of ships, boats, marine installations, floating net cages of fish farms, etc. Millions of dollars are spent to clear such fouling. We propose to rear the larvae of some of these organisms and use them as bioassay to detect natural antifouling compounds that may be present in sessile marine animals that do not get fouled. We assume that such animals are not fouled because they produce some compounds that prevent the settling of the larvae. Once this is shown to be the case using the bioassay, the compound can be purified, identified and synthesized either chemically or by recombinant DNA technology. It may then be incorporated into paint as an antifoulant.

We also plan to screen marine organisms for' antiviral, antitumor, anti-inflammatory, and antibiotic compounds. Singapore is particularly appropriate for such work because it is situated in an area which has the richest marine fauna in the world. Already we have shown that the blood cells (amoebocytes) of our local species of horseshoe crabs, like the temperate species, produce a specific antibiotic/coagulant against gram negative bacteria. This has been used to detect gram negative bacteria for clinical diagnosis and food-quality control. We are now studying the molecular biology of the antibiotic/clotting mechanism.

(3) Insect Biology and Pest/Vector Management. The emphasis is on mosquitoes and stored-products insect pests. The biology, toxicology, and management of these pests/vectors are being studied, including a study of the sense organs (sensilla) of mosquitoes with a view to developing a new approach to the control of mosquitoes through manipulations of their senses.

The study of the taxonomy and systematics of insect fauna of Singapore and the region has been an ongoing activity.

Increasing emphasis has been placed on studying the cellular and molecular aspects in the various areas of research. Molecular and cell biology of fish, marine organisms and insects offers a good opportunity for the Department to develop a niche in the scientific world, since not many laboratories in the world are working in these areas, least on tropical species. It is also building on our areas of strength — a logical extension of our studies of fish, marine and insect biology. Such studies are also likely to lead to biotechnology development.

Some of the applied aspects of our search in this decade have been reviewed by Lam (1986), Lam and Khoo (1987) and Lam (1988).

Staff and Student Profiles

From a humble beginning of only two academic staff in the 1950/1951 session, the staff strength has steadily increased to twelve by 1980 and rapidly expanded to 21 by 1989 (Table 1).

The number of Honours and postgraduate students has increased markedly in the present decade (Table 5). This reflects the expansion in teaching and research activities of the Department in the 80's.

Our graduates have found employment in teaching, government service, and industries. Some have gone on to important positions in both the public and private sectors including banking. A small percentage went on to postgraduate studies and ended up in academic or research establishments. With biotechnology offering opportunities, more of our Honours graduates are now pursuing postgraduate work.

Projections for the Future

The Department of Zoology has become a good-sized, well-equipped, modem department with well-defined research directions.1t is gaining recognition as a major centre for zoological studies in the tropics (particularly for fish, marine and insect biology).

In the next decade, we would like to consolidate this position. We would continue to build on our areas of strength, and move increasingly and steadily towards the study of the molecular and cellular aspects of fish, marine and insect biology with a view to developing appropriate biotechnology or agro-technology. In doing so, we would guard against losing sight of our roots in organismic biology and ecology. We plan to introduce postgraduate courses to attract students from around the world.

Table 5.   Student numbers in the Zoology Department, NUS.

Years	Honours	Postgraduate
1950/51–1959/60	20	13
1960/61–1969/70	95	24
1970/71–1979/80	81	26
1980/81–1989/90	190	41

Our goal for the year 2000 and beyond is a world-class department of Zoology and a major international research centre for zoological/aquatic biotechnology and high-technology aquaculture.

References

Johnson, A. (1980) Review of research effort over the past decade in the Department of Biology, Nanyang University. In "Government and University Research" (Eds. R.S. Bhathal, S. E. Chua and A. N. Rao), pp. 41–45, Sing. Nat. Acad Sci., Singapore.

Lam. T. J. (1986) Applied Research in Zoology. In "Applied Research and Its Management" (Eds. Henry Ong, Lu Sinclair and Bernard Tan), pp. 162–170, Fac. of Science, National Univ. of Singapore, Singapore.

Lam. T. J. (1988) Advances toward high technology in fish culture. In "Maximising Livestock Productivity" (Eds. Vadiveloo and C.C. Wong), pp. 35–38. Institut Pengajian Tinggi and Malaysian Society of Animal Production, Kuala Lumpur, Malaysia.

Lam, T. J. and Khoo, H.W. (1987) Towards 'high-tech' fish production. In "Agricultural Applications of Biotechnology" (Eds. A.N. Rao and H. Y. Mohan Ram), pp. 151–168. COSTED, Madras, India.

Lam, T. J. and Khoo, H.W. (1988) Fisheries Education in Singapore. In "Fisheries Education and Training in Asia", pp. 114–120. Asian Fisheries Society, Special Publication No. 2, Manila, Philippines.

Lam, T. J. and Rao, A. N. (1980) Review of research effort over the past decade in Botany and Zoology Departments, University of Singapore. In "Government and University Research" (Eds . R.S. Bhathal, S. E. Chua and A.N. Rao), pp. 46–53, Sing. Nat. Acad. Sci., Singapore.

Purchon, R. D. (1961) An Historical Review of the Department of Zoology, University of Malaya in Singapore. National Univ. of Singapore Library.

Chapter 12

Department of Biological Sciences

Paul Matsudaira, Hew Choy Leong and T J Lam

The history of the Department of Biological Science begins with the two foundings of Singapore and is inexorably linked to the development of the nation into a modern economic power, the transformation of a colonial medical school into a world-class research university, and evolution of biology science, from DNA structure to genomics and proteomics. On the one hand, the department can claim a research legacy inherited directly from Sir Stamford Raffles. On the other hand, our academic traditions originate with Singapore as a Crown Colony, through to its founding as a Republic, and the establishment of the National University of Singapore. During its brief 67-year history, Biological Sciences has kept a forward looking direction in education and research. This chapter will tell the story of the department in three parts, its constant reinvention into the department today and the people who led it, the development of the curriculum and its education mission, and finally the transformation into a research-intensive department through its Museums, Institutes, Centers, and Core Facilities.

(1949–1995) Departments of Botany and Zoology: From the University of Malaya in Bukit Timah to NUS at Kent Ridge

The formal history of the department can be traced to two biological science departments, Botany and Zoology, with the establishment of the University of Malaya in 1949. Under the leadership of the founding HoDs (Head of Departments), Profs KD Purchon (Zoology) and RE Holttum (Botany) (Fig. 1), Botany and Zoology were located in the Bukit Timah campus (Fig. 2). In this pre-DNA double helix era, both departments shared a common vision of hands-on study of the local biodiversity and a search for the mechanisms and processes dictating the genetics and development of plants and animals. Over the next three decades, a series of HoDs strengthened Botany in plant physiology, taxonomy,

Fig. 1.   The founding Heads, RD Purchon of Botany and RE Holttum of Zoology.

Fig. 2.   The Botany and Zoology Department buildings on the Bukit Timah campus of the University of Malaya.

fern studies, developmental botany as well as mycology while in Zoology, entomology, aquatic and marine biology, and physiology flourished. During this period, Singapore became independent and the National University of Singapore emerged from the merger in 1981 between the University of Singapore (previously University of Malaya) and Nanyang University and both Departments moved to the new Kent Ridge campus where they established new research capabilities at the cell and molecular level.

Botany built up research in several areas relevant to Singapore. PN Avadhani and Hew Choy Sin had established a solid foundation in plant physiology and particularly in orchid physiology and were later joined by Loh Chiang Shiong's plant tissue culture research. Their efforts were supported by AN Rao (HoD Botany 1965–85) and resulted in the organisation of the 1981 international symposium in *Tissue Culture of Economically Important Plants* held at the Botany Department. Gloria Lim (HoD, Botany 1985–91) nurtured Botany to international reputations in mycology. She served with distinction not only as Head and but later as Dean of Science. In 1991, Goh Chong Jin, who was already a Vice-Dean of Science, took over the Headship of the Botany Department and strengthened the Department with new laboratories for plant tissue culture (Fig. 3).

Meanwhile the Department of Zoology had made its name internationally in areas such as fish & marine biology and entomology. Lam Toong Jin (HoD Zoology 1981–96) had been responsible for the department's increasing regional and international prominence in fish & marine biology and in fish reproductive & developmental biology in particular. The practical applications

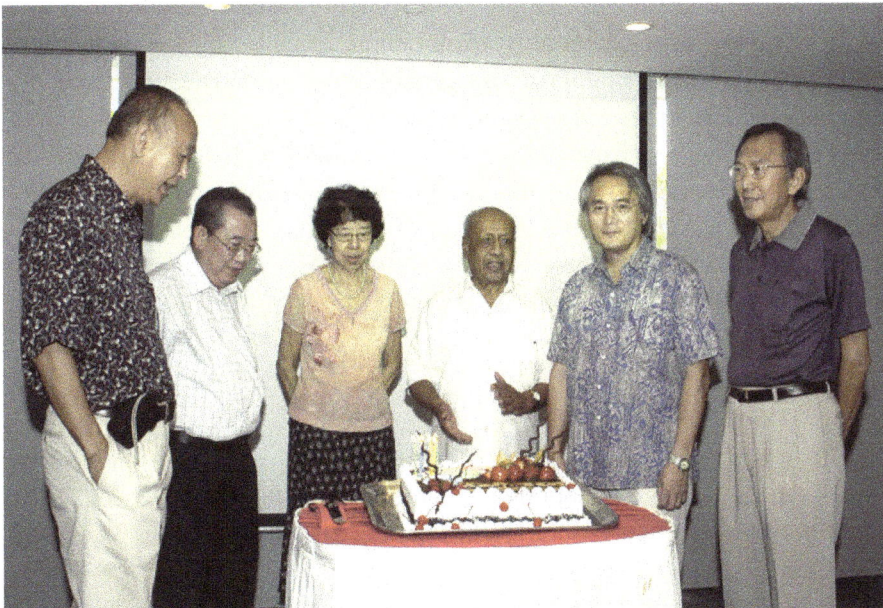

Fig. 3.   DBS HoDs at 2009 DBS 60th Anniversary celebration (from left): Hew Choy Leong, Goh Chong Jin, Gloria Lim, AN Rao, Paul Matsudaira, and Lam Toong Jin.

of such research to the commercial area of fish culture, and to the development of fish larvae in particular, made the Zoology department well-known not just in the academic sphere but also in the important food fish industry as well.

While the research continued along separate directions, the educational missions of the two Departments began to merge. In the 1969/70 academic year, Botany and Zoology collaborated with the Departments of Anatomy, Biochemistry and Physiology to offer two new subjects, Biology A and Biology B. The more traditional aspects of plant and animal sciences were combined in Biology A, while Biology B focused on medically related topics. In the early eighties, the Departments of Anatomy, Biochemistry and Physiology withdrew from the Biology B curriculum, and both Biology A and Biology B were taken over and taught jointly by Botany and Zoology. Such thinking eventually brought the two disciplines much closer together at the cellular and molecular level and eventually led to a more integrated approach for teaching and research in the Life Sciences.

Thus, in recognition of the common principles and mechanisms that underly all living organisms, the curricula underwent a major change in 1990 when Biology A and Biology B were replaced by three new subjects — Cell and Molecular Biology (CMB), Developmental and Systems Biology (DSB) and Integrative and Organismal Biology (IOB). In 1991, Cell and Molecular Biology was elevated to an Honours specialization alongside Botany and Zoology. The revised undergraduate programme also reintroduced Botany and Zoology (under Plant Biology and Animal Biology) as single subjects in the second year. This significant revision in curriculum not only provided the students with a comprehensive and up-to-date knowledge, but also brought the biological education at NUS to an internationally benchmarked level. The teaching of such cutting edge subjects made NUS a regional leader in biological science education. Furthermore, this served as the firm foundation for the step change in biology research and education at NUS.

(1996–1998) Botany and Zoology became Biological Sciences at NUS

The 90's witnessed a worldwide trend in mergers between botany and zoology departments. At NUS, the notion of a merger of the two Departments had also been discussed. After a prolonged period of negotiation and discussion between and within both departments, and with strong assurances from the Dean's Office that a combined department would be ultimately beneficial for the two disciplines, a merger was finally agreed. The new School of Biological Sciences officially came into existence as a merger of the Botany and Zoology Departments on 1 May 1996. The name "School" was deliberately chosen by the Science Dean to emphasize the merged entity's special status within the Faculty as embracing two hitherto separate disciplines. Vice-Chancellor Lim Pin launched the new School at a simple ceremony and appointed Lam Toong Jin as Director. However, two years later, the School was re-designated as a department of the Science Faculty, and Lam Toong Jin became Head of the Department of Biological Sciences.

(1998–present) Department of Biological Sciences (DBS) Gains International Stature

This period was marked by the transition of NUS from a teaching university to a research-intensive university. Much of the change in direction was fueled by the refocus of Singapore industry to a knowledge-based economy and the establishment of biotechnology as a new industrial base. During this period, the government established major research funding schemes under successive five year Science and Technology (S&T) and Research, Innovation, and Enterprise (RIE) plans managed by MOE, MOH, MTI, and the Office of the Prime Minister. R&D spending in Singapore grew from S$2.0 billion in 1995 to S$6 billion in S&T 2005 and is projected to reach S$19.0 billion under RIE 2020. From this intensive build-up in technology and industry, two areas received particular focus. Initial emphasis was placed on biomedical sciences and biotechnology when Life Sciences was earmarked for development. Later in 2010, the ecological and environmental aspects of life sciences benefited when Environmental and Water Technologies were also identified as national priorities. As a result, the department was able to strengthen fundamental disciplines such as Molecular, Cell and Developmental Biology, Structural Biology, and Ecology and Biodiversity Sciences.

Under the leadership of three HoDs (Fig. 3), DBS embarked on major changes in its research and educational missions. During his tenure as Head of Zoology and Biological Sciences Lam Toong Jin had firmly established highly respected programs in tropical fish & marine biology. The TMSI (Tropical Marine Science Institute) remains as one of his legacies and embodies the reputation for excellence in fish & marine biology. He stepped down as Head in 1999, to be succeeded by Hew Choy Leong, a graduate of Nanyang University and a distinguished fish biologist from Toronto who was widely recognised for his work on the biology of fish antifreeze proteins and transgenic fish. Prof Hew was responsible for the great leap forward in research within the department. Under his direction, the Department built up major research infrastructures in protein sciences, functional genomics, plant and fish facilities, recruitment of many young faculty members, implementation of open thematic research laboratories, establishment of major international networks with USA, China and India, and the implementation of the joint Life Sciences

Undergraduate Curriculum in partnership with several of the basic science departments in the Faculty of Medicine. In 2008, Hew Choy Leong stepped down as HoD and was succeeded for a year by Prakash Kumar, a long-serving member of the Department of Biological Sciences from the days of the Botany Department. Prakash Kumar was in turn succeeded in 2009 by Paul Matsudaira from the Massachusetts Institute of Technology, who brought with him a well-recognised research record in areas such as mechanobiology and biological imaging. With Paul Matsudaira's headship, research between faculty in DBS is now closely integrated under three research focus areas covering computational and biophysical sciences, cell and molecular biology, and ecology, biodiversity, and environmental sciences. Following the direction of the previous Heads, the department continued to expand and build up departmental resources with a new Insectary and revamped core facilities in cryoEM, CBIS computer centre, and confocal microscopy. Also, with the establishment of an internal grants office, DBS has been visibly successful in funding its research through competitive grants. By 2016, DBS consisted of 88 full time faculty, 55 staff, 1819 undergraduates, and 248 graduate students and they conduct their research and teaching in Buildings S1, S1A, S2, and S3 (Fig. 4).

Fig. 4. DBS is located entirely in buildings S3 (left), S2 (rear), S1 (low, middle), and S1A (right). Many lectures are conducted in Lecture Theatre 20 (front).

(2000–present) The Life Sciences Undergraduate Program

1998 witnessed a further development in the evolution of the biology curricula with two new degrees, the Bachelor of Applied Science in Biotechnology and the Bachelor of Science in Biology. The intent was to reflect modern trends in biology as well as its integrated character as a science discipline. In particular, the Bachelor of Applied Science in Biotechnology was part of the efforts of NUS to synergise with the nation's R&D push into biotechnology, and especially in agrobiology. The Biotechnology degree, in order to emphasize its applied and industrial focus, required students to spend a semester on industrial attachment. With these two new degrees, the three existing majors — Plant Biology, Animal Biology and Cell and Molecular Biology — were phased out. Students were allowed to read modules across the two new degree programmes. These changes provided the students with various options to choose from ranging from academic to applied angles.

In 1998, the new Biomedical Science major was also introduced in partnership with Medical School Faculty. Its aim was to train graduates for careers in medically-related areas such as biomedical research, pharmaceutical industries, and in hospital, government or industrial laboratories. This broad-based course involved 10 departments in the Faculties of Science and Medicine including Anatomy, Biochemistry, Biological Sciences, Chemistry, Community, Occupational & Family Medicine, Microbiology, Pathology (Laboratory Medicine, NUH), Pharmacology, Physiology, and Statistics & Applied Probability. The Biomedical Sciences programme underwent further developments, with the setting up of an Office of Life Sciences in NUS in 2001. A Life Sciences Curriculum Implementation Committee was set up in 2002 by the Department of Biological Sciences, Faculty of Science and Faculty of Medicine to implement the new Life Sciences curriculum to be jointly taught by DBS with the Faculty of Medicine. The aim was to train NUS undergraduates in the fundamentals in biological and biomedical sciences and to enable its graduates to contribute to various life sciences initiatives in Singapore. The Life Sciences curriculum with several specialisations eliminated duplication of courses taught among the various departments and had two major effects. First, with reduced teaching hours, the faculty members could devote more time for research activities, which has contributed significantly to enhance research output in the following years. Furthermore, elimination of duplication in teaching permitted efficient resource allocation to upgrade the Life Sciences teaching laboratories under the coordination of the Life Sciences Undergraduate Program Committee. The students benefited greatly from an integrated curriculum taught in teaching laboratories equipped with state-of-the-art facilities. Each year 450–550 students choose to read Life Sciences. The curriculum framework supports specialisations in Biomedical Sciences, Molecular Cell Biology, and Evolution and Environmental Biology with the majority of students later going on to careers in education, industry, and post-graduate schools.

Degree Programs in Computational Biology and Environmental Studies

DBS and the Faculty of Science also established degree programs with other Faculties in non-medically related areas. Computational Biology has been a joint degree with School of Computing since 2004 and trains students in bioinformatics. Environmental Sciences was launched in 2011 between the Science and Arts & Social Sciences Faculties as a truly interdisciplinary programme to reflect the growing national concern with the environment and nature.

Research Centers and Core Facilities

Throughout its history, the faculty have recognised that advanced teaching and research required investment in specialised resources. As a result, from their origins in Botany and Zoology, a number of initiatives have gained national importance and international recognition.

From Raffles to the Lee Kong Chian Natural History Museum

One of the most important possessions within the stewardship of the Faculty of Science was the Zoological Reference Collection or ZRC. This important collection of zoological specimens dates back to the time of Sir Stamford Raffles, who was a keen zoologist in addition to being the Founder of Modern Singapore. Consisting of over a million specimens, the collection was for many years a part of the Raffles Museum in colonial days, and portions of the collection were displayed to the visiting public.

When Raffles Museum became the National Museum after independence, it was decided that the collection of specimens would be removed from the Museum, and it took many years before this precious collection could find a permanent home. Under the leadership of HoD Lam Toong Jin and the Chief Curator of the collection, Yang Chang Man, the Department of Zoology had tenaciously and carefully looked after the collection and provided a proper home for the collection where its specimens could be properly stored for researchers. In 1988, the Zoological Reference Collection or ZRC was officially opened by the then Minister for Education, Dr Tony Tan. However, ZRC lacked space for a public gallery where it could present its collection for display to visitors. This deficiency was remedied on October 1, 1998 when the ZRC together with the equally precious Herbarium collection of plant specimens built up by the Botany Department was incorporated into the new Raffles Museum for Biodiversity Research (RMBR) and its new public gallery was officially opened by Minister for Education Teo Chee Hean in 2001.

The RMBR is now permanently housed and magnificently displayed in the new seven storey, 8,500 sqm Lee Kong Chian Natural History Museum, a project which was brought to fruition by Leo Tan and Peter Ng Kee Lin of the Department of Biological Sciences. On 18 April 2015, in the year of Singapore's 50th Anniversary of Independence as well as the 110th Anniversary of NUS, the Lee Kong Chian Natural History Museum was officially opened by the President of the Republic, Dr Tony Tan. Complete with three intact dinosaur skeletons, the new Museum is a fitting destination for the natural history collections inherited from ZRC and the NUS Herbarium.

Tropical Marine Science Institute

The strength of the Zoology Department in marine biology was integrated into the School of Biological Sciences, and the nature of this research demanded that the School have access to marine biology facilities beyond the limits of the NUS campus. In order to extend the scope and range of the School's work in marine biology, it became one of the major sponsors of a new research Institute established in 1997 both on the Kent Ridge campus and on St John's Island. Initially named the Tropical Marine Science Initiative (TMSI) and later upgraded from an Initiative to an Institute, TMSI brought together research groups which had close links with the tropical marine environment surrounding Singapore, including work on marine biology and aquaculture. Its St John's Island research facilities were launched in April 1998 providing the Department's marine biology and aquaculture researchers with facilities for marine animals beyond what was available on the campus. TMSI was indeed the world's first marine research institute in tropical waters and remains an important research resource for the nation.

Development of State-of-the-Art Core Technology Facilities

Under the leadership of Lam Toong Jin and Hew Choy Leong, DBS built up several high technology research core facilities. First, in 1993 the BioScience Centre was established with a grant of S$6 million from the Economic Development Board to serve as a common-core research facility for both departments as well as other biological sciences departments in the Medical Faculty. Under the directorship of Lam Toong Jin, it provided experimental facilities at the cell and molecular level and had a specific focus in the area of bioactive compounds from plants, animals and microorganisms.

With the sequencing of the human genome at the turn of the new millennium, new disciplines and technologies especially proteomics and genomics were rapidly established. In 2000, the BioScience Centre was renamed the Protein and Proteomics Centre (PPC) to complement and enhance the research and training programmes of the Life Sciences Departments in NUS, with particular emphasis on proteomics, structural biology and functional genomics. The PPC was soon followed in 2001 by the Functional Genomics Laboratories (Knowledge Enterprise, 2001) and the Structural Biology Laboratory in 2003. The three technology platforms were combined into the Structural Biology Research Corridor which was officially opened in 2003 by then Minister for Education Mr Tharman Shanmugaratnam. The Functional Genomics Laboratories were renamed the Lee Hiok Kwee Functional Genomics Laboratories in 2004 as the result of a generous donation from the Lee Hiok Kwee Family. These were the first laboratories to adopt the open laboratory and core facility concept in which many principal investigators could share a state-of-the-art facility with the latest equipment for genomics and proteomics research.

Since its founding PPC had rapidly become a powerful mass spectrometry and protein sequencing resource for local and regional researchers as it had made great progress in cancer marker discovery and cancer biology. In 2008, the PPC became home to the first Waters Centre for Innovation based on its advances in hydrogen/deuterium exchange mass spectrometry. In March 2015, the Protein and Proteomics Centre (PPC) marked the opening of the SCIEX Centre of Distinction as a new phase of partnership between SCIEX, a global leader in mass spectrometry instrumentation, and DBS.

Computational Biology at DBS

The Genome Era established the need for computer science in biology. In 2004, a Computational Biology Cluster was established by the Faculty, which bought together 13 faculty members from the Departments of Computational Science, Statistics and Applied Probability, Mathematics, Biological Sciences and Physics. This cluster led to the launch of a four-year undergraduate Computational Biology Program jointly with the School of Computing. The Comp Bio Program was complemented by a second initiative in graduate level Computational and Systems Biology (CSBi). Joint with MIT, CSBi was launched also in 2004 by the Department and the Singapore-MIT Alliance for graduate students to undergo training and research in the areas of bioimaging, bioinformatics, biocomputing and systems biology (National University of Singapore, 2004). Hew Choy Leong and Paul Matsudaira were the NUS and MIT coordinators for this joint program. In 2016, consolidation of Computational Biology was completed when the computational faculty moved to new offices and labs located in S3 Level 1.

The Mechanobiology Institute and the Centre for BioImaging Sciences

Plans for a new Research Centre of Excellence (RCE) linked to the Department of Biological Sciences were announced in February 2009. This new RCE, the fourth in NUS, was to be in the area of mechanobiology and Lasker Award winner Prof Mike Sheetz was recruited from Columbia to lead it. The MechanoBiology Institute (MBI) was officially opened on 6 October 2010, focusing on the quantitative and systematic understanding of basic biological functional processes in living cells. Paul Matsudaira was Co-Director, Hew Choy Leong as Deputy Director of the new Institute alongside Director Michael Sheetz.

On 6 December 2010, another new research centre associated with the Department of Biological Sciences was officially opened — the Centre for BioImaging Sciences (CBIS) — by Guest-of-Honour Lim Chuan Poh, Chairman of A*STAR. CBIS, whose Director was Paul Matsudaira, established high end microscopy, especially cryoelectron microscopy, and brought together an interdisciplinary group of biologists, chemists, computer scientists, physicists and engineers from the Science and Engineering Faculties, the Duke-NUS Medical School, and the Mechanobiology Institute.

The Lee Hiok Kwee Donation Fund and the SingHaiyi Donation Fund

The Department of Biological Sciences is blessed with two major donation funds that allow DBS to support several strategic initiatives and to provide financial support for needy graduate students. The estate of the late Mr Lee Hiok Kwee had generously bequeathed approximately S$15 million in shares to the Department of Biological Sciences, NUS for its research in cell and molecular biology and the transfer of technology from academic to industry with special focus on food technology. With $5 million from this donation, DBS launched 3 research programs in July 2002. The first research program established the platform technology in Structural Biology and Proteomics. The second research program on Cell and Molecular Biology focused on developing fish embryonic stem cells which will be invaluable to the study of vertebrate development, gene function and to understand how cells regenerate and propagate. Food security and safety are global concerns. Using molecular biology and gene transfer technology, this third program studies host and pathogen interactions to improve fish health, the use of plants as bioreactors to produce therapeutic proteins such as vaccines and antibodies, and the extension of shelf life of Asian leafy vegetables. The remaining $10 million of the donation and the matching funding from MOE was set up as an endowment fund to provide additional start-up funds for young faculty recruits, bridging funds for other PIs, maintenance of research infrastructures as well as the support of travel scholarships for our graduate students to attend international conferences.

The Department has a large graduate student population of about 250 coming from more than 25 countries. As most life sciences students require four to five years to complete their PhD degrees, the SingHaiyi Donation Fund of S$1 million allows the Department to provide our graduate students partial financial support during the last year of studies. As of end of 2015, more than 55 students have benefited from this Fund.

Concluding Remarks

This chapter has traced the key events in the history of the department, starting from the early efforts that were necessarily focused on training manpower until the present day research emphasis. There have been continual efforts to modernise the teaching and research activities within the department to keep pace with the rapid changes in the field as well as the evolving needs of the country. Looking to the future, the research in the department will be directed toward food security, health, and wellness as the Singapore population ages and the climate changes from global warming.

Chapter 13

Pharmacy and Pharmacy Education (1965–2015)

Go Mei Lin

> *"It was the best of times, it was the worst of times. ... It was the spring of hope, it was the winter of despair."*
>
> Charles Dickens, A Tale of Two Cities

A widely accepted symbol of pharmacy is the bowl of Hygeia with the serpent of Epidaurus entwined around its stem, its head elegantly poised above it. Hygeia is the Greek goddess of health and the daughter of Aesculapius, one of the Greek Gods of medicine. Serpents depict healing in the ancient world and from some accounts, were free-ranging in the temple dedicated to Aesculapius in the city of Epidaurus, the ruins of which remain today. This symbolism positions pharmacy squarely as a profession dedicated to maintaining and restoring health. Unlike medical doctors whose symbol also depicts a serpent entwined around the staff of Aesculapius, the healing comes from a bowl, presumably filled with a restorative potion. Thus, the root word of pharmacy is" pharmakon" which in Greek translates to a cure but ironically can also be interpreted as a poison or magical charm, depending on the context in which it is used. Cures are essentially titrated poisons, a point succinctly emphasized by the Renaissance alchemist Paracelsus ("What is not a poison? Only the dose allows anything not to be poisonous") and the 18th century English physician William Withering who elegantly juxtaposed the two when he wrote "Poisons in small doses are the best medicines and the best medicines in too large doses are poisons." As for the magical element in our cures, recall the old adage that astutely admonished healers to treat the patient and not the disease. This is an oblique reference to the placebo effect, one of the strangest phenomenon in medical science, where patients respond to their expectations of recovery, even when given an inert treatment. Clearly, there is much more to a "pill" that meets the eye. The priest-practitioners or shamans who brewed (questionable) potions for their hopeful patients appealed to their religiosity while the mediaeval apothecary with his armamentarium of henbane, opium and arsenicals cured some but killed many. Those were certainly more exciting times.

The modern day pharmacist is the learned intermediary between a prescriber and the patient. They are valued members in a health care team, with a unique role in drug therapy which revolves around optimising therapies to ensure good patient outcomes. In the community, pharmacists are often the first to make contact with the patient (or customer) with health enquiries and they contribute to the management of chronic diseases which will ultimately result in better health and savings in medical costs. Pharmacists have established niches in the pharmaceutical industry as well but in this article, the focus will be on their patient-centric roles in the hospital and the community.

Fifty years ago, the practice of pharmacy in Singapore was a far cry from what it is today. Consider the following. The now ubiquitous pharmacy in the polyclinic was not introduced until 1985. Their predecessors were Outpatient Dispensaries (OPD) of which there were 30, scattered throughout the island. One pharmacist was in charge of the dispensing activities in five OPDs and the main task was to ensure that medicines were dispensed to the never-ending queues of patients. Time was of the essence and hurried dispensing assistants had hardly any to spare except to scoop tablets from a bin (with cups, estimated to last till the next visit) and empty them into paper envelopes with directions hastily scribbled within boxes of an outline made by a rubber stamp. Fortunately, antibiotics were dispensed with greater caution and patients who required them were given just enough and asked to return the following week for a refill. For liquids and lotions, patients had to bring their own containers which were usually soya sauce, ketchup or liquor bottles. These, like other dispensed items, were passed to the dispenser through an opening in the glass panel or a grilled window (in earlier years), for filling before being returned the same way. Unfortunately, many who sought medical consultation at the OPDs were not *bona fide* patients, and were more keen on obtaining a "Medical Certificate" to justify absence from work. Discarded tablets and syrups poured into drains outside the dispensary were a regular sight.

In the late 1960s, it was not uncommon for a hospital to have only one pharmacist. The task of this pharmacist focused on the purchasing of drugs and to a lesser extent, some pharmaceutical "manufacturing". Until the late 1980s, standard medicines in government hospitals and outpatient dispensaries were prepared and supplied by the Government Pharmaceutical Laboratory and Store (GPLS). Hospitals also carried out the production of lotions, ointments, creams, extemporaneous and unstable drug preparations but on a lesser scale. For example, Eusol solution, a disinfectant containing sodium hypochlorite and used for wound irrigation and debridement, had to be freshly prepared on a daily basis in the hospital pharmacy. A lot of work went into pre-packing the bulk items received from GPLS. Liquids had to be poured from Winchester bottles into dispensing bottles (these were supplied in hospitals) and tablets packed from tins into paper envelopes of different sizes. Creams in huge jars were distributed into ice-cream cups. No wonder, pre-registration pharmacists (or pupil pharmacists as they were called then), newly graduated from the university, were left bewildered and disheartened at the prospect of devoting an entire year to "counting and pouring".

Today, a visit to the outpatient pharmacy in a restructured hospital or the pharmacy in a polyclinic will clearly illustrate how far pharmacy practice has advanced. The crowds will still be there but the patient will be given a queue number, a practice adopted from the banking sector. Queue numbers and the serving counters will flash on electronic wall displays although announcements are necessary for those who failed to respond. More likely than not, service will be provided by a pharmacist, neatly clad in a white coat with a visible name tag, across an open counter. Tablets are in blister packs and lotions or syrups in proprietary bottles. Neatly pasted on the preparation will be a printed label with date, the name of the medication, its dosage and instructions on the frequency with which it should be taken. Also listed are the name and ID of patient, cost of the medication as well as cautionary labels (where necessary). The pharmacist will first confirm the identity of the patient and then proceed to enquire if this is the first time the medication is taken and if there are known allergies. The nature of the medication is briefly explained and if it is delivered via an inhaler or child-proof device (which can be challenging to adults), a quick demonstration will follow. Payment will be made at a different counter, a subtle message that the professional services rendered by a pharmacist does not extend to the collection of money. The paradigm shift from product to patient is the most striking change to have occurred over the past 50 years.

Behind the counters, an even more startling scene is now enacted. The GPLS discontinued its manufacturing services in the late 1980s. Standard drugs are now purchased in bulk by tender and there is no manufacturing activity in hospitals except for the preparation of sterile infusions of which more will be mentioned latter. Computerisation which started in the 1980s and automation in the 2000s have changed the dispensing landscape. Computerisation improved the standard of dispensing by replacing hand-written and often illegible instructions with computer-generated labels. It provided accurate stock control through the barcoding of products, improved speed and accuracy of the dispensing process, and resulted in more productive use of human resources. The brave new world of automation is apparent today in most of Singapore's restructured hospitals. Their outpatient pharmacies could readily be mistaken for a futuristic assembly line, and indeed many ideas were adopted from state of the art processes in the automobile industry. The Outpatient Pharmacy Automation System has helped to halve patient waiting times, improve medication safety and cut-down reworking — when prescriptions have to be sent back — by two-thirds. How does it work? The local press provided a glimpse in its report on September 2014, "After getting electronic orders from doctors' clinics upstairs, machines pick and pack the correct medication for each prescription. These are put into baskets which are channeled to 3 robots via conveyor belts and sorted into shelves. ... Now, four in ten (patients) arrive at the pharmacy to find their medications ready for collection. The rework rate has also dipped from 30% to around 5%. ... Only 13 pharmacy technicians are needed in the main pharmacy instead of 32 previously. Extra staff have been moved to clinics, where they go through prescriptions with patients face-to-face."

Meeting patients face-to-face has elevated the visibility of the hospital pharmacist. From a tentative start in 1985 when ward pharmacy services were provided, it was progressed to pharmacists taking part in ward rounds with the medical team (1987), and has now gained a steady momentum with pharmacists providing a slew of patient-centered services for oncology, diabetes, asthma, cardiac conditions and those on anticoagulant therapy. Pharmacists are found in intensive care units while others are responsible for providing a more responsive drug information service. In 2007, the Pharmacists Registration Act made provisions for the accreditation and registration of pharmacy "specialists". The register which lists 25 specialists (Annual Report 2013, Singapore Pharmacy Council) has created additional opportunities for pharmacists to excel in those patient care areas where specialist input is relevant and sought after.

One aspect of pharmaceutical "manufacturing" that has been retained in hospitals is the specialised compounding of sterile parenteral feeds (total parenteral nutrition, TPN), enteral nutrition and extemporaneous sterile preparations. Of these, TPN manufacturing was the first to be introduced and as recounted by the pharmacist who introduced the service to the Singapore General Hospital, the need for it became apparent when no intravenous nutrition in a suitable formulation or pack size could be found to support the first bone marrow transplant in the hospital. The pharmacist, Mr Liak Teng Lit, described his early attempts at TPN preparation "Having prepared the three-litre solution, I was at a loss as to how to seal the ports of the plastic bag. I had to hurriedly seal it with a rubber band !" Such is the "can-do" and "will do" spirit of the remarkable band of hospital pharmacists of the 1970s and 1980s who propelled the practice to what it is presently.

Other than hospitals and primary healthcare, the presence of the pharmacist is most apparent in community pharmacies. In 2015, almost 500 of the more than 2,500 pharmacists work in the community, with the majority in retail pharmacies. The community pharmacist is in the unique position of being the first point of contact with the general public, not all of whom are necessarily unwell. Getting to know the "customer-patient" in an informal and unhurried environment, they are well placed to counsel on medication–related issues and disease prevention. As chronic diseases become more common in an ageing population, their services may extend to the provision of medication services to home bound senior citizens, filtering those cases that warrant medical attention from those that are less urgent, or simply to provide a helpful word of advice on health related matters. Like James Herriot, the well-loved veterinarian of Yorkshire in *All Creatures Great and Small* and their uplifting sequels, their interactions with the community may well go beyond that of professional counselling to that of listener and friend.

Unlike their hospital counterparts, community pharmacists face an "identity" problem which is peculiar to Singapore. Walk into a community pharmacy, and you can be forgiven for thinking you are in just another shop selling toiletries, cosmetics and health products. To be fair, this is not a problem of the profession's making. Prescribing and dispensing of medicines are undertaken by medical practitioners, a practice which is unlikely to change in the near future. Only a small fraction of prescriptions are routed to the community pharmacy, and these are generally for the occasional medicines that are not found in the doctor's dispensary. Under these circumstances, community pharmacies are left with few options if they are to survive and serve the community.

In the 1960s and 1970s, community pharmacies came under intense criticism.[2] The recurring rebuke was that a graduate pharmacist was not required to run a shop. Strong words came from no less than the Vice-Chancellor of the University of Singapore, the late Dr Toh Chin Chye. Writing in the Pharmacy undergraduate publication *Pharmaceutica* in 1974, he said "The traditional role of pharmacists as retail chemists has changed over the last 25 years. In North America, the drug store is the place where patent medicines are sold together with toiletries and other consumer goods. The drug store has also become a meeting place where people can talk over an ice cream or a hamburger. Does it require a university education to run a drug store? There is a need therefore for us to re-examine the role and training of pharmacists in Singapore so that graduates will be able to meet the new demands of the profession." Yet in 2015, the Minister of Health speaking at a Pledge taking ceremony for new pharmacists, had this to say "Community pharmacists have to work with medical practitioners and other healthcare providers to care for patients with chronic diseases, optimising medication therapy management and achieving good patient outcomes." Clearly, community pharmacies have thrived in the intervening years, notwithstanding bleak prospects and poor public perception, to warrant this recognition.

A pharmacist who is widely acknowledged to have brought the practice of community pharmacy to a higher level in Singapore is Mrs Pauline Ong. Mrs Ong graduated from the University of Singapore in 1966. She completed her pre-registration training at a retail pharmacy (Federal Dispensary) which was then the premier retail pharmacy. After working in pharmaceutical marketing and the banking sector for several years, Mrs Ong joined Guardian, a retail chain presently run by Dairy Farm Holdings of Hong Kong. It was not long before Mrs Ong was tasked with setting up a new Guardian outlet at Mount Elizabeth Hospital, Singapore. This was what was written of her "Because of her motivation and running the practice like her own, people thought she owned the pharmacy. Her principle was that if you didn't make a difference, who would want to come into your pharmacy? Even with 10 prescriptions a day, she made a difference by providing excellent service complete with professional pharmaceutical care." Mrs Ong made sure that each Guardian pharmacist (there were about 60 of them) worked as a manager of an outlet, to emphasise the point to the non-pharmacy higher management that there was added value in employing a pharmacist. Furthermore, "if the media referred to Guardian as a store, she would immediately call … to tell them that it is was a pharmacy, not a store." Credit should go to the small but determined group of community pharmacists, who like Mrs Ong, persevered in promoting the profile of the profession.

Fifty years ago, pharmacy education was provided by the School of Pharmacy at the University of Singapore. The degree conferred was the Bachelor of Pharmacy. Today, there is still one institution providing university level education in pharmacy. This is the Department of Pharmacy at the National University of Singapore. The intervening fifty years have seen several watersheds that have shaped pharmacy education to what it is today.

In 1965, the School of Pharmacy was inaugurated. Prior to this, it was a Department within the Faculty of Medicine. Dr Ngiam Tong Lan, then a pharmacy undergraduate, had this to say of the event: "It was the greatest event for pharmacy over half a century. The occasion was graced by the Minister of Health, Mr Yong Nyuk Lin who made an important speech. We had a tea party at the quadrangle outside the School to celebrate the occasion." In his "important" speech, the Minister announced an administrative decision of the Ministry of Health that only registered pharmacists would be allowed to hold wholesale licences under the Poisons Act. Although not specifically spelled out in the Act, that decision committed the entire chain of control for the distribution of medicines containing poisons to the pharmacist. The underpinning rationale for this decision was that an errant pharmacist would be sanctioned by prevailing laws as well as by his profession. Rebuke by his peers together with personal shame may prove to be a stronger deterrent than a cursory fine. This concession has supported many careers and livelihoods of pharmacists at a time when patient-care careers were far and few between.

The elevation to a School allowed pharmacy to reclaim its first year, which was previously spent on subjects like physics, chemistry, botany and zoology which were taught by the Faculty of Science. The leadership of the School pondered on how the curriculum should be restructured to reflect its new status. Finally the decision was made to implement the following changes. The first was to introduce anatomy, biochemistry and physiology into the curriculum to provide a foundation for the emerging biological slant in the pharmaceutical sciences. Second, biopharmaceutics, better known today as drug disposition or pharmacokinetics, and a staple in US and UK pharmacy curricula, was included. Curiously, the School held back on introducing clinical pharmacy which was then the buzz word in pharmacy education, following the introduction of US-styled PharmD programs in the 1960s. Instead, the School opted for Food Science and Technology on the grounds that it would provide employment opportunities for pharmacy graduates. Many processes employed in the pharmaceutical industry were indeed employed in the food industry as well and it was felt that this would hasten the acquisition of the necessary skillsets for the new discipline. Lastly, a basic course in business administration (taught by the Department of Business Administration) was deemed necessary because a large proportion of graduates found employment in the wholesale pharmaceutical sector.

The school took several years to prepare for the new curriculum. Young faculty members were sent to UK to undertake higher degrees in Food Science and Technology and some existing faculty members changed their research directions to embrace the new discipline. The curriculum was finally introduced in 1970 and along with it, a change in the degree conferred which became Bachelor of Science (Pharmacy).

Sadly, the School of Pharmacy was short-lived. In 1974, it reverted to its previous status as a department but hosted by the Faculty of Science. This decision came in spite of strong objections raised by the faculty members and students of the School. The reasons for the change are possibly only known to those who were instrumental in effecting it. The oft cited reason was that it would better reflect the Bachelor of Science degree that pharmacy graduates were already receiving. Significantly, this move necessitated another revamp in the curriculum, this time reverting back to the pre-1965 era where only 2nd and 3rd years were dedicated to pharmacy subjects. The loss could have been greater if not for the courageous stand taken by the then Director of the School, Dr Rosalind Tao, who insisted that her colleagues as pharmaceutical educators should be given the major say as to how the course and training of pharmacy students should be run. Dr Tao resigned shortly after Pharmacy was transferred to Science and migrated to the United States. She never returned to academia and passed away about six years ago.

In the new system, students who wanted to read pharmacy were admitted as students in the Faculty of Science. They were required to read Chemistry and Biology B (comprising biochemistry and human physiology) in their first year and to re-apply for admission to pharmacy at the end of their 1st year. The 4th (Honors) year which was optional, generally had a low uptake rate. Effectively, this meant that most students had only two years of rigorous pharmacy education before exiting the program for pre-registration training. It is a tribute to the caliber of the pharmacy students that very few failed and notwithstanding, went on to form the core of "young Turks" who change pharmacy practice in ways described earlier.

The two-year pharmacy program formally continued till 1994 when the modular system was introduced in NUS. In the wake of this watershed event, the 1st year was reinstated in pharmacy. In the intervening years before 1994, incremental changes have been made to increase the pharmacy content in the curriculum. In 1986, an introductory Pharmacy course was introduced in the 1st year, alongside the science-based disciplines. In 1988, Science students were streamed into pharmacy on admission, without having to wait until the end of their 1st year to exercise their choice. However, pharmacy was not listed as a direct entry program in the admission prospectus until 1995.

With the return of the 3-year pharmacy course, the department made up for lost time by actively recruiting faculty members to boost the teaching of pharmacy practice and clinical pharmacy. But with the clinical pharmacy tsunami cresting in pharmacy schools all over the developed world, more was needed. In 1997, the pharmacy curriculum was extended to 4 years. Pharmacotherapy modules were introduced for the first time and preceptorship programs were implemented. These programs which were undertaken in the summer vacations and spanned 12 weeks in all, aimed at giving undergraduate students practical exposure to the practice of pharmacy in community, hospitals and industrial settings. In view of their structured nature, the program was accepted by the Singapore Pharmacy Council as a component of pre-registration training, thus effectively reducing the latter from 12 to nine months.

By 1999, the Department had recruited 3 practice faculty members — two of whom were Pharm Ds. With a core group of expertise on hand, it decided to introduce a Masters program in Clinical Pharmacy, with the aim of bringing up to par the clinical knowledge and skills of pharmacists who had graduated before the implementation of the new undergraduate curriculum. From the start, the writing on the wall was that once the "backlog" was cleared, demand for the program would diminish. That came 2009 when the Pharm D program was introduced by the department. In part, the demise of the Masters program was accelerated by the fact that many of the hospitals were more keen to sponsor their pharmacists for "distance-learning" Clinical Pharmacy Master programs run by universities in US and UK, than the one offered by NUS. Nonetheless, 20 pharmacists went through the program, and their qualifications served them well in their careers.

By 2000, the Pharm D was the sole entry-level degree awarded to all pharmacy graduates in the US. In Singapore, the paradigm shift in practice standards has seen an increasing demand and acceptance by physicians to include a pharmacist as part of the patient management team. As such, there is a need to enhance the clinical knowledge and skills of pharmacists if they are to assume a more patient-oriented role in their patient care areas. The Pharm D program was initiated in 2009, with strong support from the Ministry of Health. 34 Pharm D degrees have been awarded as of 2015.

In the meantime, the pharmaceutical sciences component of the curriculum, which comprises the core disciplines underpinning pharmacy practice — was not neglected and equal effort was made to revitalise its contents. In 2011, a minor in pharmaceutical sciences was launched for students in Science and Engineering.

Another important development was the introduction of the Masters program in Pharmaceutical Sciences and Technology in 2009. In 2005, the Economic Development Board (EDB) Biomedical Sciences Division approached the Department of Pharmacy to initiate discussions on developing a course on Formulation Science and Pharmaceutical Technology. The objective was to ensure a supply of competent manpower to pharmaceutical companies which have set up or plan to set up manufacturing and pharmaceutical development bases in Singapore. To date, the degree has been conferred 60 graduates, many of whom have first degrees in Science and Engineering.

Yet another change was made to the curriculum in 2014. This was to incorporate part of the preregistration training into the program. Thus, students in their final year will spend a semester in training. Didactics will be limited to 3½ years, with a compulsory research project to be undertaken by all students. Within that time, students will have to comply with the General Education Requirements of the university as well. Thus, pharmacy modules have to be tightly structured and a major revamp was again made to the curriculum.

Taken together, the last 50 years of pharmacy practice and education have seen tumultuous changes. The article started with a quote from A Tale of Two Cities by Charles Dickens. It would surely strike a chord with those who had witnessed these changes. Fifty years on, we have gone through the deepest winter of despair. If the present generation of pharmacists have glimpsed light on the horizon, it is because they stand on the shoulders of giants — those remarkable men and women in hospitals, community pharmacies and the university who have worked selflessly for the betterment of their chosen profession.

Acknowledgements

The views reflected in this article are those of the author's alone. Discussions with Dr Ngiam Tong Lan are warmly acknowledged.

References

[1] 75 Years of Our Alumni. Edited by Dr Lim Kuang Hui. Published by the Alumni Association, 1998.

[2] Pharmacy in Singapore: A Journey Through the Years. Edited by Tan Shook Fong. Published by Ministry of Health, Republic of Singapore, 2001.

[3] From the Alumni Perspective: Centenary of Tertiary Education 1905–2005. Edited by Dr Lim Kuang Hui. Published by the Alumni Association, 2005.

Chapter 14

The School of Computing

Leong Hon Wai, Chan Sing Chai, Hsu Loke Soo and Thio Hoe Tong

The Budding Stage (1968–1974)

Computer Center at Nanyang University

In October 1969, Prof Hsu Loke Soo proposed to establish a Computer Center at the then Nanyang University with the following objectives:

- Offer computer related courses to students of all departments.
- Help computerise administrative processes.
- Support the research work of academic staff.

The Lee Foundation was approached for funding the computing equipment. A sum of S$280,000 was given.

The Center was located at the basement of the Nanyang University Auditorium. It has enough space to house a machine room, two key punch rooms, and a number of offices for teaching and administrative staff. In the machine room, an IBM 1130 computing system was installed.

Due to shortage of space, courses were conducted at the Science and Commerce faculties.

The Center was officially opened on 8 Jan 1970, and named Lee Kong Chian Computer Center.

Computing Courses

From 1972 onwards, two types of courses were offered:

- Programming courses for students.
  FORTRAN and COBOL programming courses were offered as elective courses to students from the Science and Commerce faculties. They were very popular. Students rushed to apply for admission to these courses, and even crashed the gate of the auditorium that housed the Computer Center!

- Programming courses for the public.
  During term breaks, similar but shorter courses were offered to the public. These were very popular too. The oldest student was a retired civil servant aged 80 years old who was very interested in inventing new methods for Chinese character input. The Center awarded a certificate for those that passed an examination.

Teaching Unit at the Computer Center

By 1973, the Center had acquired enough teaching experience, and started to offer other non-programming computer courses and the University agreed the Computer Centre should conduct courses to all students of the then Nanyang University.

The Expansion Stage (1975–1983)

The Computer Science Department

On 1 April 1975, the Computer Science Department was set up. With a new grant from the Lee Foundation, the IBM 1130 computer was replaced by an IBM System 3 computer. Many new faculty members were recruited. The Department was authorised to award degrees.

The Department expanded rapidly and the student population topped that of all other departments in the Science faculty. The graduates were well received by employers and were placed in key positions in many organisations.

With the increase in faculty members, the Department offered many more courses such as Assembly Language Programming, Data Structures, Algorithmic Analysis, Systems Analysis and Design, Computer Organisation, and Computer Architecture, etc.

Installation of A Time-Shared Computer System

In 1976, the Department solicited the donation of a computer system which was time-shared and allowed concurrent terminal access from its users. With the donation of the Totalisator Board, a PDP 11/34 computer system was installed with 16 terminals and was completely card-less. Students and faculty members were very excited to be able to do computing on this amazing machine.

The Returned Scholars' and the Teachers' Program

In 1975, the Department was invited by the Ministry of Defence (MINDEF) to develop a Post-Graduate Diploma Course for MINDEF's returned scholars. The purpose was to ensure that the scholars were fully aware of the rising computer technology in order to enhance their work in the Government. The impact of these scholars to the various ministries of the Government was tremendous. The courses were jointly taught by staff of the then Nanyang University, the then University of Singapore and MINDEF. There were many outstanding students who attended the course. Among the lecturers from MINDEF was Mr Lee Hsien Loong, our current Prime Minister of Singapore.

In response to a request by the Curriculum Development Institute of Singapore (CDIS), the Department offered Diploma in Computer Science courses to selected teachers from 1979 to 1981 to meet the immediate demand for Computer Science teachers.

Exchange Research Programme

During the period of 1975–83, the Department established two Research Exchange Programmes with financial aid from the Singapore French Embassy and the Japanese Research Foundation.

A regular research exchange program was with LAAS–CNRS Toulouse, France. Professors from LAAS (Laboratory for Analysis and Architecture of Systems) came to teach courses on a regular basis for a period of two weeks to six months. Academic staff were sent to be attached to the Laboratory for about six weeks. This arrangement continued for more than five years. In addition, LAAS also offered scholarships for graduates to pursue their Doctorate degrees.

The Japanese Research Foundation also established the exchange pro-gramme with the Department in the same period. Scholarships were offered to graduates to pursue higher degrees in renowned universities in Japan. The professors from Tokyo University also came to teach in the Department. Likewise the academic staff were sent to Japan to do research work. This research exchange programme enriched the staff and also widened their hori-zons in the computer research arena.

Move to Kent Ridge

After three years of the joint campus in both Bukit Timah and the then Nanyang University, the Department finally moved to its new home at Kent Ridge on

1 July 1980. At the same time, Nanyang University and Singapore University were officially merged to form the National University of Singapore. The Department shared a four-storey building with other departments of the Science faculty. Teaching equipment was upgraded to a then widely used system — the IBM 360. Each year, 200–300 good students were admitted with high admission cut-off points. The faculty size increased to more than 20, together with more than 10 teaching assistants to assist in conducting and marking tutorials.

The Changing and Fast Expanding Stage (1983–1985)

The Department developed a good relationship with the computer industry. Computer-user companies and statutory boards became the potential employers of the graduates. In order to equip the students with the practical experience of real life in the computer industry, the Department sent students for internships to many computer-related companies and statutory boards. In addition, the Department also invited potential employers to give career talks to the final year students. They allowed the students to be acquainted with their companies and told them about job opportunities. More relationships were developed during tea breaks. With the student internship and industry career talks in the campus, as a result, many of the graduates were employed before graduation. Others were employed within three months of their graduation.

The rapid increase in student population resulted in the shortage of faculty members. In addition to regular recruitment exercises, the Department also gave scholarships to bright graduates to study at renowned universities aboard. They signed bonds that required them to come back and serve as faculty members. This helped to solve the faculty shortage problem and also provided momentum to the development of the Department. During this period, the student population increased to about 500.

The Department of Computer Science in NUS took note of the new requirements to make changes to its curriculum and to expand the Department rapidly. It needed to offer an academically sound Computer Science curriculum with substantive research content for its students. It had to be software biased with some basic hardware content. The curriculum had to grow with time to mature into a comprehensive curriculum that could compete well with the other world-class computer science departments. However, the demand of the local industries then did not need such content with strong bias towards academia. The research component at that point in time was not needed by the local industry. Nevertheless, having a strong research orientation in the curriculum did help the Department in its recruitment effort for overseas academic faculty. The industry needed applications know-how to computerise different sectors of the industry, such as banking, airline reservation, trading, retail transaction processing and others. An Information Systems (IS) curriculum had to be developed as an option for students.

The critical success factors in the development of a new information systems curriculum had many aspects:

- Industry Orientation.
- Integrated Business Content.
- Development of a Sound and Comprehensive Curriculum.
- Teaching Resources from the Academia and the Practitioners.
- Strong Support from the University Administration.

The Department made a conscious effort to partner with the industry. Its key and senior faculty members were teamed up with the industry partners to develop the IS curriculum. Two key industry partners, which represented over 80% of the IT professionals in Singapore were selected:

- Data Processing Managers Association (DPMA) which represented most of the IT management personnel in the industry. It is presently renamed as ITMA.
- Singapore Computer Society (SCS) which represented both the practitioners as well as the IT management personnel.

These two key industry partners provided the required feedback on the needs of the industry. A steering advisory committee was set up and chaired by Mr Johnny Moo, a founding member of SCS. It had members from the senior academia as well as from DPMA and SCS. It provided a very clear direction to develop the Information Systems (IS) curriculum, leading to the establishment of an IS component in the Department's curriculum.

A strong business content was also added to the IS curriculum to provide a basic business domain expertise. With the cooperation of the Business School and the Department of Economics of the University, the Department was able to offer a number of basic business courses, such as Principles of Economics, Financial Accounting, Cost Accounting, and Marketing.

Teaching resources were crucial to the successful delivery of the IS curriculum to the students. Faculty members were recruited from the academia and the industry. In the early stage, there were problems in getting high-caliber professionals from the industry who needed to have at least a Master degree from recognised universities with at least ten years of industry experience. To integrate them with the CS academia was also a problem which the Department's management had to resolve.

Strong support from the University Administration was obtained to convince the opposition viewpoint in order to accept this new component in the Department. The Department was renamed as the Department of Information Systems and Computer Science (DISCS). The University Administration generously provided funding for the resources required by the new Department and committed to expand the number of academic faculty required with the projected number of students that the new component (IS) and the old component (CS) would be able to attract. At its peak, the Department was able to have a faculty strength of over one hundred and an annual intake of new CS and IS students of over 800.

The Department then show cased in the middle of 1984 its CS and IS curricula as well as facilities and manpower resources to support this integrated curriculum. Strong support was given by DPMA and SCS to this approach. The Department managed to push its way ahead and implemented the IS curriculum with a small yet cohesive IS faculty recruited from UK, US and the local industry.

A Steady State Department (1986–1997)

Starting from about 1995, the Department (DISCS) and later the School of Computing (SoC) engaged Visiting Committees of external experts. The first Visiting Committee consisted of Profs. JD Ullman (Stanford) and HT Kung (Harvard), and the committees have evolved with time. The committees were asked to look critically at *all* aspects of the School of Computing, to critically appraise how the school was doing and to make recommendations for critical changes, where necessary. Initially, there was one DISCS International Review Committee for all aspects of both CS and IS components of DISCS. After the formation of the SoC in 1998, with two departments CS and IS, it was later decided that there should be one IRC for CS (CS International Review Committee) and one for IS (IS International Visiting Committee).

Starting from the 1980's the DISCS and then SoC always had an Industry Advisory Committee (IAC) consisting of top executives from the industry and national bodies. Through the years, the SoC IAC has provided counsel on the strategic direction of the School, and helped strengthen the partnership between the School, national research institutes and industry.

Starting in 1993, the department joined the university-wide initiative to move from a year-based system, to a more flexible, semester-based, modular curriculum. The DISCS curriculum committee took the opportunity to go beyond that and adopt an ACM-style curriculum, following guidelines published by the latest ACM Curricula 1991 for CS and IS. With this change, the curriculum was able to adopt a flexible core-elective structure, with additional free electives to allow students to customize their program of study to their interests. It allowed greater flexibility in the length of study, allowing some students to accelerate, and other students to take a slower pace to complete their study. This curriculum was first launched in the 1994/1995 academic year. Since then there had been various modifications and changes made to the curriculum to move with technological changes in the field, but the overall flexible core-elective structure has remained intact. New focus areas for specialisations have been created and have evolved over time, demonstrating the flexibility afforded by this overall structure.

School of Computing (1998–now)

The formation of the School of Computing (SoC) was approved (on the third attempt) and officially started in 1998, with two departments, the Department of Computer Science, and the Department of Information Systems. It happened at the height of the dot-com craze happening around the world and also in Singapore, with many IT startups springing up in Singapore. And it sent a message about the key importance of Computing, not just for CS and IS, but also in all disciplines in the university and beyond.

With the formation of the school, there is even stronger emphasis on pursuing the creation of greater opportunities for students, on pursuing excellence in all areas, including research, teaching and service. The school held several retreats to identify core strengths and gaps in all areas: student programmes (both undergraduate and graduate), student experiences, the teaching programmes, our many excellent and passionate educators, our teaching facilities and resources, our research programmes, key areas of research strength, our research facilities, research funding, and also our support staff (in administration, computing and building services).

Degree Programmes Offered by the School

The School offered three degree programmes:

- More Diverse Undergraduate Degree Programmes.
- Double Degree and Concurrent Degree Programmes.
- Graduate Degree Programmes.

More Diverse Degree Programmes

With the formation of the School of Computing in 1998, graduates received the Bachelor of Computing (BComp) degree for the first time. The School of Computing started with its two trademark degree programmes, namely, BComp (CS) and BComp (IS).

As an independent faculty, the SoC had greater flexibility to adjust the degree programmes offered to meet the rapidly changing landscape and needs of the IT industry in Singapore and worldwide. So, over time, new and more diverse degree programmes were added to meet the needs of the industry (e-Commerce, Communication and Media, Computational Biology, Business Analytics, Information Security). Some of these specialised degree programmes were later discontinued as things evolved — either their content was incorporated into the main curriculum as they became mainstream or they became electives as their popularity dropped over the years. Currently, SoC offers the following degree programmes:

- BComp (CS) — Bachelor of Computing in Computer Science.
- BComp (IS) — Bachelor of Computing in Information Systems.
- BComp (InfoSec) — Bachelor of Computing in Information Security.
- BEng (CEG) — Bachelor of Engineering in Computer Engineering (jointly with the Department of Electrical and Computer Engineering).
- BComp (CB) — Bachelor of Computing in Computational Biology (in collaboration with the Department of Life Sciences).
- BSc (BA) — Bachelor of Science in Business Analytics (interdisciplinary programme in collaboration with the Business School, Faculty of Arts and Social Sciences, Faculty of Science, and Faculty of Engineering).

Double Degree and Concurrent Degree Programmes

The NUS *Double Degree Programmes (DDPs)* allow students to pursue and graduate with two different degrees in two disciplines (either from with the same faculty or in different faculties within NUS). Specially customised DDPs allow students to complete the two degrees within five years, fulfilling the requirements of both degrees (with some specially approved double counted modules). In addition, students can also design *their own* double degree programmes between

Computing with another discipline in NUS. In summary, the SoC now has the following DDP:

- DDP for CS + Math/Applied-Math.
- DDP for CS/IS + Biz-Ad/Accountancy.
- DDP for SoC + Another NUS Faculty Course.

The NUS *Concurrent Degree Programmes* (*CDPs*) allow students to graduate with a BComp and a MSc degrees in 5 years with a partner institution or faculty/school in NUS.

- CDP with Brown University   (BComp + MSc in Computer Science).
- CDP with Brown University   (BComp + MSc in Computational Biology).
- CDP with CMU (Media)   (BComp + MSc in ETC).
- CDP with NUS-Biz School   (BComp (CS/IS) + MSc in Management).

Graduate Degree Programmes

As an independent faculty, the SoC has more flexibility to design the graduate degrees offered.

Apart from the trademark PhD degree programmes, SoC also offers several graduate degrees that are in demand. Hence, the current graduate degree offered are:

- PhD (CS), PhD (IS).
- MComp (CS), MComp (IS).
- MComp (Information Security).
- MSc (Business Analytics).
- MTech (IT Leadership).

The School also offers special programmes with other University facilities and outside bodies.

- University-wide Special Programmes.
- SoC Special Programmes.

University-wide Special Programmes

SoC students can also enjoy a wide variety of university-wide *special programmes*. Some of these are listed below.

University Scholars' Programme (USP)

The SoC is a partner faculty of the USP (University Scholars' Programme). USP is university-wide programme designed to help some of the best NUS students to fully develop their potential with a broad-based multi-disciplinary approach to teaching and learning. USP equips students to work across disciplines and cultures and provides SoC students with interaction with outstanding peers in many disciplines.

Student Exchange Programme (SEP)

The NUS Student Exchange Programme (SEP) provides students with the opportunity to study in an overseas partner university, usually for a semester or two, to enhance their learning experience. This new experience provided by SEP is a once-in-a-lifetime opportunity that will enrich the life of the student forever.

NUS Overseas Campus (NOC)

The NUS Overseas Campus (NOC) is another special programme that is very popular among SoC students. NOC is a one-of-a-kind *internship programme* where a student spends a *full year* working as full-time intern in a startups in top entrepreneurial hubs around the world, and takes classes at nearby universities (such as Stanford). Students also interact with famous startup founders, angel investors and other inspiring role models to learn the secrets of their success.

SoC Special Programmes

SoC students are also given a vast array of rich educational opportunities to broaden their knowledge, exposure, and experiences.

Turing Programme (TP)

The Turing Program (TP) for CS majors is named after Alan M Turing (1912-1954), the multi-talented computer scientist, mathematician, logician, and cryptanalyst, most famous for his proposed model of computation, the universal Turing Machine model. TP is designed for students who love to pursue fundamental work, to take bold new directions, and to make concrete contributions to the world of computing. It aims to nurture students who aspire to engage in a research career in computing.

The von Neumann Programme (vNP)

The von Neumann Programme (vNP) for CS majors is named after John von Neumann, the computer scientist and mathematician who is most famous for his proposed *von Neumann architecture* (that is now *the standard* computer architecture). vNP is designed for students who love to solve complex real-world problems and develop software systems to solve them. It aims to nurture students who aspire to translate theory to practice, build software solutions, and make contributions to the computing world.

Undergraduate Research Opportunity Programme (UROP)

Students who are keen in pursuing research as a career can enroll in the UROP. They are given a chance to undertake a research project (working with a SoC supervisor) and receive credit for two research modules. They get to have a first-hand experience with doing CS research work. Most of these students continue to pursue research work as a career, either in academia or in industry.

Orbital (Summer project experience)

SoC's Orbital is a self-driven programming experience designed to give first-year students the opportunity to self-learn, build something useful, and pick up software development skills over the summer. The Orbital framework helps students stay motivated and driven to complete a project *of their own design*, by structuring peer evaluation, critique and presentation milestones over the summer period.

Internships for Students

SoC also provides internship opportunities for all students. Internship provides students with experiential learning, an opportunity to apply their computing knowledge to real projects, and hone their competencies in a real-world environment. In a world where employers are looking for graduates with experience, doing an internship can provide them with the experience that would give them an edge over others. SoC provides several types of internship: (a) a 6-months Advanced Technology Attachment Programme (ATAP) internship with students earning credit for three modules, and offered twice every year, (b) a 3-month summer Student Internship Programme (SIP) with students earning credit for 1.5 modules, and (c) other *ad hoc* internship opportunities for students that are done without course credit.

For students enrolled from AY2014/15 onwards, industry attachment is compulsory for all.

Innovation and Startup Activities

SoC actively nurtures innovation and entrepreneurship among the staff, students and alumni. Various steps have been taken to promote innovation and to support faculty and students who wish to embark on an entrepreneurial track. These include:

- Awards for undergraduates with outstanding innovations.
- Incubation facilities and support, including guidance and networking.
- Exposure to industry specialists and entrepreneurs, particularly those working in regional and global markets, through seminars and workshops.
- Targeted inclusion of adjunct faculty with entrepreneurial track records.

At any given time the Furnace has up to 10 startups at various stages. The SoC incubator has seeded more than 25 startup companies, including some that have successfully exited via acquisition.

FYP/UROP Innovation Awards

The FYP/UROP Innovation Award recognises SoC students who have developed innovative, practical and commercialisable ideas with the potential of substantial real-world impact. It provides the opportunity for students to demonstrate their individuality and inspiration, independence and originality, innovativeness and impact on the real world. The successful awardee will win a cash prize up to S$2,000.

VaSCo Award

The *VaSCo* (*Validating Startup Concept*) award is named after intrepid explorer Vasco da Gama. It recognises SoC students who have innovative ideas for the digital markets. It is awarded on a project-basis to student teams and gives support of up to S$10,000 to develop ideas with good commercial potential. The main applicant should be a SoC student but the project teams can come from different disciplines. Applications are judged on (a) differentiated innovation, (b) commercial potential, (c) clearly defined target market, and (d) quality of prototype/demo.

The Furnace (Incubation Centre):

The furnace is SoC's own early-stage startup incubator. The Furnace offers infrastructure and management support to foster an entrepreneurial climate and help bring ideas through development to commercial fruition. The goal is to nurture young businesses and help them to survive and grow during the start-up period when they are most vulnerable. All SoC students, staff and alumni are eligible to apply for residency with their business plans (with innovative, IT related technology based products or services). For successful applicants, Furnace provides physical space (lockable rooms) and network and telecom connectivity, etc for a period of half a year, with renewal options.

Executive Education

SoC started from 2009 onwards active effort to provide executive education to the infocomm sector and workforce.

STMI (Strategic Technology Management Institute): (http://stmi.nus.edu.sg/)

The *Strategic Technology Management Institute* (*STMI*) was jointly appointed in 2009 by the Singapore Workforce Development Agency (WDA) and InfoComm Development Authority of Singapore (IDA) to provide training that is intended to best reflect the present and emerging needs of the Infocomm industry. The STMI approach combines the rigour of the Singapore Workforce Skills Qualifications (WSQ) system with cutting-edge research by the NUS School of Computing. STMI adopts a strong industry orientation to ensure that its trainees are imparted with the relevant competencies, best practices and necessary job skills to keep pace with the rapid developments in the Infocomm sector.

Centre for Health Informatics (CHI): (http://chi.nus.edu.sg/)

The mission of the Centre for Health Informatics is to train healthcare and IT professionals to use health informatics to find new ways of delivering healthcare to achieve better patient care and patient satisfaction. CHI aims to serve as the leader provider and thought leader in Health Informatics. It runs seminar and short courses on various topics in Health Informatics.

Business Analytics Centre (MSBA): (http://msba.nus.edu/)

The Business Analytics Centre is a collaboration between SoC and Business School (Biz) which offers a MS Business Analytics degree. Leveraging on the domain expertise of SoC and Biz, MSBA offers a unique holistic BA degree that blends business modelling and analytics, industry domain knowledge, and IT expertise. The Centre works with other schools and faculties with specific domain expertise. One example is with the NUS Saw Swee Hock School of Public Health that has domain expertise in Biostatistics, Epidemiology and Health Systems & Behavioural Sciences. The degree programme is based on also integrated leading academic and industry players (such as IBM) in implementing business analytics solutions.

Teaching Excellence

Teaching has always been a priority for DISCS and for SoC. And especially over the past decade, SoC has built a good reputation (among the students in NUS) as a school with excellence in teaching, with many excellent and cool educators in its midst.

Teaching Innovations

Many educators in SoC also innovate by building tools for teaching. In particular, two of these stand out — Coursemology (A platform to turn course exercises into an online game. This gamification platform can support any course subjects), and VISUALGO (A series of configurable visualisation for data structures and algorithms, from simple sorting algorithms to complex graph data).

(See the complete list at http://www.comp.nus.edu.sg/about/depts/cs/teach/innovations/)

Teaching Luncheon Seminar Series

SoC also organise a "Teaching Luncheon Seminar" every semester where winners of teaching award share their teaching secrets and/or other topics related to teaching with colleagues in the school.

Undergraduate TA Scheme

Since 1994, DISCS initiated the idea of engaging top final year undergraduates who are passionate about teaching as *undergraduate TAs* to assist in the teaching of some of the introductory courses in the department. The practice continues in SoC and has evolved over time.

Undergraduate Discussion Leaders (UDL) Scheme

More recently, SoC started the *Undergraduate Discussion Leader (UDL)* scheme, where talented students (not necessarily in their final year) with passion for teaching and who have good communication skills are employed to lead discussion sessions for freshmen classes. The UDLs first attend a teaching assistant workshop (conducted by staff of SoC) to further sharpen their teaching skill before teaching their first classes as UDLs. The scheme has worked very well and has been beneficial not only to the students, but also the UDLs as well.

Excellence in Research

Since the mid-90s, the department (DISCS) had moved toward strengthening research while maintaining excellence in teaching. At around this time, DISCS had the first International Visiting Committee and the report and recommendations also helped dovetail and support DISCS's effort to strengthen research. During the period when SoC was formed, there was also a university-wide effort to move towards quality research.

Hence, many important measures were taken by the management of DISCS/SoC to help to strengthen research. Some of these were:

- Engaging an International Visiting Committee.
- Better support for research students.
- Innovating the PhD programme to support stronger research work.
- Further strengthening recruitment of excellent faculty members.
- Providing startup grants for new faculty members.
- Recognising the unique importance of conferences in CS and IS.
- Better support for conference travels (from research grants).
- Adopting international best practices in promotion and tenure.
- Encouragement to collaborate and apply for larger research grants, and so on.

Research Excellence all Around

Today, SoC is strongly committed to research excellence in all its dimensions: searching for fundamental results and insights, developing novel computational solutions to a wide range of applications, building large-scale experimental systems and improving the well being of society. SoC seeks to play an active role both internationally and locally in the core and emerging areas of Computer Science and Information Systems.

SoC is mindful of the vital interplay that needs to exist between teaching and research at the undergraduate and graduate levels of education. SoC also constantly endeavors to forge closer collaboration with other faculties, research institutes and industry.

Research Area Grouping

With a large faculty size, SoC faculty members engage in research in almost all the major areas of research in CS and IS. To make it more manageable, the research areas were consolidated into the following groupings:

Research Grouping for Computer Science:

Artificial Intelligence	Computational Biology
Database	Media
Prog. Languages & Software Engg	Security
Systems and Networking	

Research Grouping for Information Systems:

Data Science & Business Analytics	Economics of IS
Social Media & Digital Business	Healthcare Informatics
IS Development & Management	Digital Innovation in the Service Economy

Research Centres in SoC

NeXT Search Centre (NUS-Tsinghua Extreme Search Centre), from May 2010, $10M over 5 years from MDA IDMPO (Interactive Digital Media Programme Office).

COSMIC (Centre of Social Media Innovations for Communities), from Sep 2010, $10M over 5 years from MDA IDMPO

FCI (Felicitous Computing Institute), from Jan 2012 Seed funding from SoC and Office of the Deputy President (Research & Technology).

SeSaMe (Sensor-enhanced Social Media) Centre, from July 2012 $10M from MDA IDMPO

TSUNAMi Centre (on software and system security), from Oct 2014
$6.1M from NRF

NCL (National Cybersecurity Research & Development Laboratory), from Nov 2015

World-Class Standing of the School

The School, after 40 years of highly intensive teaching and research, with industrial and worldwide university collaborations, has achieved a world-renowned standing based on the latest surveys by these two esteemed organisations:

- In the QS World University Ranking 2015/2016 for Computer Science and Information System, SoC was ranked 9th in the world and 1st in Asia.
- The Times Higher Education (THE) World University Ranking 2015/2016 ranked SoC as 13th in Engineering and Technology in the world, and 1st in Asia.

Chapter 15

A New Garden of Science in Singapore: Science at Nanyang Technological University

Ling San

A Seed Sown

As the world ushered in the new millennium on 1 January 2000, Nanyang Technological University (NTU) in Singapore was given a "millennium challenge".

At the launch of NTU's Millennium Celebration in the first evening of the new millennium, then Singapore's Deputy Prime Minister and Minister for Defence, Dr Tony Tan Keng Yam, said in his speech:

> "Singapore is well placed to benefit from the life sciences revolution. World-class organisations, ranging from healthcare giants such as Merck, Glaxo Wellcome and Baxter to research universities such as Johns Hopkins, have set up significant establishments in Singapore. The output of life sciences companies manufacturing in Singapore is expected to double within the next five years.
>
> "Recognizing the potential of the life sciences area, world-renowned technical universities overseas are rapidly building strong capabilities in the life sciences …
>
> "NTU has made a start in building a capability in the life sciences area through its collaboration with the Singapore General Hospital (SGH) in the field of biomedical engineering. This is a good start but more can be done. To realize the potential of the life sciences area, it is timely for NTU to set up a task force to consider how NTU can expand and strengthen its collaboration with SGH and other hospitals in Singapore to build a world class capability in the life sciences to enable Singapore to take advantage of the opportunities which will abound in this rapidly growing field in the coming years. In building such a world class capability in the life sciences, NTU will face new challenges, create new opportunities and open new horizons for tertiary education in Singapore in the new millennium."[1]

The First Sprout

NTU was inaugurated on 1 July 1991, incorporating its predecessor, Nanyang Technological Institute (NTI), and the National Institute of Education (NIE), a new institution formed from the merger of the Institute of Education and the College of Physical Education. NTI had only engineering, accountancy, and applied science (essentially computer science) at the point of the merger, while two new undergraduate degree programmes — Bachelor of Arts with Diploma in Education and Bachelor of Science with Diploma in Education — were introduced when NIE was established. With this latter programme, in some sense Science education had begun its presence in NTU. However, it was only available within NIE, strictly for students training to become teachers.

Following the challenge issued by Dr Tony Tan, NTU, under the leadership of its President Dr Cham Tao Soon, explored seriously the feasibility of starting a medical school, and concluded that the first step would be to establish a School of Biological Sciences. In November 2000, the University revealed its plans to establish a College of Life Sciences, comprising the School of Biological Sciences and the Biosciences Research Centre.

In July the following year, NTU's search for the Founding Dean of the School of Biological Sciences (SBS) ended with the announcement of the appointment of Professor James P Tam, a world-renowned expert in peptides and protein chemistry from Vanderbilt University, to head the new School. NTU's venture into the life sciences was also given a boost by a generous S$10 million gift from Dr Lee Hiok Huang, in the name of his late father Mr Lee Wee Nam. This was the biggest private sector donation to NTU's endowment fund up to that point in time, and it blessed the new School with a new endowed professorship — the Lee Wee Nam Professorship in Life Sciences — right from the beginning. In July 2002, 100 undergraduate students and 23 PhD students commenced their studies at the SBS. The undergraduate programme in biological sciences was (and still is) a Direct Honours programme, where the students did not have to qualify for an extra year of Honours programme only after completing a basic Bachelor's degree.

Growing Landscape

Ten months after the SBS welcomed its first students, in May 2003, the Committee to Review the University Sector and Graduate Manpower Planning (USR), chaired by Dr Ng Eng Hen, then Minister of State (Education and Manpower), recommended that "NTU expand into a full-fledged, comprehensive university to include disciplines in the physical sciences, humanities and design & media".[2] This announcement gave a sudden change to the landscape of Science originally envisaged for NTU. Instead of focusing solely on the life sciences, it would now also introduce and grow other disciplines in Science.

This new challenge landed on Dr Su Guaning, who had just taken over as President of NTU on 1 January 2003. Plans were rapidly put in place to prepare for the establishment of a new School of Physical Sciences by 2005, which was to include chemistry, mathematics, and physics. After an extensive global search, Professor Lee Soo Ying, a well-known physical chemist with rich academic leadership experience from the National University of Singapore (NUS), joined NTU in October 2004 as Dean-Designate of the new School of Physical & Mathematical Sciences (SPMS), and was officially appointed as its Dean in January 2005. The careful reader would have noticed a subtle difference between the name proposed earlier for the new School and the name it eventually took. This was the outcome of a recommendation from Professor Lee, who explained that the new name would be a more accurate reflection of the nature of the School — while the term 'physical sciences' would normally refer to disciplines including chemistry and physics, mathematics is usually not among them.

The SPMS was to have three disciplines — chemistry, mathematical sciences, and physics — under a single School, quite unlike most of the other Schools already existing in NTU. It would be organised according to these disciplines into three Divisions — Division of Chemistry & Biological Chemistry, Division of Mathematical Sciences, and Division of Physics & Applied Physics.

There was no time to be wasted. The SPMS was to open its doors to students by July 2005. A team of faculty and support staff was quickly assembled and swung into action — curriculum, space, facilities, recruitment, outreach, etc. In July 2005, 183 undergraduate freshmen matriculated as the pioneer cohort in the SPMS. They joined eight other PhD students in chemistry and three in physics, who had joined the SPMS earlier with their supervisors, and another 18 newly enrolled PhD students in chemistry and seven in physics, who began their PhD studies also in July 2005 in the SPMS. The PhD programme in mathematical sciences was only launched the following year, with seven students in its pioneer batch. Like the SBS, the undergraduate programmes at the SPMS were (and still are) Direct Honours programmes.

As the SPMS was busy getting ready in its operations, its elder sibling — the SBS — was growing furiously. As soon as the SBS was formed, a purpose-built home was planned to house its educational and research activities. In May 2004, students, faculty, and staff of the SBS finally moved into the spanking new 30,000 m² building, located across the Quad from the Nanyang Auditorium, where Dr Tony Tan's "millennium challenge" was issued. This new building, the first in NTU for Science, was officially opened by Mr Lee Hsien Loong, Prime Minister of Singapore, on 29 August 2005. This occasion also saw the birth of the Institute of Advanced Studies (IAS) at NTU.

The IAS has been instrumental in bringing eminent scientists, especially Nobel laureates and Fields medalists, to NTU through the numerous activities it has organised.

But the growth of the SBS was not just in its physical space. In late 2004, the School announced that an agreement of collaboration had been reached with the Beijing University of Chinese Medicine (BUCM), and with effect from July 2005, a new innovative Double Degree Programme in Biomedical Science and Chinese Medicine would be offered, where students would spend the first three of five years at NTU for the Biomedical Science programme, followed by two years at the BUCM for the programme in Traditional Chinese Medicine. This unique "East-meets-West" programme welcomed its first batch of 63 students in July 2005.

The timing of the completion of the SBS building was perfect. The temporary space it was previously occupying was then freed up, just in time to house the new kid on the block, the SPMS. It was now the SPMS's turn to plan for its permanent home. With inspiration drawn from the likes of Oxford's Chemistry Research Laboratory and the Isaac Newton Institute at Cambridge, UK, a sprawling 38,000 m² complex, comprising both teaching and research facilities, was ready for use by early 2008. The official opening of the new complex a year later, on 21 July 2009, was graced by Dr Ng Eng Hen, then Minister for Education and Second Minister for Defence, who had earlier chaired the USR committee whose 2003 report had led to the birth of the SPMS.

Addition of Further Species

The year 2009 was a very exciting one for Science at NTU. The Earth Observatory of Singapore (EOS), a Research Centre of Excellence (RCE) with S$150 million funding from the Singapore National Research Foundation (NRF) and the Ministry of Education (MOE), and the first RCE hosted at NTU, was officially launched at the beginning of the year. Led by its Founding Director, Professor Kerry Sieh, who hailed from the California Institute of Technology (Caltech), the EOS conducts research on hazardous natural processes that pose threats to the region, including tsunamis, volcanic eruptions, earthquakes, and climate change.

The presence of the EOS has given NTU the opportunity to significantly broaden and deepen its research capabilities in natural hazards. It also brought about new possibilities in education. A new academic home for the Earth scientists, the Division of Earth Sciences (DES), was hence created in 2010 and housed under the SPMS. A new PhD programme in Earth sciences was also launched. This marked a new milestone for Science in Singapore.

Major changes to university governance were introduced to public universities in Singapore in 2006, where NTU, NUS, and SMU became autonomous universities, with greater flexibility to strategise and innovate, "to respond to the opportunities and challenges of a more competitive landscape, and to achieve global excellence".[3] The academic organisational structure of NTU was changed in August of the same year, with the entire University organised into four Colleges: Engineering, Science, Business, and Humanities, Arts & Social Sciences. Each of the existing Schools came under a College, led by a Dean. The academic head of the School was re-designated as the Chair (previously the Dean). The College of Science then comprised both the SBS and the SPMS. Professor Lee Soo Ying was appointed the first Dean of the College of Science. The office of the Provost was also instituted at NTU. Professor Bertil Andersson, an eminent biochemist who had been the Chief Executive of the European Science Foundation and Rector of Linköping University in Sweden, joined NTU as its first Provost, on 1 April 2007.

Both Schools in the College of Science continued to grow rapidly during the first decade of the new millennium. The pioneer batch of undergraduate students in the SBS graduated in 2006, while the School also produced its first PhD graduates the following year. A year after it took in its first undergraduates, the SPMS introduced a new undergraduate combined major programme in Mathematics & Economics, in collaboration with the Division of Economics in the School of Humanities and Social Sciences. This was the first time such a programme, popular in some universities in the UK and the US, was made available in Singapore. While the majority of the pioneer cohort of SPMS undergraduates, who matriculated in 2005, graduated in 2009, seven amongst them completed their studies a year ahead of their peers and graduated in 2008. The School also produced its first PhD graduate in 2007.

Both the SBS and the SPMS made the decision right from the beginning to be research-intensive, and made great efforts to put in place the proper research infrastructure and to hire promising scientists and established researchers. Very soon, prominent research groups became entrenched, in areas such as structural biology, synthetic organic chemistry, applied discrete mathematics, and condensed matter physics. Excellence begets more excellence. When the Singapore NRF introduced, in late 2007, a new NRF Fellowship scheme — a competitive programme to attract, recruit, and root outstanding young scientists from around the world to conduct independent research in Singapore — four of the 10 winners in the inaugural class of NRF Fellows chose to join the College of Science as their host institution, one each in biology and physics, and two in chemistry. Not a bad haul for an institution less than a decade in age. To date, more than 30 of the recipients of the NRF Fellowship have joined NTU's College of Science. In 2008, Provost Andersson also introduced NTU's own in-house analogue of the NRF Fellowship — the Nanyang Assistant Professorship (NAP) scheme. The College of Science has also benefited from the NAP scheme, through which more than 20 promising young scientists have been recruited since the scheme's inception.

While the first 10 years of the new millennium saw phenomenal growth in Science at NTU from virtually nothing, further exciting development continued to unfold in the second decade of the 21st century.

In September 2011, a second RCE — the Singapore Centre for Environmental Life Sciences Engineering (SCELSE) — opened in NTU, with S$120 million of funding from the NRF and the MOE. Headed by renowned microbiologist Professor Staffan Kjelleberg, SCELSE focuses on microbial biofilm communities and related environmental issues. Together, the two RCEs — EOS and SCELSE — provided synergistic opportunities for NTU to build up new capabilities in environmental, ecological, and Earth systems sciences. Tapping on the expertise in these two RCEs, an innovative interdisciplinary undergraduate major programme in Environmental Earth Systems Science was launched in 2014, the first of its kind in Singapore. A class of 27 freshmen embarked on this new programme in August 2014.

It was time for the DES, initially created to house the Earth scientists, to morph into a full-fledged School, as a multidisciplinary School focused on the environment. The College of Science saw the birth of its third baby on 1 January 2015. The Asian School of the Environment (ASE) was officially established, while the SPMS returned to its original state with three divisions. The undergraduate programme in Environmental Earth Systems Science also came under the administration of this new School.

Hot on the heels of the establishment of SCELSE, in early 2012, a team from NTU, led by Professor Nikolay Zheludev, became the first in Singapore to win an Academic Research Fund (AcRF) Tier 3 grant (approximately S$10 million), newly introduced by the MOE. The success of this proposal on photonics led rapidly to the growth of photonics research at NTU. The Centre for Disruptive Photonic Technologies (CDPT) was soon launched. Later in October 2014, The Photonics Institute at NTU, comprising the CDPT and four other centres, was officially opened. Photonics research at NTU had come of age.

A year after the success of Professor Zheludev's team, a group led by renowned structural biologist Professor Daniela Rhodes, FRS, of the SBS, went on to receive the first S$25 million AcRF Tier 3 award at the subsequent grant call, this time for the study of telomere dynamics and genome functions. This award enabled NTU to further strengthen its excellence in structural biology. The following year, yet another S$25 million AcRF Tier 3 grant was awarded to a team of scientists, led by Professor Stephan Schuster, of the SBS and SCELSE, to study urban air microbiomes. Life sciences at NTU had come a long way since the humble beginnings barely more than a decade earlier.

A New Shoot from The First Seed

What started as a precursor to the aspiration of starting a medical school grew to become a full-fledged presence of Science at NTU. Yet, the initial plan was never forgotten. With the growth of Science, the time had come to revisit the idea of the medical school. A new chapter began in July 2010, when NTU signed a memorandum agreement with Imperial College London to jointly set up a medical school. Following a significant gift in 2011 by the Lee Foundation, the medical school was officially named after the local philanthropist Tan Sri Dato Lee Kong Chian — the Lee Kong Chian School of Medicine (LKCMedicine). With government matching, the gift, which amounted to S$400 million, remains to date the largest single gift to any institution in Singapore. Housed both at NTU's main Yunnan campus and the new campus at Novena (next to Tan Tock Seng Hospital, a member of the National Health Group, LKCMedicine's primary clinical partner), LKCMedicine welcomed its pioneer intake of 54 students in August 2013.

With the birth of SCELSE and LKCMedicine, NTU had built up a significant presence in the life sciences. In February 2015, NTU launched the NTU Institute of Structural Biology (NISB) to bring together related expertise to address important problems in biology and medicine. Seven months later, an ambitious cluster, the NTU Integrated Medical, Biological & Environmental Life Sciences (NIMBELS) Cluster, which features the Singapore Phenome Centre, the first of its kind in Southeast Asia, was formed. This new initiative brought together, under a single umbrella, the SBS, LKCMedicine, SCELSE, and NISB, to promote interdisciplinary and collaborative approaches to address challenges in healthcare and environmental sustainability.

A Flourishing Garden

The seed planted by Dr Tony Tan, in the "millennium challenge" given on the first day of the new millennium, has flourished into a garden that has in turn enriched the landscape of Science both in Singapore and worldwide. NTU is now a significant contributor internationally to the creation of new scientific knowledge and a key player in preparing future generations for the knowledge-driven economy, through its plethora of research and educational programmes in Science. Today, the College of Science alone has nearly 200 faculty members, about 500 research staff, 3600 undergraduate students, and more than 700 graduate students. LKCMedicine, which enjoys an intimate relationship with the College of Science, continues to expand. In the meantime, the NIE, which began offering a Bachelor of Science with Diploma in Education at its establishment, had also grown its Academic Groups in Mathematics & Mathematics Education and Natural Sciences & Science Education. While NIE's mission remains focused on teacher education and educational research, collaboration with the College of Science in education and research has also been increasing steadily. NTU, including its various scientific disciplines, has seen a meteoric rise in various international rankings of universities or on research publications (though rankings really aren't everything (!), the trajectory of an institution is nonetheless useful information).

Several factors have contributed to this rapid growth of Science at NTU. The determined focus and unrelenting support that the Singapore government has given to research and education has not gone unnoticed by the international community and remains the envy of many academics. Bold and visionary leadership within the University cannot be taken for granted either. These have led to the recruitment of outstanding individuals passionate in research and education, resulting in a vibrant environment where scientists constantly endeavour to tread into uncharted territories and push boundaries.

The story of the garden of Science at NTU certainly does not end here. The world continues to evolve, and new challenges arise that demand new thinking and knowledge to tackle them. Many interesting fundamental problems also remain elusive to date, and will not cease to capture the imagination of scientists. With all these challenges come opportunities, and this garden of Science at NTU will no doubt continue to flourish further amidst this fertile landscape and open new horizons.

Acknowledgements

The author thanks the staff of the NTU Corporate Communications Office, Student & Academic Services Department, and College of Science, especially Chan Kwong Lok, Vivien Chiong, Choy Kam Luen, Ronald Anthony Lin

Linxiong, and Eileen Tan Leng Hwee, for their assistance in providing some of the facts and figures used in this article.

References

[1] Speech by Dr Tony Tan Keng Yam, Deputy Prime Minister and Minister for Defence, at the NTU Launch of Millennium Celebration and Official Opening of Nanyang Auditorium held on Saturday, 1 January 2000 at 7:30 pm.

[2] Restructuring the University Sector — More Opportunities, Better Quality: Report of the Committee to Review the University Sector and Graduate Manpower Planning, May 2003, Ministry of Education, Singapore.

[3] NUS, NTU, SMU to become Autonomous Universities: Press Release by the Ministry of Education, Singapore, 12 April 2005.

Chapter 16

The Institute of Advanced Studies at Nanyang Technological University: Ten Years of Achievement

K K Phua

In 2015, Singapore celebrates its 50th birthday. At the same time, the Institute of Advanced Studies (IAS) at the Nanyang Technological University (NTU), established in July 2005, celebrates its ten years of achievement. IAS at NTU has been formed to identify and push strategic research areas at NTU with appropriate research programs supplemented with high-level international conferences and workshops. These conferences and workshops are organised with the help of eminent scientists to provide NTU's science and technology initiatives with a "Nobel boost": establishing the hallmarks of science at the highest level. Since then, IAS has also organised several successful graduate schools and outreach activities. To shape the strategic directions of the Institute, IAS is advised and guided by a committee of world-renowned scientists, including 11 Nobel Laureates and a Fields Medallist. Mr George Yeo, Former Minister for Foreign Affairs, is the Patron of IAS.

Since its establishment in 2005, IAS at NTU is generally regarded as one of the best Institutes of Advanced Studies in the Asia Pacific region. The Institute hosts a significant number of Nobel Laureates and Fields Medallists every year, and organises a large number of high impact, international conferences, workshops and schools across many disciplines in the Asia Pacific region. The Institute has helped to forge interdisciplinary research and close collaboration between NTU and major centres of research around the world. It has also inspired numerous talented youths through its many programs to move into a scientific career. Moreover, these events have allowed NTU faculty members and students to interact closely with eminent visitors and speakers from around the world. The multidisciplinary topics that IAS oversees include physics, chemistry, engineering, biomedical imaging, materials science, maths, liberal arts, urban planning, and so forth.

Although IAS is based at NTU, the close rapport with many active scientists and distinguished scholars from all disciplines abroad that has been established through the years has also helped to promote Singapore as a hub for international research and development. This successful model has encouraged the National Research Foundation to engage IAS with the initial planning and organisation of the well-known Global Young Scientists Summit (GYSS) each year in Singapore. The GYSS is modelled closely after the Lindau conferences organised by the Nobel Foundation each year.

IAS received generous financial support from the Lee Foundation. For the past ten years, IAS has organised numerous conferences, workshops, symposiums and graduate schools. Many of these events were co-organised with the National University of Singapore, Tan Kah Kee International Society, National Institute of Education, Ministry of Health and Ministry of Education, particularly through activities in secondary schools and junior colleges. Some of the key events are:

100 Years of Physics Symposium in Singapore (10 to 12 August 2005)

This event was held as part of the celebration of the World Year of Physics 2005. Besides reaching out to the next generation of scientists and engineers, the symposium, organised jointly by IAS and the newly formed School of Physical & Mathematical Sciences at NTU, was also intended to reach out and capture a wider non-physics community.

We were glad that Mr Tharman Shanmugaratnam (current Deputy Prime Minister of Singapore) was able to grace the opening ceremony on 10 August 2005. Keynote speakers invited to the conference included Prof Robert Blaughlin (Nobel Laureate in Physics 1998), Prof Alex Pines (Wolf Prize Winner 1991), Dr Akira Tonomura (Imperial Prize Winner 1991) and Prof Paul Davies (Templeton Prize Winner 1995).

(from left) Mr Tharman Shanmugaratnam (current Deputy Prime Minister of Singapore) had a cordial conversation with Prof Eng-Chye Tan and Prof Soo-Ying Lee.

An educational forum and a regional cooperation panel discussion took place during the symposium. Both sessions, particularly the panel discussion, was intended to pave the way for better future cooperation between institutions from different countries.

Conference in Honour of Nobel Laureate Professor C N Yang's 85th Birthday: Statistical Physics, High Energy, Condensed Matter and Mathematical Physics (31 October to 3 November 2007)

The special conference in celebration of Prof Chen-Ning Yang's pioneering contributions to physics on the occasion of his 85th Birthday was held at the Swissotel Merchant Court.

In 1967, he attended an International Conference at Nantah's Department of Physics ("Nantah" is the Chinese abbreviation of the Nanyang University, a precursor to the current NTU). In 1971, Prof Yang was appointed as External Examiner for the Department of Physics. He also accepted IAS's invitation to be a member of the International Advisory Committee in 2005. When NTU organised the *37th International Physics Olympiad* in July 2006, Prof Yang delivered two inspiring lectures to students and staff at NTU. During the event, he also launched the NTU's C N Yang Scholars Programme designed for talented students with a deep passion in Science, Technology, Engineering, Mathematics and Research.

(from left) Mr Tharman Shanmugaratnam (current Deputy Prime Minister of Singapore), Mrs C N Yang, Prof C N Yang, Dr Guaning Su (then President of NTU), Prof Bertil Andersson (current NTU President) and Prof Phua Kok Khoo (Director of IAS).

Prof C N Yang and Mr S R Nathan (then President of the Republic of Singapore) at the conference banquet.

We were also happy that the opening ceremony was graced by Mr Tharman Shanmugaratnam (current Deputy Prime Minister of Singapore) on 31 October 2007. The three-and-a-half day programme included presentations by more than 60 invited eminent speakers, local and overseas researchers. The Nobel Laureates speakers were Prof Walter Kohn (Nobel

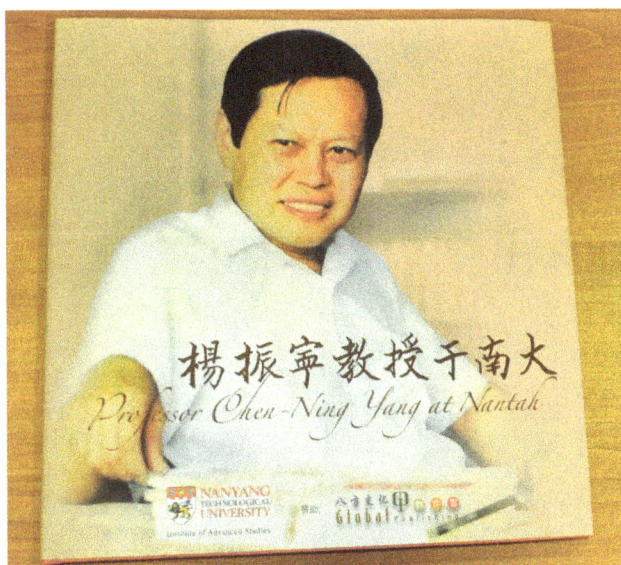

The special book titled *Professor Chen-Ning Yang at Nantah* was presented to Prof Yang at the conference.

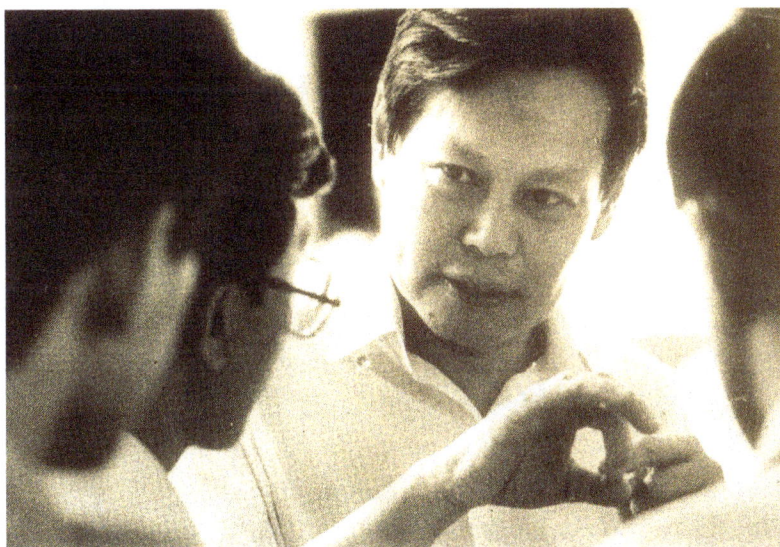

Prof C N Yang exchanged views with young students during a visit to Nantah in 1976.
Reprinted with permission from *Professor Chen-Ning Yang at Nantah*. Copyright 2007, Institute of Advanced Studies NTU, p. 16.

Laureate in Chemistry 1998), Prof Martin Perl (Nobel Laureate in Physics 1995) and Prof Claude Cohen-Tannoudji (Nobel Laureate in Physics 1997).

A series of other events were held in conjunction with the unique conference. These included C N Yang's Nobel Laureate room dedication, signing of

Prof C N Yang held a cordial talk with Mr Lee Hsien Loong (current Prime Minister of Singapore) (second from left) at the *25th International Conference on High Energy Physics* in August 1990. On the left is Prof Phua Kok Khoo, Chairman of the Organising Committee.
Reprinted with permission from *Professor Chen-Ning Yang at Nantah*. Copyright 2007, Institute of Advanced Studies NTU, p. 36.

Memorandum of Understanding between NTU and Tsinghua University, the Science Education Symposium, the Asia-Pacific Meeting on Frontiers in Plasma Physics, as well as a public lecture by C N Yang and renowned Chinese artist, Fan Zeng.

The conference banquet was held on 1 November 2007 at the Ballroom of Swissotel Merchant. Mr S R Nathan, then President of the Republic of Singapore, was the Guest-of-Honour for the evening.

Annual International Science Youth Forums (ISYF)
with Nobel Laureates in Singapore
(19 to 23 January 2009, 18 to 22 January 2010, 11 to 15 January 2011,
15 to 19 January 2012, 20 to 24 January 2013, 19 to 23 January 2014,
18 to 22 January 2015 & 17 to 21 January 2016)

The annual ISYF is an outreach program organised jointly with Hwa Chong Institution each year. The Forum is conceived as a platform for talented high school students to learn and enhance their interests in science through

ISYF 2015 panel discussion at NTU on *Becoming a Modern Scientist*. (from left) Nobel Laureates Sir Tim Hunt and Sir Andre Geim, Prof Bertil Andersson (NTU President) and Prof Lars Brink (former chairman of the Nobel Physics Committee).

ISYF 2014 participants engaging in an experiment on the embryonic *in ovo* of a chick at the NTU Lee Kong Chian School of Medicine.

intellectual dialogues with leading experts in various scientific fields: Physics, Chemistry, Biology, Mathematics and Engineering. Each year, around five to six Nobel Laureates are invited to share their insights and passion in science with top science students from premier schools worldwide. The Forums are sponsored by the Temasek Foundation (2009 to 2011) and the Agency for Science, Technology and Research (A*STAR, 2012 to 2016) as well as the Ministry of Education.

The residential camp provides an excellent platform for high school students to bond with like-minded high achievers. It is also a rare opportunity to interact with Nobel Laureates and eminent scientists. The Forums help to inspire the students and provide them with a better understanding of the importance of broad-based knowledge, diverse interest, deep passion and tenacity necessary in their pursuit of science excellence.

Les Houches School of Physics Summer Session in Singapore (29 June to 24 July 2009)

IAS organised the first "Les Houches" Session in Asia.

The one-month summer school held at the Nanyang Executive Centre of NTU was organised by IAS in collaboration with the Les Houches School of Physics in France and the Centre for Quantum Technology. The Les Houches School of Physics, widely known as one of the premier summer and winter schools in the world, is traditionally held in the Rhone Alps in south-eastern France. Session 91 was in fact the first time the prestigious French school was held outside of France since its establishment in 1951.

His Excellency Mr Pierre Buhler, then Ambassador of France to Singapore, graced the opening ceremony. Prof Leticia Cugliandolo (Director of École de

Opening ceremony of the Les Houches School of Physics in Singapore. Seated in the front row from left: Prof Martial Ducloy (CNRS), Prof Leticia Cugliandolo (Director of École de Physique des Houches), Prof Phua Kok Khoo (Director of IAS), His Excellency, Mr Pierre Buhler (then French Ambassador of Singapore) and Dr Guaning Su (then President of NTU).

Participants enjoying a discussion with Nobel Laureate Prof Anthony Leggett.

Physique des Les Houches) gave a presentation on the Les Houches School of Physics in France.

The Les Houches School in Singapore aimed to attract young graduate students and researchers into these very active fields of research frontier. It offered an overview of the latest developments and allowed the participants to share the excitement and challenges of the community with experts in the fields of Ultracold Gases and Quantum Information. More than 110 postdoctoral fellows and graduate students applied for the School and 67 participants were eventually admitted.

Conference in Honour of the 80th Birthday of Professor Murray Gell-Mann (24 to 26 February 2010)

A festival of lectures themed "Quantum Mechanics, Elementary Particles, Quantum Cosmology and Complexity" was held to celebrate Nobel Laureate Prof Gell-Mann's revolutionary contributions to physics. Many renowned scientists attended the conference, including three Nobel Laureates, Professors C N Yang, Kenneth Wilson and Gerard 't Hooft. Chaired by Prof Harald Fritzsch and Prof Phua Kok Khoo, the conference included parallel sessions on Particle Physics, Quantum Mechanics and Complexity. The event was sponsored by the Lee Foundation. More than 80 teachers attended the Physics

Nobel Laureates Professors Gerard 't Hooft and Murray Gell-Mann enjoying the conference lectures.

Education workshop which was held in conjunction with the conference. Most of the teachers who attended the workshop thoroughly enjoyed the captivating talks and hands-on activities.

International Workshops on Photosynthesis (18 to 20 August 2010, 11 to 13 June 2012, 11 to 14 June 2014 and 21 to 24 March 2016)

The biennial International Workshops on Photosynthesis were jointly organised by IAS, School of Biological Sciences, School of Material Sciences and Engineering, and Energy Research Institute at NTU, and sponsored by the Lee Foundation.

Chaired by Prof James Barber (Imperial College London), the workshops were attended by many leading experts, including Prof Alan Heeger (Nobel Laureate in Chemistry 2000), Sir Harold Kroto (Nobel Laureate in Chemistry 1996), Prof Rudolph Marcus (Nobel Laureate in Chemistry 1992), Sir John Walker (Nobel Laureate in Chemistry 1997) and Prof Michael Grätzel (Millennium Technology Prize 2010).

In 2016, the *4th International Workshop on Solar Energy for Sustainability: Photosynthesis and Bioenergetics* was specially dedicated to the celebration of the 75th Birthday of Nobel Laureate Sir John Walker and Prof Leslie Dutton FRS, as well as in memory of the outstanding contributions made to this area of research by Prof Joan Mary Anderson FRS, who passed away on 28 August 2015.

Prof James Barber (Imperial College London), the chief architect of the series of workshops on photosynthesis and bioenergetics.

Local and overseas speakers and participants of the 2014 Photosynthesis Workshop.

IAS-CERN Schools on Particle Physics and Cosmology and Implications for Technology (9 to 31 January 2012 and 2 to 6 February 2015)

In January 2012, with the enthusiastic support of CERN, a special three-week long school on Particle Physics, Cosmology and Implications for Technology was organised by IAS. Her Royal Highness, Princess Maha Chakri Sirindhorn of Thailand, was the Patron of the School.

The schools aim to bring together interested researchers to join in the scientific exploration in the exciting fields, such as frontier topics in particle physics and cosmology, and the latest technologies in accelerator and detector physics.

Singapore Deputy Prime Minister Teo Chee Hean launched the first School at the opening ceremony on 9 January 2012. "The IAS School takes the collaboration between NTU and CERN to a higher level, and helps to position NTU as the Asian research hub in the global network of scientific institutions that focus on high energy physics," said DPM Teo. "To stay ahead in the highly competitive arena of scientific research, Singapore must continue to support and recognise high-quality research, forge new links and partnerships, develop new niches of excellence and blaze new trails."

The Straits Times, page B6, 11 Jan 2012

S'pore students on hunt for God particle

IT HAS been called the building block of the universe, but has eluded the grasp of scientists around the world.

Now, students in Singapore are being given a chance to help European scientists find the Higgs Boson particle – or so-called God particle - an ultra-small particle believed to give all matter its mass.

They began doing so this week under a programme run by the Nanyang Technological University in conjunction with the European Organisation for Nuclear Research (Cern), which is at the forefront of the hunt for the particle.

The three-week programme, to be held every two years, will focus on the physics of particles and include lectures and work-shops by experts in the field.

This batch of participants will also help Cern scientists analyse data from its laboratories to prove that the Higgs Boson particle exists.

The Large Hadron Collider at Cern, a machine that collects information on particles, generates about 15 million gigabytes of data each year.

About 120 students, academics and researchers from Singapore and the region were picked for this year's programme.

Deputy Prime Minister Teo Chee Hean, the guest of honour at the opening ceremony on Monday, said the programme, called a "winter school", will give Singapore students "a high-level platform to keep up to date with the frontiers of particle physics and technology".

Such short-term schools, traditionally held in the US and Europe during the winter and summer breaks, are meant to help scientists keep pace with developments in their fields and learn about trends in related sciences.

Professor Ignatios Antoniadis, the theory division's head of physics at Cern, said the institute could in the future also collaborate with other Singapore schools and agencies, such as the National University of Singapore and the Agency for Science, Technology and Research.

DPM Teo said the Ministry of Education will send teachers to Cern to enhance their science-teaching methods.

The programme's patron is Thailand's Princess Maha Chakri Sirindhorn, who visited Cern in 2009 and initiated a collaboration between her country and the organisation.

Thailand holds its second workshop with Cern in April.

FENG ZENGKUN

The Straits Times article on the 1st IAS-CERN School (11 January 2012).

The cheerful delegates in a group photograph at the 1st IAS-CERN School.

Deputy Prime Minister Teo Chee Hean (centre) at the launch of the School, accompanied by NTU President, Prof Bertil Andersson (right) and Prof Phua Kok Khoo (Director, IAS).

The success of the School encouraged IAS, together with support from CERN, to organise the 2nd School from 2 to 6 February 2015, as well as two workshops in 2013 and 2014.

International Conference on Pan Shou: A Centennial Commemoration (31 March 2012)

The conference and exhibition in honour of the 100th anniversary of a famous Chinese literary figure, Pan Shou, was jointly organised by IAS and the College of Humanities, Arts, and Social Sciences, and supported by the Tan Kah Kee International Society. An all-rounder in the poetic and academic realms, the late Pan Shou is fondly remembered as the best calligrapher and Chinese poet in Singapore.

Mr Heng Swee Keat (then Minister for Education) graced the conference as the Guest-of-Honour. Mr George Yeo (IAS Patron and former Minister of Foreign Affairs) was unable to attend but he kindly sent a message in support of Pan Shou's important contributions to Singapore's education, cultural and arts identity. Mr Yeo, an ardent supporter of the arts, said that Singapore would have been poorer culturally if not for Pan Shou's literary works. It was therefore meaningful to remind people of Pan Shou's monumental contributions, because "without collective memories, there is no strong community."

Mr Heng Swee Keat (then Minister for Education, centre front row) was the Guest-of-Honour.

International speakers hailing from China, Hong Kong and Malaysia, including expert Chinese calligraphers and poets, attended the conference.

Global Young Scientists Summits (GYSS) (20 to 25 January 2013, 19 to 24 January 2014, 18 to 23 January 2015, 17 to 22 January 2016)

GYSS 2016 speakers discussed the trends and issues facing the scientific community at a panel discussion. (Photo: National Research Foundation Singapore)

The GYSS is an annual gathering of close to 300 young scientists and researchers from all over the world, who gather in Singapore to interact with eminent scientists and technology leaders including recipients of the Nobel Prize, Fields Medal, Millennium Technology Prize and Turing Award. It is a multidisciplinary summit, covering topics ranging from chemistry, physics, medicine, mathematics, computer science and engineering. Over a five-day programme comprising plenary lectures, panel discussions and small group sessions, the young scientists and researchers are inspired and encouraged to pursue their scientific dreams. Public talks are also held island-wide where the prize winners engaged the public, students and researchers. These public talks stimulate interest and inject excitement in science, technology and innovation.

GYSS was launched in 2013 by Dr Tony Tan Keng Yam, President of the Republic of Singapore and Patron of GYSS. GYSS is organised by National Research Foundation Singapore. IAS helped to invite some of the speakers including Nobel Laureates and Fields Medallists.

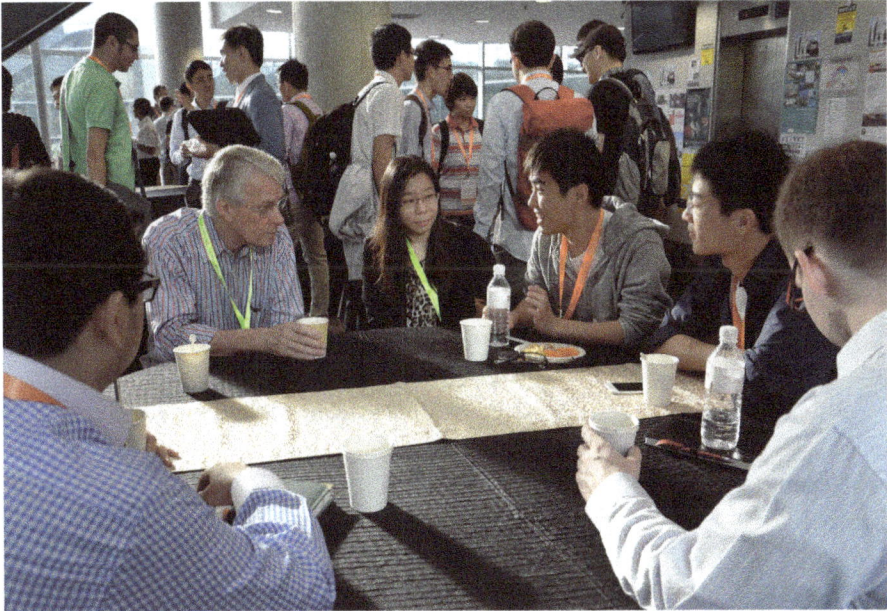

Sir Richard Roberts (Nobel Laureate in Physiology or Medicine 1993) interacting with GYSS 2015 participants over coffee. (Photo: National Research Foundation Singapore)

GYSS 2014 participants get up close with Prof Michael Grätzel, Millennium Technology Prize (2010). (Photo: National Research Foundation Singapore)

Conference in Honour of the 90th Birthday of Professor Rudolph Marcus: Fundamentals in Chemistry and Applications (22 to 24 July 2013)

The conference in honour of Prof Rudolph Marcus' contribution to chemistry was held from 22 to 24 July 2013 with sponsorship from the Lee Foundation. His groundbreaking theory on electron transfer won him the Nobel Prize in Chemistry in 1992. The Marcus theory has since become the dominant theory describing electron transfer reactions that are ubiquitous in chemistry and biology. The presence of many international participants amongst the locals added much diversity to the special occasion.

Prof Bertil Andersson (NTU President) graced the opening ceremony as Guest-of-Honour. Among the 19 international speakers were Nobel Laureate Prof Yuan-Tseh Lee and Millennium Technology Prize Laureate Prof Michael Grätzel.

A joyful Prof Marcus receiving his birthday gift from Prof Bertil Andersson at the conference banquet held at Tanglin Club.

Conference in Honour of the 90th Birthday of Professor Freeman Dyson (26 to 29 August 2013)

To celebrate Prof Freeman Dyson's illustrious career in physics, mathematics, astronomy, nuclear engineering and climate change, IAS organised the

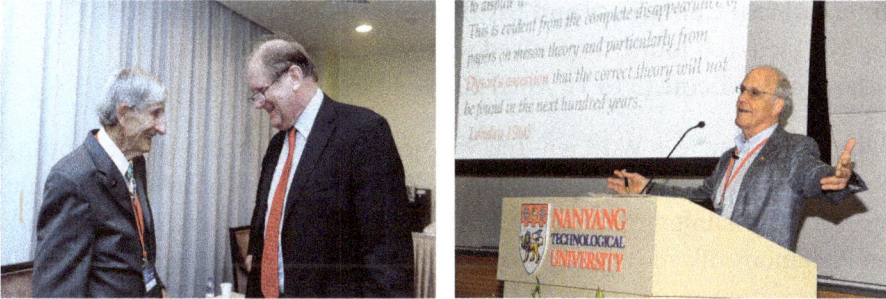

(left) Prof Dyson had a nice chat with Prof Bertil Andersson (NTU President).
(right) Nobel Laureate Prof David Gross delivering his presentation on Quantum Field Theory.

Prof Dyson cutting his birthday cake. Cheering him on were (from left) Mrs Dyson, Prof David Gross (Nobel Laureate in Physics 2004), Prof Shou-Cheng Zhang (Stanford University and Tsinghua University) and Prof Phua Kok Khoo (Director of IAS).

Conference in Honour of his 90th Birthday from 26 to 29 August 2013 with the support of the Lee Foundation. The Guest-of-Honour for the event, Prof Bertil Andersson (NTU President), addressed over 160 speakers and participants who came from different parts of the world to join in the festivities.

The distinguished speakers included Prof David Gross (Nobel Laureate in Physics 2004), Prof Shou-Cheng Zhang (Stanford University), Prof Xiao-Gang Wen (Perimeter Institute), Prof Kazuo Fujikawa (RIKEN), Prof Molin Ge (Nankai University) and many others. There were also public lectures by popular science writers Prof Lawrence Krauss and Prof Phillip Schewe dedicated to Prof Freeman Dyson.

International Conference and Exhibition on Plague Fighter Dr Wu Lien-Teh jointly organised with the Lee Kong Chian School of Medicine (5 to 12 April 2014)

"Plague Fighter" Dr Wu Lien-Teh was widely considered as the Founding Father of modern medicine for China. He saved millions of lives in the fight against the pneumonic plague of China over hundred years ago. For his "Work on Pneumonic Plague and for identifying the role of the Tarbagan marmot in the disease transmission", he was nominated for the Nobel Prize in Medicine in 1935.

The conference was organised by IAS and Lee Kong Chian School of Medicine at NTU, the Singapore China Friendship Association and Dr Wu Lien-Teh Society, Penang. A well illustrated and informative photograph exhibition on the life and work of Dr Wu was also on display for one week. Nine distinguished speakers presented interesting anecdotes covering a wide range of aspects of Dr Wu's life and work. Minister for Health Mr Gan Kim Yong, graced the occasion as Guest-of-Honour. The Ambassador of the People's

The Minister for Health, Mr Gan Kim Yong (third from right) graced the occasion as Guest-of-Honour and officiated the opening of the exhibition.

Republic of China to the Republic of Singapore, H.E. Duan Jielong, was the VIP guest.

Workshop on the Chemistry of Energy Conversion: From Molecular Design to Advanced Materials & the 6th MRS-S Conference on Advanced Materials (22 to 24 July 2014)

IAS, the College of Science and the School of Materials Science and Engineering at NTU, as well as the Materials Research Society of Singapore jointly organised the workshop and conference in 2014. Chaired by Prof Peidong Yang of the University of California, Berkeley, it focused on the importance of advancement of new materials design and development with efficient energy conversion, so as to provide for a sustainable energy future that does not rely on fossil fuels. The event brought forth many distinguished speakers to share their views on this topic and also showcased the recent development in the field. The opening ceremony was graced by Prof Freddy Boey (Provost of NTU), a materials scientist himself.

(from left) Prof Peidong Yang (University of California, Berkeley), Prof Phua Kok Khoo (Director of IAS), Prof Chowdari (Materials Research Society of Singapore) and Prof Freddy Boey (Provost of NTU).

28th General Assembly of the International Union of Pure and Applied Physics (IUPAP) (5 to 7 November 2014)

The IUPAP 2014 Executive Council at the 28th General Assembly in NTU, Singapore. (from left) Prof Bruce McKellar (current President), Prof Cecilia Jarlskog (Immediate Past President), Prof Sukekatsu Ushioda (Past President), and Prof Stuart Palmer (Immediate Past Secretary General).

The International Union of Pure and Applied Physics (IUPAP), held its 28th General Assembly at NTU in Singapore from 5 to 7 November 2014. NTU President Prof Bertil Andersson was the Guest-of-Honour at the Assembly.

IUPAP was established in 1922 at Brussels with 13 member countries. The Union is composed of members representing identified physics communities. Adhering bodies act through their Liaison Committees. Delegates from these Committees meet in the General Assemblies of the Union, held every three years. Currently, the Union, which is made up of 60 member countries, is governed by its General Assembly, which meets every three years in a different country. The aim of IUPAP is to support and promote physics and the learning of physics worldwide.

The secretariat and administration work for IUPAP had been hosted by the Institute of Physics in London, England for the last six years since 2009. From 1 January 2015, secretariat responsibilities were transferred to Singapore and hosted by NTU. The move of the IUPAP office to Asia acknowledges and supports the strong growth of physics activity in this region. Observers from

ASEAN countries, which are not yet IUPAP members, attended the General Assembly in NTU. IUPAP hopes to build fruitful relations with them and that they will eventually become members of the Union.

2nd Singapore Sustainability Symposium (15 to 17 April 2015)

IAS and the Sustainable Earth Office at NTU jointly organised the 2nd Singapore Sustainability Symposium (S3) at the Regent Hotel. The Ministry of Health, Ministry of National Development and the Ministry for Environment and Water Resources, Singapore lent their support and collaboration and made the event a success.

The theme for S3 2015 was *Sustainable City Design*. The Guest-of-Honour, Permanent Secretary (Environment and Water Resources) Mr Choi Shing Kwok presented some key objectives from Singapore's new sustainability plan. He outlined the important areas of development and change that are being put in place for a more sustainable future. S3 has now become an international platform with a Singaporean flavour for sustainability and urban solutions with a special interest in the challenges faced by cities and city decision makers.

Panellists of the roundtable discussion on water recycling with the Guest-of-Honour Choi Shing Kwok (third from left), Permanent Secretary, Ministry for Environment and Water Resources.

Conference on 60 Years of Yang-Mills Gauge Field Theory: C N Yang's Contributions to Physics (25 to 28 May 2015)

During the last six decades, Yang-Mills theory has increasingly become the cornerstone of theoretical physics. It has widespread applications in statistical physics, condensed-matter physics, atomic and nonlinear optics and nonlinear systems. In May 2015, IAS held a major conference on *60 Years of Yang-Mills Gauge Field Theories* with sponsorship from the Lee Foundation, attracting more than 180 participants from around the world. Many eminent speakers attended the conference, including Nobel Prize winners Professors C N Yang and David Gross, Royal Medallist Michael Fisher, National Medal of Science winner Paul Chu, and many more.

In addition, Professors C N Yang, David Gross and Michael Fisher gave public lectures on their personal perspectives in physics. The conference also included a roundtable discussion on the role of regional labs and hubs in promoting collaboration in theoretical and fundamental physics. Various suggestions were made during the discussion, including strong arguments for creating an Asian version of CERN, where Asian countries could work with scientists from the rest of the world to uncover new levels of fundamental physics.

Roundtable discussion on international collaboration chaired by Prof Ngee-Pong Chang.

During the conference, the NTU C N Yang Scholars also had the privilege of engaging in an informal discussion with Prof C N Yang on 27 May 2015. The scholars were enlightened by his unique philosophy of research work and life that he shared with the scholars with some humour.

2nd Pan Asia Liberal Arts Education Conference (28 to 29 October 2015)

Jointly organised by IAS and the College of Humanities, Arts, and Social Sciences at NTU, the conference aimed to explore concepts and development directions of liberal arts education around the world, and to address issues that universities and liberal arts colleges face in this modern era. The 1st Liberal Arts Education Conference was held in April 2014 at the University of Nottingham Ningbo China. NTU was pleased to be the host of the 2nd Liberal Arts Education Conference which featured eight presidents of renowned universities — Professors Fujia Yang (President, University of Nottingham Ningbo China), Steve Kang (President, KAIST), Chia-Wei Woo (Founding President, HKUST), Pericles Lewis (President, Yale-NUS College), Frank Chang (President, National Chiao Tung University), Peihua Gu (Provost and Vice President, Shantou

Prof Da Hsuan Feng (standing) chaired the roundtable discussion on *Liberal Arts Education in Asia in the 21st Century*.

University), Joseph Sung (President, The Chinese University of Hong Kong) and Wei Zhao (Rector, University of Macau).

Around 120 attendees attended the conference, including high school students, undergraduates and interested members of the public. There was also a roundtable discussion on *Liberal Arts Education in Asia in the 21st Century* which provided useful insights on the state of liberal arts education in different parts of Asia. The 3rd conference will be hosted by Fudan University.

Memorial Meeting for Nobel Laureate Professor Abdus Salam's 90th Birthday (25 to 28 January 2016)

Co-chaired by Prof Lars Brink (Chalmers Institute of Technology), Prof Michael Duff (Imperial College London) and Prof Phua Kok Khoo (Director of IAS), the memorial meeting was held to commemorate and celebrate Prof Salam's numerous achievements and pioneering contributions to physics. Among the speakers were four Nobel Laureates, Professors David Gross, Anthony Leggett, Carlo Rubbia and Gerard t' Hooft. NTU President Prof Bertil Andersson delivered a welcome address at the opening ceremony. The event was sponsored by the Lee Foundation.

The late Abdus Salam was the first Muslim to be awarded the Nobel Prize in science. He was one of the most prolific and exciting scientists of the second half of the last century. Salam believed that "scientific thought is the common heritage of all mankind" and that the developing world should play its part, not merely by importing technology but by being the arbiter of its own scientific destiny. In 1964, he founded the ICTP in Trieste, where thousands of scientists from developing countries have been trained.

Professor Gerard 't Hooft
Nobel Prize in Physics (1999)

Professor Carlo Rubbia
Nobel Prize in Physics (1984)

Professor Sir Anthony Leggett
Nobel Prize in Physics (2003)

Engaging talks by Nobel Laureates at the Memorial Meeting of the late Prof Abdus Salam.

Workshop on the Evolution of Cells, Genomes and Proteins
(1 to 6 February 2016)

The workshop was co-organised with the Royal Swedish Academy of Sciences and the NTU School of Biological Sciences (SBS). The event is an outgrowth of the 2014 meeting in Stockholm to celebrate three centuries of science supported by the Royal Science Academy. It seeks to introduce to the Singapore community of molecular cell biologists the newly emergent molecular phylogenomic methods as well as their applications to studies of evolution at the hierarchical levels of cells and protein molecules.

The workshop was chaired by Prof Charles Kurland (Uppsala University, Sweden), and co-chaired by Prof Lars Nordenskiold (SBS, NTU) and Prof Phua Kok Khoo (IAS, NTU). Among the speakers were two Nobel Laureates, Prof Michael Levitt (Nobel Laureate in Chemistry 2013) and Prof Sydney Brenner (Nobel Laureate in Physiology or Medicine 2002).

Speakers and participants shared camaraderie moments at the Workshop on Evolution of Cells, Genomes and Proteins.

Appendix

Key Events for the Past 10 Years	
2005	
100 Years of Physics Symposium in Singapore	10–12 Aug 2005
2006	
International Workshop on Spintronics	8–12 May 2006
International Workshop on Multiscale Analysis and Applications	18–22 Dec 2006
2007	
International Workshop on New Trends in Biomolecular Modeling: From Protein Folding to DNA Compaction	15–16 Mar 2007
International Workshop on Plasma Applications in Nanofabrication and Photovoltaic Solar Cells	5–6 Jul 2007
4th Asia Pacific Workshop and 3rd Asia Pacific Conference on Quantum Information Science	30 Jul–2 Aug 2007
Conference in Honour of the 85th Birthday of Nobel Laureate Prof CN Yang: Statistical Physics, High Energy, Condensed Matter and Mathematical Physics	31 Oct–3 Nov 2007
Science Education Symposium	2 Nov 2007
2008	
Conference on Particle Physics, Astrophysics and Quantum Field Theory: 75 Years since Solvay	27–29 Nov 2008
NTU-Tsinghua Joint Workshop on Discrete Mathematics and Theoretical Computer Science	12–14 Dec 2008
2009	
1st International Science Youth Forum @ Singapore with Nobel Laureates	19–23 Jan 2009
Tsinghua-NTU Joint Workshop on Nanoscience	26–28 Feb 2009
Les Houches School of Physics Summer Session in Singapore	29 Jun–24 Jul 2009
Conference on Spare Representation of Multiscale Data and Images	14–17 Dec 2009
2010	
2nd International Science Youth Forum @ Singapore with Nobel Laureates	18–22 Jan 2010
Conference on Recent Development in Chinese Herbal Medicine	25–26 Jan 2010
Conference in Honour of the 80th Birthday of Prof Murray Gell-Mann	24–26 Feb 2010
Tsinghua-NTU Joint Workshop on Life Sciences	14–17 May 2010

(Continued)

(Continued)

Workshop on Physics with Ultra Cold Atoms	21 Jul 2010
Satellite Meeting of the 15th International Photosynthesis Congress	18–20 Aug 2010
IAS-Julian Schwinger Foundation Joint Workshop: Spontaneous Energy Focusing Phenomena and Multiscale Physics	30 Aug–3 Sep 2010
International Conference on Flavour Physics in the LHC Era	8–12 Nov 2010
NTU-Tsinghua Joint Workshop on Nanoscience	26–27 Nov 2010

2011

3rd International Science Youth Forum @ Singapore with Nobel Laureates	11–15 Jan 2011
5th Asia-Pacific Workshop on Quantum Information Science 2011 in conjunction with the Festschrift in honour of Vladimir Korepin	25–28 May 2011
IAS-MIT Joint Workshop on Emergence in Field Theory	5–8 Aug 2011
1st Workshop on Standardisation of Chinese Physics Terminology	11–12 Dec 2011

2012

1st IAS-CERN School on Particle Physics and Cosmology and Implications for Technology	9–31 Jan 2012
4th International Science Youth Forum @ Singapore with Nobel Laureates	15–19 Jan 2012
International Conference on Pan Shou: A Centennial Commemoration	31 Mar 2012
Tsinghua-NTU Joint Workshop on Number Theory, Discrete Mathematics and their Applications	25–27 May 2012
Singapore School of Physics Session I: Strong Light-Matter Coupling: From Atoms to Solid State Systems	21 May–8 Jun 2012
2nd International Workshop on Photosynthesis: Natural and Artificial Photosynthesis, Bioenergetics and Sustainability	11–13 Jun 2012
International Symposium on Theory and Evidence in Acupuncture	21–22 Aug 2012
2nd Workshop on Standardisation of Chinese Physics Terminology	23–26 Nov 2012

2013

5th International Science Youth Forum @ Singapore with Nobel Laureates	20–24 Jan 2013
1st Global Young Scientists Summit	20–25 Jan 2013
Complexity Conference	4–6 Mar 2013
International Workshop on Determination of the Fundamental Parameters of QCD	18–21 Mar 2013
1st IAS-CERN Workshop on Particle Physics and Cosmology: Status, Implications and Technology	25–27 Mar 2013

(Continued)

(Continued)

5th APCTP Workshop on Multiferroics	22–24 May 2013
9th Singapore-China Joint Symposium on Research Frontiers in Physics	28–29 Jun 2013
Conference in Honour of the 90th Birthday of Nobel Laureate Prof Rudolph Marcus: Fundamentals in Chemistry and Applications	22–24 Jul 2013
Conference in Honour of the 90th Birthday of Prof Freeman Dyson	26–29 Aug 2013
3rd Workshop on Standardisation of Chinese Physics Terminology	22–25 Nov 2013
2014	
6th International Science Youth Forum @ Singapore with Nobel Laureates	19–23 Jan 2014
2nd Global Young Scientists Summit	19–24 Jan 2014
IAS-CERN Novice Workshop on Higgs Boson, Particle Physics and Cosmology	7 Feb 2014
International Conference on Flavor Physics and Mass Generation	10–14 Feb 2014
Complexity Conference	3–5 Mar 2014
International Conference and Exhibition on Plague Fighter Dr Wu Lien-Teh jointly organised with the Lee Kong Chian School of Medicine	5–12 Apr 2014
Berge Fest Conference on Quantum Information, Quantum Optics and the Foundations of Quantum Mechanics	22–25 Apr 2014
3rd International Workshop on Photosynthesis: Advances in Solar Fuels and Photovoltaics in Honour of the 70th Birthday of Prof Michael Grätzel	11–14 Jun 2014
The OCPA8 International Conference on Physics Education and Frontier Physics	23–27 Jun 2014
Workshop on the Chemistry of Energy Conversion and the 6th MRS-S Conference on Advanced Materials	22–24 Jul 2014
8th Asian Science Camp With Nobel Laureates & Eminent Scientists in Singapore	24–29 Aug 2014
28th General Assembly of the International Union of Pure and Applied Physics (IUPAP)	5–7 Nov 2014
International Workshop on Exceptional Symmetries and Emerging Spacetime	10–12 Nov 2014
Workshop on Quantum Effects in Biological Systems	2–5 Dec 2014
2015	
7th International Science Youth Forum @ Singapore with Nobel Laureates	18–22 Jan 2015

(Continued)

(Continued)

3rd Global Young Scientists Summit	18–23 Jan 2015
International Workshop on Polyelectrolytes in Chemistry, Biology and Technology	26–28 Jan 2015
2nd IAS-CERN School on Particle Physics and Cosmology and Implications for Technology	2–6 Feb 2015
International Conference on Massive Neutrinos	9–13 Feb 2015
2nd Singapore Sustainability Symposium	15–17 Apr 2015
Conference on 60 Years of Yang-Mills Gauge Field Theories: CN Yang's Contributions to Physics	25–28 May 2015
2nd Pan Asia Liberal Arts Education Conference	28–29 Oct 2015
International Workshop on Higher Spin Gauge Theories	4–6 Nov 2015
2016	
8th International Science Youth Forum @ Singapore with Nobel Laureates	17–21 Jan 2016
4th Young Scientists Summit	17–22 Jan 2016
IAS-ICTP School on Quantum Information Processing	18–29 Jan 2016
Memorial Meeting for Nobel Laureate Prof Abdus Salam's 90th Birthday	25–28 Jan 2016
Workshop on the Evolution of Cells, Genomes and Proteins	1–6 Feb 2016
Conference on New Physics at the Large Hadron Collider	29 Feb–4 Mar 2016
4th International Workshop on Solar Energy for Sustainability: Photosynthesis and Bioenergetics	21–24 Mar 2016

Chapter 17

Institute for Mathematical Sciences: A Dream Come True*

Louis Chen

The setting up of the Institute for Mathematical Sciences (IMS) at the National University of Singapore (NUS) is a dream come true for the mathematical community in Singapore. The idea of setting up a mathematical institute in Singapore dates back to the 1980's following the enormous successes of two mathematical institutes established in the United States with funding from the National Science Foundation. These were the Mathematical Sciences Research Institute (MSRI) at Berkeley and the Institute for Mathematics and Its Applications (IMA) at the University of Minnesota at Minneapolis.

The main functions of these institutes were to promote and support mathematical research through organizing year-long programs and bringing together mathematicians who share common interests from all over the world to interact and do research with one another. Unlike other institutes, these new institutes did not hire in-house researchers. Instead, interested researchers would submit proposals to these institutes and they would organize programs under some specific themes at these institutes if their proposals were accepted. These programs would then be funded by the host institutes. Mathematicians, some by invitation, would visit the institutes for various lengths to participate in the programs.

Inspired by the successes of these institutes, a group of mathematicians led by Peng Tsu Ann, who was then Head of the Department of Mathematics at NUS, started talking about setting up a similar institute in Singapore. This topic was to become a constant conversation piece during lunches and coffee breaks in the following years. It was thought that such an institute would bring huge benefits to the mathematical community in Singapore and could also become a foremost mathematical institute in this region, if not in Asia.

These thoughts finally crystallized into a proposal for the establishment of a mathematical institute, which we wrote and submitted to the University for

---

*In this article, all Chinese names are written with surnames first.

funding in 1991. It was revised and, with the support of the University, we submitted it to the National Science and Technology Board (predecessor of A*STAR) for funding in 1996. Both attempts were unsuccessful. In 1998, with strong support from NUS, we rewrote the proposal and submitted it to the Ministry of Education (MOE) for funding. This time, our proposal was approved and we received generous start-up funding from MOE and NUS for the first five years. In addition, the University granted the new institute the use of two colonial houses in Prince George's Park.

In retrospect, the time was right for us to have a mathematical institute. There were three reasons. First, by 1998 the mathematical community in Singapore had grown substantially both in size and strength. Second, NUS was in the process of transforming itself from a teaching university to a research university and then to a world-class university. Third, Singapore was in a transition to a knowledge-based economy and was moving towards a first world status. Mathematics has permeated into almost every scientific discipline. Both fundamental research in mathematics as well as its applications to the various scientific disciplines cannot be separated from national development.

The Institute for Mathematical Sciences (IMS) was formally established in July 2000 as a University-level institute. I was fortunate to be appointed as the director. As the founding director, I was immediately faced with two challenges. First, the physical infrastructure needed to be in place quickly. Second, good quality programs had to be planned and solicited with only a year of lead time and in an unfavorable geographical position of being far away from the major academic institutions in Europe and the US.

The Scientific Advisory Board was appointed to provide advice and guidance on the scientific development of IMS and to assist the director in evaluating and selecting proposals for the Institute's programs. The Board was chaired by Roger Howe of Yale University. The other members were Jacques-Louis Lions (1928–2001) (Collège de France), Keith Moffatt (University of Cambridge), Hans Föllmer (Humboldt University of Berlin), Avner Friedman (Ohio State University), Lui Pao Chuen (Ministry of Defence) and David Siegmund (Stanford University). They all agreed to serve on the Board without hesitation. It was clear that they all took a personal interest in this new endeavor and were determined to ensure that the Institute would be a successful one. It is sad to mention that Lions passed away shortly before the Institute was formally inaugurated.

The Management Board was also formed with Chong Chi Tat as the Chair to oversee the running of the Institute. Other members came from the Office

Founding Scientific Advisory Board, December 2003: (from left) Sun Yeneng, Louis Chen, Chong Chi Tat, Hans Föllmer, Lui Pao Chuen, Avner Friedman, Keith Moffatt, David Siegmund, and Roger Howe.

of Research, Faculty of Science, Faculty of Engineering, Nanyang Technological University, Defence Science & Technology Agency and industry.

In the first year of its inception, the Institute was housed in the Department of Mathematics while the two colonial houses were being renovated. I recall that my first deputy director, Chen Kan, was helping me with the planning of the IT infrastructure of the Institute and the hiring of support staff. I was working very closely with the architect from PWD Consultants on the concept and details of the renovation. For example, we had to explore where to make an extension for a seminar room, which walls to knock down, how many toilets to keep, what colors to use for the walls and what tiles to use for the floors and the toilet walls.

Three six-month programs were planned and the themes decided after consultation with the Scientific Advisory Board. The first three programs were *Coding Theory and Data Integrity* (July–December 2001), *Post-Genome Knowledge Discovery* (January–June 2002) and *Representation Theory of Lie Groups* (July 2002–January 2003).

We moved to the renovated houses in June 2001. The Institute was officially opened on 17 July 2001 by the guest of honor, Rear Admiral Teo Chee Hean, then Minister for Education. In his speech, the Chair of the Scientific

IMS buildings, 2001: two renovated colonial houses in Prince George's Park.

Official Opening, 17 July 2001: (from left) Louis Chen, NUS President Shih Choon Fong, Minister for Education Teo Chee Hean.

Advisory Board, Roger Howe, charted the paths for IMS's development. He argued that the establishment of IMS was an inevitable undertaking because Singapore was "a full participant in the information revolution" and "mathematics is deeply and broadly engaged with information technology". But due to the small size of the Singapore scientific community, IMS would have to engage in the whole spectrum of mathematical activity, ranging from research on "subjects chosen for their lively internal agendas" to strengthening "the ties between mathematical research and other sectors in Singapore". Thus, the establishment of IMS was also an audacious undertaking. He ended his speech with a word of encouragement, saying that he had high hopes for IMS's success.

The first program started in mid-July 2001, followed by the other two programs in 2002. The Fields Medalist, Jean-Pierre Serre of Collège de France, was invited to the first program. He delivered the inaugural lecture on "Codes, curves and Weil numbers". Serre first visited Singapore in February 1985 and has made several return visits, most recently in 2009. As the number of visitors increased, the seminar room became too small for seminars and workshops, and with the support of the President and Provost of NUS, an auditorium with a seating capacity of 80 was built in 2003.

Jean-Pierre Serre, Inaugural Lecture, *Coding Theory and Data Integrity* (Jul–Dec 2001).

From the beginning, it was decided that the programs at the Institute should cover all of the mathematical sciences and their applications. This direction has proved to be quite fruitful, particularly for a small country like Singapore. But as time moved on, it became clear that six-month programs were difficult to sustain, due to the small size of the scientific community in Singapore. Gradually, we reduced the program duration to two months and we reached the steady state of holding programs lasting from one to three months.

Through the years, many local and overseas researchers submitted proposals to organize programs. By December 2012, IMS had supported 47 programs,

47 standalone workshops, conferences and symposia, 16 summer/winter/ spring (seasonal) schools, 43 public lectures and many colloquium lectures. Many fields were covered. Broadly classified, these were the various branches of pure mathematics, scientific computing, imaging science and digital media, biological and medical sciences, mathematical physics, fluid dynamics, turbulence, hydrology and climate change, probability and statistics, and economics and finance. These activities highlighted local developments, such as a breakthrough in the study of braids, and brought new results from overseas to local scientists.

Charles and Margaret Stein, *Stein's Method and Applications: A program in honor of Charles Stein* (28 Jul–31 Aug 2003).

Many programs and workshops were co-organized or co-funded by other research institutes, faculties and departments in NUS and also by other universities, national research institutes, government ministries and industrial and commercial bodies in Singapore. A few were co-organized and co-funded by foreign mathematical institutes such as the Pacific Institute for the Mathematical Sciences (PIMS) at Vancouver and the Research Institute for Mathematical Sciences (RIMS) at Kyoto University. Programs such as the following are related to Singapore's strategic interests: *Financial Mathematics* (2 November–23 December 2009); *Mathematical Imaging and Digital Media* (5 May–27 June

Eric Maskin, *Uncertainty and Information in Economics* (9 May–3 Jul 2005).

Facing camera (from left) Stanley Osher, Tony Chan, Shen Zuowei, *Mathematical Imaging and Digital Media* (5 May–27 Jun 2008).

2008); *Data-driven and Physically-based Models for Characterization of Processes in Hydrology, Hydraulics, Oceanography and Climate Change* (6–28 January 2008); *Mathematical Modeling of Infectious Diseases: Dynamics and Control* (15 August–9 October 2005); *Post-Genome Knowledge Discovery* (January–June 2002).

The programs and workshops were attended by many mathematicians and scientists as well as graduate students from Singapore and overseas. Among the participants and visitors to IMS were Nobel Laureates and Fields Medalists. The IMS activities have provided much stimulus to research in various fields. Applications of mathematics to other disciplines are better understood. Many research collaborations were also forged, which in many cases have led to discovery of significant results, such as in representation theory, logic, multiscale modelling, wavelets and frames, mathematical economics, and probability.

In 2009, the John Templeton Foundation endowed a gift of about S$1.4 million (approximately US$1 million) to establish the program *Asian Initiative for Infinity* (AII) at IMS. The AII was initiated and conceived by Theodore Slaman and Hugh Woodin of the University of California at Berkeley and Chong Chi Tat. It was a three-year program with a special focus on innovative research on "Infinity" and creating a community of researchers on Infinity in Asia. This initiative came at a time when the subject Infinity was on the brink of major breakthroughs and there was an increasing interest in the subject in Asia. Both Slaman and Woodin were among the organizers of the two-month program *Computational Prospects of Infinity* in 2005 and had been conducting annual summer schools in logic at IMS since 2006. During the three-year period of the AII program, more workshops were organized and the scope of the summer schools enlarged with more invited speakers and graduate students from other parts of Asia supported. The annual summer school in logic continues till today even after the AII program ended.

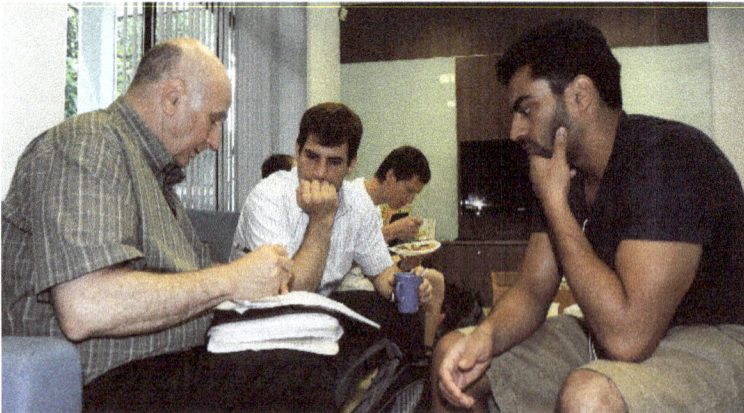

Menachem Magidor (far left), *Asian Initiative for Infinity (AII) Graduate Summer School* (28 Jun–23 Jul 2010).

Asian Initiative for Infinity (AII) Graduate Summer School (20 Jun–17 Jul 2012).

The gift of S$1.4 million was matched dollar-for-dollar by MOE. While the Templeton Foundation fund was used for the activities under the AII program, the matching fund from MOE went into an endowment fund for IMS. This was the first time IMS has set up an endowment fund.

In addition to the lectures conducted at the seasonal schools, many of the programs at IMS contain tutorial lectures for graduate students and those who are interested in learning the subject. In order to make the notes for the tutorial lectures and seasonal schools available to a wider audience, IMS publishes a lecture notes series. This series also occasionally includes special lectures and workshop proceedings organized wholly or jointly by the Institute. Each volume is edited by the program or school organizer(s) and the series is published and distributed by World Scientific. So far, 32 volumes of the lecture notes series have been published. To give an idea of the wide spectrum of fields covered in the series, here is a sample of the volumes published. Vol 32: *Mathemusical Conversations*; Vol 27: *E-Recursion, Forcing and C*-Algebras*; Vol 22: *Multiscale Modeling and Analysis for Materials Simulation*; Vol 21: *Environmental Hazards*; Vol 19: *Braids*; Vol 16: *Mathematical Understanding of Infectious Disease Dynamics*; Vol 12: *Harmonic Analysis, Group Representations, Automorphic Forms and Invariant Theory*; Vol 10: *Gabor and Wavelet Frames*; Vol 7: *Markov Chain Monte Carlo*; Vol 4: *An Introduction to Stein's Method*.

As part of its outreach program, the Institute organizes public lectures given by prominent mathematicians. It also publishes a widely circulated newsletter called *Imprints* which contains news of the Institute's activities, invited articles of special interest and interviews with some eminent visitors to the Institute. Two issues of *Imprints* are published every year with World Scientific as the printer. Our colleague Leong Yu Kiang served as the founding editor of *Imprints* and has been consistently the sole interviewer of those mathematicians and scientists. Yu Kiang and I jointly conceptualized the newsletter.

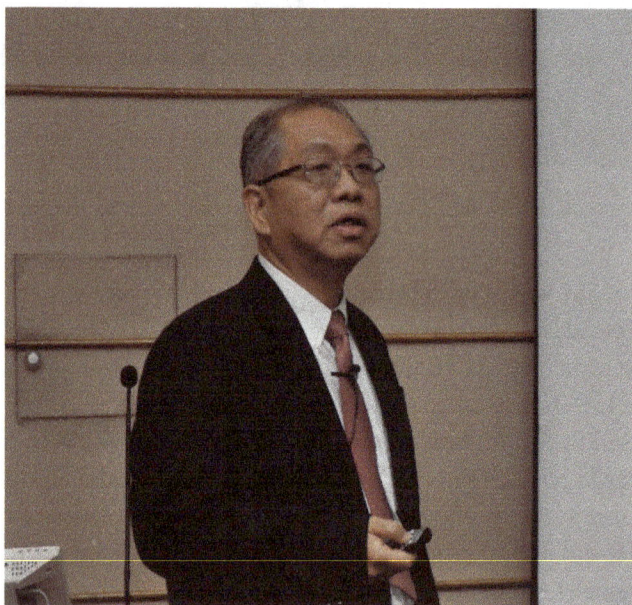

Yau Shing-Tung, public lecture, Jan 2011.

On 24 June 2010, IMS celebrated its tenth anniversary with the NUS President, Tan Chorh Chuan, as the guest of honor. To commemorate the occasion, a volume entitled *Creative Minds, Charmed Lives* featuring interviews of 38 mathematicians and scientists was published. It is a collection of the interviews that had appeared in *Imprints* up to the time of the publication of the volume. World Scientific is also the publisher of this commemorative volume. During the celebration, our guest of honor, Tan Chorh Chuan, was presented with a copy of the volume.

Tenth Anniversary Celebration, 24 June 2010: Louis Chen (right) presenting the book *Creative Minds, Charmed Lives* to NUS President Tan Chorh Chuan.

Our colleague from the Department of Physics, Bernard Tan, specially composed a piece of music *Remembrance* for flute and harp. The title of the piece befitted the occasion on which we reflected on the development of the Institute over the past 10 years. The music was performed by two young budding musicians during the celebration. A video highlighting the achievements of the Institute was made with the help of the NUS Centre for Instructional Technology and was also shown during the celebration.

In addition, a commemorative booklet entitled *Celebrating 10 Years of Mathematical Synergy* was published with the help of World Scientific. Among the features in the booklet are articles by Chong Chi Tat and by Roger Howe, an interview with the director, reminiscences by the deputy directors and thoughts of program organizers and participants.

The afternoon of the day of celebration was devoted to three lectures delivered by Tony Chan, President of Hong Kong University of Science and Technology, Hugh Woodin of University of California at Berkeley and our colleague Sun Yeneng. It was a happy and memorable occasion.

I served as director of the Institute until the end of 2012. Chong Chi Tat succeeded me and has been the director since January 2013. At the same time Ho Teck Hua became the Chair of the Management Board. He was succeeded by Lai Choy Heng in September 2015. The University continues to provide funding for IMS after the five-year start-up period ended.

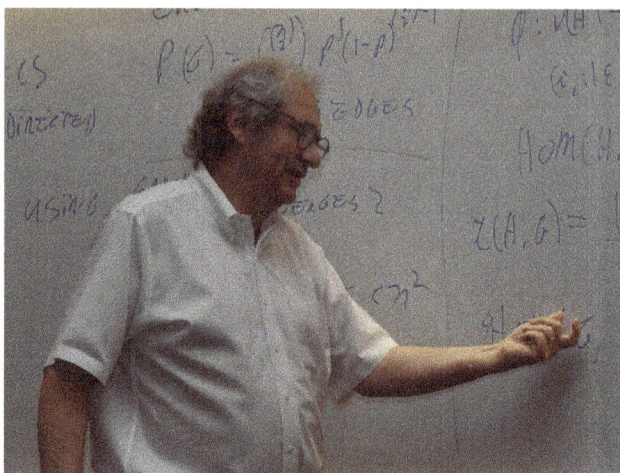

Persi Diaconis, *Probability and Discrete Mathematics in Mathematical Biology* (14 Mar–10 Jun 2011).

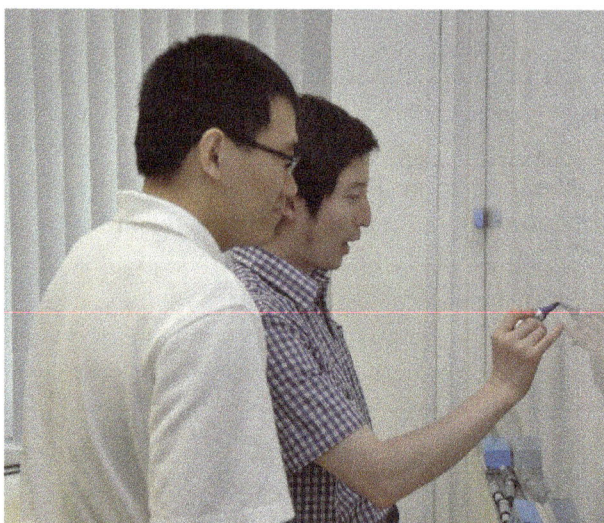

Li Jiyou (right), *Coding, Cryptology and Combinatorial Designs* (15 May–11 Jun 2011).

The Chair of the current Scientific Advisory Board is Siu Yum-Tong of Harvard University, who succeeded Roger Howe in 2010. The other members of the Board are Alice Chang (Princeton University), Louis Chen (NUS), Wolfgang Hackbusch (Max Planck Institute for Mathematics in the Sciences), Iain Johnstone (Stanford University), Lin Fang-hua (New York University), Jill Pipher (Brown University), Quek Tong Boon (DSO National Laboratories,

Multiscale Modeling, Simulation, Analysis and Applications (1 Nov 2011–20 Jan 2012).

Singapore) and Hugh Woodin (Harvard University). The following had also served on the Board for various periods of time: David Mumford (Brown University), Olivier Pironneau (Université Paris VI), Fan Jianqing (Princeton University) and Douglas Arnold (University of Minnesota).

During my term as director, I was assisted by the following deputy directors: Chen Kan, Sun Yeneng, Denny Leung, Leung Ka Hin, Tan Ser Peow and To Wing Keung, each serving for one to three years. The current deputy director is Choi Kwok Pui.

As we look back, the establishment of IMS and its development is very much the result of collective efforts. The Ministry of Education, the University, Scientific Advisory Board, Management Board, benefactors, my deputy directors, support staff, program organizers, colleagues, and last but not least, the visitors and participants of programs, all played crucial roles in contributing to the growth of the Institute into what it is today.

Since Chi Tat took over the directorship, two new successful schemes have been introduced. The first is the IMS Long-term Visitor scheme for young researchers who visit the Institute for at least a month or the entire program period. The second is the IMS Distinguished Visitor scheme under which prominent senior scientists, who could raise the quality and prestige of a program, are appointed as Distinguished Visitors for visits of at least two weeks. Among the Distinguished Visitors appointed was the Fields Medalist, Vladimir Voevodsky, of the Institute for Advanced Study, USA, who participated in the program *Combinatorial and Toric Homotopy* (1–31 August 2015).

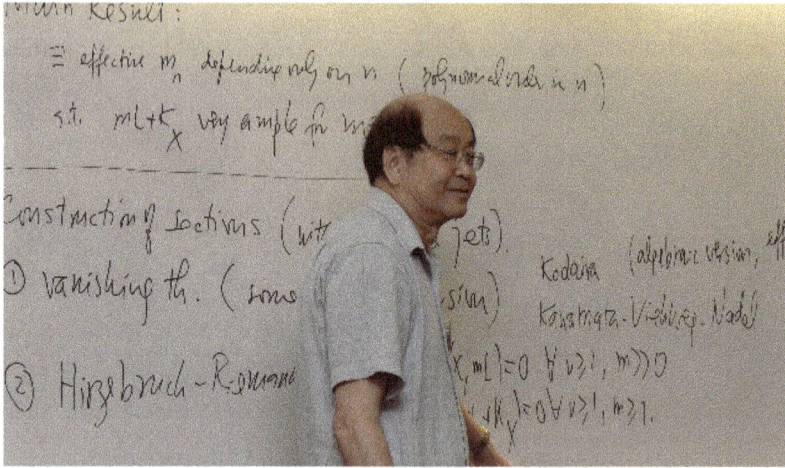

Siu Yum-Tong, *Complex Geometry* (22 Jul–9 Aug 2013).

From left: Loke Hung Yean, Zhu Chengbo and Gan Wee Teck, *New Developments in Representation Theory* (6–31 Mar 2016).

The Institute has also signed agreements with the Korean National Institute for Mathematical Sciences and the Vietnam Institute for Advanced Study in Mathematics for the cooperation on scientific exchanges and development of areas of the mathematical sciences of mutual interest.

The IMS public lecture series is now called the Ng Kong Beng Public Lecture Series, named after the late father of Ng Kok Lip and Ng Kok Koon, who jointly made a generous donation of S$250,000 to the Institute's endowment fund. The Institute has received a dollar-for-dollar matching grant from MOE. The interests from the total endowed sum are used to provide funding support for the public lectures and the seasonal schools.

In addition, the Institute organizes jointly with the Department of Mathematics the Oppenheim Lectures, a distinguished lecture series named after Sir Alexander Oppenheim, who was the first Head of the Department of Mathematics and later became the Vice Chancellor of the University of Malaya (predecessor of the University of Singapore, which in turn is a predecessor of NUS). The lecture series, which began in 2015, is held annually. Each year, an eminent mathematician is invited to give lectures at IMS and the Department of Mathematics. The Fields Medalist, Ngo Bao Chau, of the University of Chicago gave the inaugural Oppenheim Lecture in 2015.

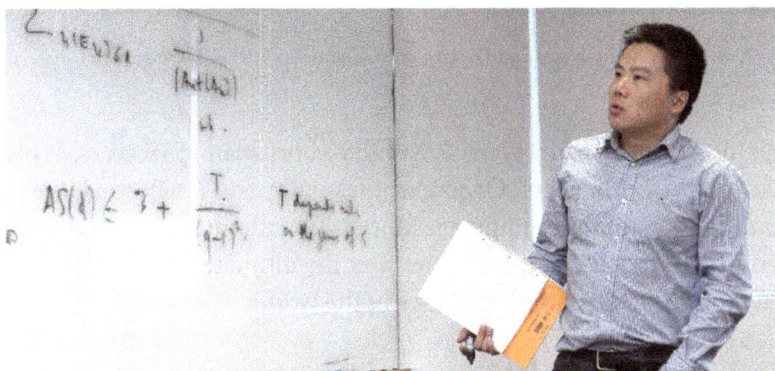

Ngo Bao Chau, Inaugural Oppenheim Lecture, Jan 2015.

The Institute has also expanded the scope of its programs. It now hosts more than 700 mathematical scientists every year from Singapore and from other countries worldwide. The activities of the Institute have helped create a research vibrancy at a level unseen before in Singapore. The Institute for Mathematical Sciences has become an important part of the research landscape at NUS. It brings top mathematicians and well-known researchers from other disciplines to NUS and connects local mathematicians to the international scientific community. The visibility and reputation it has acquired and developed over the years is a competitive advantage for the Department and NUS as a whole.

Geometry, Topology and Dynamics of Moduli Spaces (1–19 Aug 2016).

The Institute has made its mark within a short span of 16 years. What will it be like in the next 16 years? Of course, the quality and intellectual level of the programs will continue to improve. More top mathematicians and scientists will come to participate in its activities. But the ultimate aim is to become one of the foremost mathematical institutes in the world.

Acknowledgements

The author is indebted to Chong Chi Tat and Leong Yu Kiang for their very helpful comments, corrections and suggestions. Thanks also go to Emily Chan, Stephen Auyong and Jolyn Wong of IMS for their help in finding suitable photographs from the Institute's photo archive for this article. The author would like to thank IMS for its kind permission to reproduce the photographs in this article.

References

[1] Institute for Mathematical Sciences website http://www2.ims.nus.edu.sg/.

[2] Chong Chi Tat, "Fifty Years of Mathematics in Singapore: A Personal Perspective" in *50 Years of Science in Singapore*, World Scientific, 2016.

Chapter 18

The Singapore Synchrotron Light Source (SSLS) — A Personal Account of its Founding

B T G Tan

The Idea of a Synchrotron for Singapore

The idea of the acquisition of a synchrotron X-ray source for the National University of Singapore (NUS) may have originated when I was a member of a delegation from Singapore visiting various scientific research parks and facilities in Europe in September 1990. This was organised by the Science Council of Singapore and I was a member of the delegation as a representative of NUS. The trip was to prepare the ground for the establishment of what today is the One-North science and technology research precinct adjacent to Kent Ridge.

One of the key places which we visited was Grenoble in France, and I clearly remember being brought to the site of what was to become the largest synchrotron radiation facility in Europe — the European Synchrotron Radiation Facility or ESRF (now one of the big three in the world, the other two being SPRING-8 in Japan and the Advanced Photon Source or APS at Argonne in the US).

In general, synchrotron radiation facilities are built around a large machine known as an electron synchrotron storage ring,[1] which is basically a particle accelerator much like the famous machine known as the Large Hadron Collider or LHC based at CERN, the European Centre for Nuclear Research. The LHC is a very large ring spanning a circumference of 27 km (the ESRF storage ring circumference is 844 m), within which the fundamental particles which constitute matter, such as protons, can be accelerated by electromagnetic forces to extremely high energies. This enables collisions amongst such particles which may result in the creation of hitherto unobserved particles, the most famous such particle being the recently observed Higgs Boson.

In synchrotron storage rings, electrons are usually the particles being accelerated, not to produce collisions as in the LHC, but to produce various types of electromagnetic radiation as the electrons travel round and round in the storage rings. This radiation can be extremely intense and hence this is an excellent method of producing intense beams of electromagnetic radiation such as X-rays, ultra-violet radiation, visible light, and infra-red radiation, all of which can be extremely valuable tools for scientific research and technological applications.

Experimental Physics Research at NUS

Under the leadership of the then Vice-Chancellor Lim Pin, the National University of Singapore (NUS), which came into being in 1980 from a merger of the then University of Singapore with Nanyang University, rapidly developed from what was mainly a good teaching university designed to train manpower for colonial (and later independent) Singapore into a research university gaining international attention for the scope and quality of its research. At around the time when the University of Singapore was transitioning to become NUS, the Science Faculty of the University moved from the old Bukit Timah campus to the new Kent Ridge campus. This gave the Physics Department a chance to expand the scope and depth of its experimental research facilities, and new laboratories in key areas such as ion beam analysis using nuclear particle accelerators like the Van der Graaf accelerator, and surface science analysis using X-ray fluorescence machines were set up, greatly strengthening our experimental research capabilities.

By the 1990s, the Physics Department had developed a credible experimental research reputation, alongside its already strong theoretical physics credentials. There is no doubt that the progress of physics is ultimately dependent on leading-edge experimental research facilities, and for the Physics Department to attain credibility in the international physics research community, it is imperative that our experimental physics facilities be comparable to those in internationally renowned physics departments.

However, much leading-edge physics research is now heavily dependent on big and expensive machines like particle and synchrotron accelerators such as the LHC or ESRF, SPRING-8 and APS. It is obviously well outside the financial capabilities of a small country like Singapore to acquire such large machines. Was there a comparable experimental facility which would be within the bounds of possibility for the physics community in Singapore to acquire?

The Helios Synchrotrons

Sometime in 1994 or so, a Visiting Professor in the NUS Materials Science Department, Antony Bourdillon, told me that a synchrotron of compact dimensions, but having the capabilities of a machine considerably larger in size, was available for a price which was very much lower than that of a large synchrotron. This was a machine named Helios-2, which had been designed and constructed by a relatively small but well-known company named Oxford Instruments, renowned for the design and production of compact superconducting magnets. The company had been built on the pioneering superconducting expertise of Martin Wood who was already well-known for founding the company when I was doing my DPhil at Oxford in the late 1960s.

Oxford Instruments had already designed and built Helios-1[2] which had been installed at the IBM semiconductor fabrication complex at East Fishkill in New York state in the US. Helios-1 was designed specifically for the next generation of semiconductor lithography with far finer resolution and device density than was possible for the current generation of semiconductor chips. The present technology using visible light for the photographic lithography process needed to form the patterns on the chips was already capable of resolutions well below 1 micron (i.e. 1 micrometre or one-millionth of a metre or 1,000 nanometres or nm).[3]

As resolutions became even finer, at some point the limiting factor would be the wavelength of visible light, which could go not much below about 380 nm. To get around this limiting wavelength, light at ultra-violet or UV wavelengths from gas discharge lamps and later on, deep UV light from excimer lasers with wavelengths below 200 nm, enabled resolutions (or "feature sizes") to go down to 50 nm. Other techniques such as immersing the optical apparatus in water to decrease the UV wavelengths were also being employed.

To reach even smaller feature sizes, it was believed that lithography had to be done with electromagnetic radiation of even shorter wavelengths than deep UV, and one approach was to use X-rays with wavelengths below 10 nm for lithography. However, the generation of X-rays with traditional X-ray sources would not produce X-rays of sufficient intensity for the lithographic process, and therefore synchrotron storage rings which could generate X-rays of intensity many orders of magnitude greater than the most powerful X-ray tubes were seen as the ideal source of the X-rays needed for lithography.

Oxford Instruments had designed Helios-1 such that the high magnetic fields required to bend the electron beams circulating in the storage rings were generated by superconducting magnets which could produce magnetic fields much stronger than normal magnets. IBM was a pioneer in the use of X-rays for wafer lithography, and had already demonstrated that this new lithographic technology could produce integrated circuits of very small feature size.[3] Oxford Instruments had built a second synchrotron similar to Helios-1, which was called Helios-2, and while it had been built expressly for the purpose of X-ray lithography, it was also highly usable as a synchrotron for research, both basic and applied.

Because Helios-2, like Helios-1, was designed to meet the stringent cost and space requirements of the semiconductor wafer industry, it was much more compact, with a circumference of 10.8 m, than a non-superconducting synchrotron of comparable electron beam energy capability. The powerful Helios superconducting magnets bent the circulating electron beams into a radius much smaller than possible by non-superconducting magnets. The smaller size of the Helios machines also resulted in other cost savings, and hence made the acquisition and siting of Helios-2 by NUS within the bounds of possibility, though it was going to take a lot to win the support of the funding authorities.

The shortest wavelength of the electromagnetic radiation which a synchrotron storage ring can usefully produce is defined by its characteristic or critical photon energy, which in turn depends on the storage ring's electron energy and the magnetic field strength bending the beams.[1] The electron beam circulating within the storage ring of Helios-2 could attain an energy of 700 MeV or 700 million electron volts which was sufficient, with the superconducting magnetic fields, to produce electromagnetic radiation with a characteristic photon energy of 1.47 KeV or 1.47 thousand electron volts, which gave a wavelength of 0.845 nm.

For comparison, visible light has a wavelength extending from about 380 nm for violet light to 750 nm for red light. Beyond violet light are the shorter wavelengths of ultra-violet or UV light, and beyond that lie the very short wavelengths of X-rays ranging from about 0.01 to 10 nm. The shorter the wavelength, the "harder" and more energetic the X-rays. To produce harder X-rays, the storage ring electron energy has to increase for a given magnetic field strength.

Helios-2 could produce useful amounts of X-rays up to a photon energy of about 10 KeV, but very little beyond that. However, for photon energies less than the characteristic energy and longer wavelengths, Helios-2 could produce radiation of undiminished intensity. So a wide spectrum of electromagnetic radiation from 0.845 nm X-rays downwards in energy to the entire range of UV, visible, infra-red and microwave radiation could be produced by Helios-2.

Of course, lasers can produce radiation of greater intensity and coherency than that from a synchrotron, but lasers are limited to single fixed frequencies or wavelengths while a synchrotron can produce electromagnetic radiation over a wide spectrum range to its shortest wavelength limit. For some wavelengths such as for X-rays, lasers are not really a viable option at present.

In comparison, a large synchrotron storage ring like SPRING-8 has a maximum electron energy of 8 GeV or 8 giga electron volts (a giga electron volt being equal to 1,000 million electron volts), and hence could produce X-rays of photon energies much greater than possible from Helios-2, for example of more than 300 KeV. The resultant very hard X-rays are extremely useful for research on the structural characteristics of complex biological molecules, a task beyond the capabilities of Helios-2. For further comparison, the synchrotrons in Switzerland (the Swiss Light Source), UK (Diamond), Australia (Boomerang), Taiwan (Taiwan Photon Source) and Shanghai (SSRF) have electron beam energies of 2.4 GeV, 3 GeV, 3 GeV, 3 GeV and 3.5 GeV respectively.[4]

A Synchrotron Storage Ring for NUS?

The late former Head of the Physics Department, Arthur Rajaratnam who had retired by then, visited the NUS Physics Department around this time and remarked that to take the Physics Department into the future, a synchrotron storage ring would be a significant boost to our experimental capabilities. This was another confirmation that the acquisition of a synchrotron was a step in the right direction for the Physics Department. I therefore asked Antony Bourdillon to prepare the ground for a formal proposal to the main national research funding agency at that time, the National Science and Technology Board or NSTB.

There was ample evidence that a synchrotron storage ring such as Helios-2, though much less powerful than large machines like ESRF, would still provide a significant resource for basic and applied research. It would not have enough electron beam energy to generate the X-ray wavelengths needed for the kind of structural analysis of biological molecules which made up much of the justification for the large synchrotrons. But it would still be the most powerful experimental research facility ever installed in Singapore, and would be a major step forward for our scientists and engineers, particularly in materials science and allied areas of work.

Important as Helios-2 could be for scientific and engineering research, from the purely research angle it would be difficult to justify a machine which at that time was going to cost well in excess of 25 million pounds sterling. The main driver for research funding in Singapore then was not so much the building up of a first class research infrastructure for basic and applied research, important as that was (and still is), but the tangible economic benefits which could result from such R & D.

This meant that the synchrotron proposal had to firmly couple its research justification with some key industry application. It was not difficult to work out that the most appropriate industrial application of Helios-2 would be precisely what it had been designed for in the first place: X-ray lithography for semiconductor wafer fabrication. The Physics Department was already in contact with Singapore's pioneering wafer fabrication corporation, Chartered Semiconductor, and they would be the obvious industrial partner for the synchrotron proposal. The then CEO of Chartered Semiconductor, Tan Bock Seng and his senior colleagues (whom we met at a lunch on 29 November 1994), were extremely technically competent and their technological vision was more far-sighted than that of many other industry players.

While X-ray lithography was certainly a front-runner for next generation lithography, there were many other competing lithographic technologies such as extreme ultra-violet or EUV lithography. One of the significant challenges in X-ray lithography was the need to fabricate much finer masks for the lithographic process than was necessary with competing techniques. However, as IBM had already demonstrated the practical feasibility of X-ray lithography, it was then seen by many as the front-runner candidate for the next generation lithographic process.

It was also important that we try to arouse interest amongst the scientific and engineering research community in synchrotron radiation as a powerful tool for R & D, as well as for industrial applications. We thus organised a series of seminars targeted at the science and engineering research community in academia as well as in industry to acquaint them with the capabilities of synchrotron radiation and its wide variety of applications in research and industry. Indeed, a synchrotron, powerful as it might be, was merely a tool which had to be made use of by researchers who understood its capabilities and versatility.

Applying to NSTB for Funding

The drafting of the formal proposal to NSTB gradually took shape as we started to introduce the concept of synchrotron radiation and what it could do to a wide range of scientists and engineers both from academia and from industry. We were very careful in our proposal to emphasise the importance of the synchrotron to basic and applied research as well as its potential uses in industry. We were aware that this would be the largest ever proposal for a single piece of research equipment, and getting the proposal through the complex procedure for research funding was not going to be easy, to put it mildly.

One of the major difficulties in our proposal which we had to overcome was that we were not making a proposal for a research grant in the normal sense, i.e. the proposal was not for a specific research objective which would yield certain desired results. We would be essentially asking for a very large piece of research infrastructure which was not for a specific scientific experiment or even for a specific group of experiments. The versatility of the synchrotron storage ring would lend itself to a very wide range of experimental research and industrial applications in many different fields, so we would have to try to justify the request by appealing to its infrastructural significance both for research and industrial applications.

The Helios-2 machine was designed to provide up to 20 beam ports, each of which could output electromagnetic radiation within the very wide range of wavelengths the synchrotron could provide. In most synchrotron storage ring facilities, a single beam port would normally support not just one experimentalist but was often shared by a number of experimentalists interested in the same wavelength range. Therefore, in theory, Helios-2 could support up to 20 groups of experimentalists, so that in full operation and with all the ports used, hundreds of scientists and technologists could be accommodated on a time-sharing basis (as is normal for all synchrotron storage rings).

Hence our proposal to NSTB concentrated on the acquisition of a synchrotron storage ring, and would not actually request for the associated equipment for specific experiments which the beam ports could support. One option would have been to request for funding to equip some of the 20 beam ports with beam lines to receive the radiation from the ports, ready for use by experimentalists. However, in practice the particular requirements of experimentalists for their beam lines are usually quite specific depending on the actual type of research to be conducted with the beam line, so it may have been wiser not to have included the beam lines in the formal proposal.

Indeed, including any beam lines in the proposal would have considerably enlarged the size of the funding required. I was very well aware that the synchrotron proposal would probably be by far the largest funding request ever made from bottom-up to a research funding agency in Singapore. Of course, larger amounts had been disbursed for projects such as the setting up of new research institutes, but such proposals had usually originated top-down from higher policy-making levels. The synchrotron, not being such a proposal, would hence require a great deal of support from the policy-makers if it was to have any chance of success.

Here I have to acknowledge the strong and unstinting support of Lim Pin for the synchrotron project from its very inception, which was undoubtedly a key factor in our eventual success. Lim Pin was the driver in building NUS from a mainly teaching university into a leading research university, and hence he completely understood the significance to the research capabilities of NUS of acquiring such a major research facility.

The Executive Director of the NSTB was then Vijay Mehta, a highly competent administrator who had been seconded from the Ministry of Defence. Vijay understood very well the need for both basic and applied research in transforming the research landscape of the country. One of his major tasks at NSTB was to oversee the disbursement of research funding, as NSTB was the most important funding agency for R & D in Singapore. Understandably, one of NSTB's principal thrusts was to ensure that the research funding being disbursed would make a real contribution to the economic progress of Singapore.

NSTB was the successor to Singapore's first Government science agency charged with the shaping of the country's science and technology policy, the Science Council of Singapore. Founded in the early years of the country's independence, the Science Council under its founding Chairman, Lee Kum Tatt, had laid down the foundations for the young nation's scientific development and organization under the then Minister for Science and Technology, Toh Chin Chye, who later became the Vice-Chancellor of the then University of Singapore. The Science Council was responsible for early science promotion initiatives such as the Science Centre and the Science and Industry Quiz.

In Singapore's short research history, the availability of research funding in non-trivial amounts really began only after about 1980, when the National University of Singapore was established and began its transition from a teaching to a research university. The Science Council of Singapore was transformed into the NSTB in the early 1990s and began to play a key role in the research scene, being charged with the disbursement of research funds much more substantial than hitherto. University researchers were able to apply for funding in larger amounts than had been available, but their applications had to be supported by strong arguments not only for the excellence of the research but also for their potential contribution to the country's economic development.

The Importance of both Basic and Applied Research

The importance of R & D in a country's economic development is now well recognised, and it was understood that R & D had to be well funded in order for it to contribute significantly to a nation's progress. However, the relative importance of basic research, applied research and development to a nation's technological and economic development is not always as well understood as it should be. The idea that nations can ignore basic research and reap economic benefits by concentrating only on applied research and development is a seriously flawed one. In fact the relationship between basic and applied research is highly complex and dynamic. What may be deemed to be basic research could very quickly become applied research in the light of rapid changes in industry and technology.

A small country clearly cannot support basic experimental research on the scale of the LHC and CERN or of SPRING-8, ESRF and APS. Nevertheless, for a country which seriously wishes to progress significantly in its understanding and mastery of advanced technology, participation in basic research, including experimental as well as theoretical research, is essential. Science and technology cannot be divided into salami slices from which one can pick the supposedly most cost-effective pieces. The advancement of science and technology is a highly complex and holistic enterprise which requires the deep understanding of both basic theory as well as leading-edge experimentation, all of which requires the engagement of the best minds. Research talent of the highest quality will only be attracted if the funding of research is based on the excellence of the work, and not just on its perceived economic benefits.

Therefore, it was to Vijay Mehta's great credit that once the scope of our synchrotron storage ring proposal had been explained to him, he quickly grasped its potential not only for its possible economic benefits, but also for its importance in attracting the best research minds to work in Singapore in a wide variety of basic and applied fields. He quickly became a strong supporter of the synchrotron proposal and was undoubtedly a key factor in its eventual success. I attribute this to Vijay's clear vision for NSTB's long-term objectives as well as his superb administrative and organisational abilities which helped us navigate the proposal through the complex funding process. His tragic early death in October 1997 was a great loss as he could have contributed even more to the country's research advancement.

Preparing the Preliminary Report for NSTB

In preparing the formal proposal, one of the NSTB's senior officers who was a tremendous help and a major factor in our success was Yeo You Huan. He guided us through the complex process and certainly helped to make our proposal as strong as possible. The proposal had the key support of its then Chairman, Teo Ming Kian and its Deputy Chairman, CC Hang. In the preparation of the proposal, we also worked closely with Oxford Instruments, who helped us to shape the technical proposal and to adapt the Helios-2 to our specific purposes.

The two key people in Oxford Instruments with whom we worked closely were Peter Williams, Chairman of Oxford Instruments, and Alistair Smith, its Managing Director.

An NUS committee was then formed, called the Committee on Advanced Materials and Electro-Optics, to prepare a draft preliminary proposal for NSTB which would lay down the rationale for a synchrotron storage ring and the scope of research and applications work which could be done with such a machine. By 16 January 1995, a memo from Antony Bourdillon to the members of the Committee (listed as R Chu, Ang HG, Chan SO, Hor TS, Shen ZX, Tan KL, Tang SH, Ji W, A Huan CH, Chong TC, I Novak and S Jaenicke) makes it clear that this draft was in an advanced stage.[5] A memo from Antony to me dated 27 February[6] indicates last minute changes to the draft, and hence this preliminary proposal was probably submitted to NSTB within the next few months or so. Entitled "Advanced Materials — An Eye For the Future" and subtitled "Strategic Areas in Research and Development", it was submitted as "A Preliminary Report for NSTB."[7]

The preamble to the Report was an Introduction and Objectives in Section 1, and the Background in Section 2. Part I of this Report, entitled "Characterization and Processes of Advanced Materials" included the following sections:

Section 3: Benefits of synchrotron radiation.
Section 4.1: Research targets in the strategic areas of X-ray lithography, materials development and devices, surface analysis, electro-optics and spectroscopy, micromachining and molecular biology & agrobiology.
Section 4.2: Research targets in strategic areas with established expertise including basic surface science research, nanotechnology, defect characterization in semiconductors, atomic-scale imaging of advanced materials by transmission electron microscopy and electro-optics & spectroscopy.
Section 5: Facility.
Section 6: Experience.
Section 7: Conclusion and recommendations.
Section 8: Selected bibliography.

Part 2 consisted of just Section 9: synthesis and properties of advanced materials, which covered polymers, catalysis, materials derived from MOCVD and ceramics.

The Preliminary Report was thus clearly focused on the vast potential of synchrotron radiation in advanced materials research. While industrial applications were an integral part of the report, it was apparent to any physicist or chemist, and in particular those engaged in materials related work, that the real power of the synchrotron storage ring lay in its ability to produce highly intense beams of electromagnetic radiation for the specific purpose of materials analysis.

The Advanced Materials Committee, in order to build up support for a synchrotron storage ring radiation facility, organised a seminar at NUS on X-ray lithography on 22 September 1995, as part of the NSTB Tech Month and also as part of the NUS 90th Anniversary celebrations that year.[8] It was a whole day seminar with speakers from IBM, NTT, Oxford Instruments, Karl Süss, Leica, Chartered Semiconductor, NUS, NTU and Warwick University, and sponsored by NSTB, Chartered Semiconductor and SGS Thomson.

A memo dated 6 October 1995 from Antony to me and my then Vice-Dean in charge of research, Lee Soo Ying, made it clear that after the submission of the preliminary proposal, we were looking at the relative merits of other X-ray synchrotron storage rings as compared to Helios-2, attaching a list of existing storage rings in the world.[9] The memo also mentioned the need to obtain from NSTB a request from us for a full proposal.

Our Visit to IBM at Burlington and East Fishkill

NSTB, now seriously considering our Preliminary Report, then organized a trip for the key people involved in the proposal to visit IBM's X-ray lithography facilities in the US, and in particular to visit IBM's East Fishkill facility in New York state where Helios-1 was located. From the very outset, our proposal was conceived as a national synchrotron radiation facility which would be accessible by all potential users in Singapore, and hence the group going on the trip included not just representatives from NUS (Antony Bourdillon and myself), NSTB (Yeo You Huan and Ong Beng Thiam) and Chartered Semiconductor (CK Lau), but from Nanyang Technological University (NTU) as well (Lee Sing and Tan Hong Siang).

The trip from 15–21 May 1996 first brought us to Burlington in the state of Vermont, where IBM's mask-making facilities were located, and where we were briefed by Patrick Hughes and Kurt Kimmel of IBM. The masks which define the patterns to be fabricated by X-ray lithography on the semiconductor wafers are extraordinarily fine, and IBM had one of the most advanced mask-making facilities in the world at Burlington.

We then went on to East Fishkill, where we saw Helios-1 and were joined by Alistair Smith of Oxford Instruments. Here we had a clearer idea of the machine we wanted to acquire and the supporting facilities it would require, such as liquid helium cooling for its superconducting magnet coils. We were briefed by Jeff Kristoff, Jerome Silverman and George Gomba of IBM, who also showed us around the state of the art X-ray lithography and wafer fabrication facilities. The trip concluded with a visit to NTT's synchrotron facilities in Japan, where we were briefed by Yasuyoshi Sakai, Haruo Tsuzuyaki and others from NTT.

Having actually seen Helios-1 at close quarters, our proposal could now be put into sharper focus, and we then worked very closely with NSTB and with Yeo You Huan in particular, to fashion a formal paper which could be submitted to the senior management and the Board of NSTB for approval. This meant that much work now had to be done on defining the specifications of Helios-2 as a research tool much more precisely than in our preliminary draft. It was absolutely necessary that the paper for NSTB had to include strong justification and clearly defined goals in accordance with Singapore's overall research strategy.

The Budget for the Proposal

The paper for the NSTB also had to include a budget, with cost figures for all major items in the synchrotron and its ancillary equipment. This would include the cooling systems for the superconducting coils which used liquid helium (certainly not an inexpensive item). Fortunately, liquid helium was available from industrial gas suppliers in Singapore, but we had to include a system for the transfer and storage of the liquid helium used for the cooling of the superconducting magnet coils.

Of course, the paper also had to include the cost of the building in which Helios-2 was to be housed. This meant that we had to make an estimate as to what building costs per square metre would be like at a future time, which was not going to be very accurate as such costs are always difficult to predict. There was very little land left in the NUS Kent Ridge campus available for a new research facility, but we were quite determined to locate the synchrotron facility on campus, where it could be easily accessible to the many physicists, chemists and engineers who would be potential users of the facility. We therefore identified a location next to the Institute of Materials Research and Engineering or IMRE, which was an excellent site since many of IMRE's research staff would be potential users.

We thus had to work with the NUS Estate Office and the campus planners to determine a realistic budget for the proposed synchrotron building. A balance had to be struck between making the building large enough to accommodate all the beam lines as well as the researchers who would be work-ing at these lines, while keeping the budget for the building within reasonable limits. A circular-shaped structure was envisaged to house Helios-2 at its core and the beam lines coming out of the shielding around the machine. Indeed, one of the key design issues was the vault and shielding needed to protect researchers at the beam lines from harmful radiation generated by Helios-2 while it was in operation.

The Paper for the NSTB Board Meeting

As we neared the end of 1996, the text of the submission to the NSTB Board came close to its final form. We had many discussions with Oxford Instruments on the final technical specifications for Helios-2 including the necessary essential spares for maintenance. This included getting an idea of the costing for every item, since we would have to include the budget for the entire project in the submission. It was at this stage that Oxford Instruments dropped the price of Helios-2 offered to NUS from an original figure of 25 million pounds sterling to S$25 million, a significant reduction which indicated how keen they were to complete the sale to us.

We would also need to strongly justify why we were recommending one particular synchrotron storage ring design, since other synchrotron storage ring designs from other manufacturers were available. Hence we had made a careful study of all synchrotron storage ring designs of a similar capability to justify our final choice of Helios-2. All of the alternative designs used non-superconducting magnets for the synchrotron storage rings. We compared Helios-2 with the Aurora 2D ring from Sumitomo, the Melco ring from Mitsubishi and the HECTOR and MAX II rings from Scanditronics. In techni-cal terms, HECTOR and MAX II were the best rings but not so appropriate for X-ray lithography, and were more expensive and would require a much larger space than Helios-2.

Since we had already submitted the preliminary report to NSTB which had broadly justified the acquisition of a synchrotron storage ring for research and industry, the paper to be submitted to the NSTB Board made a case for the syn-chrotron from a national strategic perspective. With inputs from NUS, the paper which was skillfully crafted by Yeo You Huan cogently argued that the synchro-tron storage ring would be a major addition to our research capabilities and would also contribute to industry at the cutting-edge level.[10]

In Part I of the Paper, its objective was stated as the seeking of the NSTB Board's approval for the "establishment of synchrotron radiation user facility for economically relevant and applied research." A sum of S$32.1 million was sought being 80% of the total project cost of S$39 million, with the remaining 20% to be sought from the Ministry of Education's Academic Research Fund.

Part II gave the background to the proposal, largely drawn from our Preliminary Report. Part III entitled "Support for Industry", gave the main strategic justification for the proposal, linking it to major industrial growth areas such as:

Semiconductor Wafer Fabrication
Microelectronics (which included the key area of MEMS or micro-electromechanical systems which could be fabricated using X-ray lithography-related technology)
Chemicals
Biotechnology/pharmaceutical
Materials

In Part IV, the Technology Infrastructure i.e. the synchrotron-based technology capabilities to be developed were described, and in Part V the Manpower Development details were given. In Part VI, Helios-2 was briefly described. Part VII contained the all-important budget, which requested a sum of S$39 million spread out over three years. The electron storage ring which made up Helios-2 cost S$25 million, and the remainder of the costs were made up by the ancillary equipment for the ring, the spares for maintenance, the liquid helium systems, the building, and manpower. The facility was to have a small manpower complement of just 5 RSEs (research scientists and engineers) to run it. The building costs were estimated at S$4.8 million for a 1,600 sq metre space i.e. S$3,000 per sq metre which was based on the construction costs for IME (the Institute of Microelectronics) and IMRE. The annual running costs were estimated to be around S$5 million.

The R & D programmes for the first two or three years after commissioning were to be in the following areas:

X-ray lithography for semiconductor fabrication
MEMS and sensor technology
Life science and biotechnology
Biomaterials
Failure analysis and process development in microelectronics
Biochemistry
Catalysis
Surface science

(As it eventually turned out, only some of these listed areas became actual beam lines dedicated to research or industry projects.)

Part VIII gave a positive overall assessment of the proposal which stated that "the proposed facility has been endorsed by the NSTB Strategic Research Planning Committee." Part IX thus sought NSTB Board's endorsement for the objectives stated by the paper and also laid out the expected deliverables.

The paper had four Annexes:

Annex 1 gave the Semiconductor Industry Association's Roadmap.
Annex 2 showed the proposed layout of the Helios-2 facility.
Annex 3 listed the main constituent components of Helios-2.
Annex 4 was a timetable for the milestones for the entire project.

The paper was submitted to the NSTB Board at its meeting on 12 December 1996, but no decision was made then about its approval, as this was obviously a major item which would have required a certain amount of due diligence on the part of the Board. Thus far, we had not directly involved IME in the planning for the synchrotron storage ring, and so it seemed logical for NSTB to seek IME's opinion on the relevance of the machine particularly with respect to the microelectronics industry.

IME's response, in the form of a letter from Robert Tsai to David Lim, NSTB's Head of Electronic Components and Systems, was somewhat less than enthusiastic.[11] His assessment was that the introduction of X-ray lithography in the wafer industry had been set back for a few generations due to continuous improvements in optical lithography and that the synchrotron should be justified not just by X-ray lithography for wafer fabrication, but also for applications in medicine, physics and engineering. My own response to Robert Tsai's points in a letter to Gong Wee Lik (who had succeeded Vijay Mehta) dated 31 March 1997 was that while X-ray lithography remained a major candidate for next generation lithography for wafer fabrication, the synchrotron would be essential for R & D into other key areas such as micromachining, MEMS and LIGA (an emerging technology for microfabrication) as well as state-of-the-art R & D in materials science.[11]

Approval of the Paper by NSTB

The paper was discussed by the NSTB Board at its meeting on 13 March 1997, and approval was given at that meeting. However, this was only the first stage of a complex process before definitive government approval of the project was obtained. The NSTB was part of the Ministry of Trade and Industry (MTI), and the paper did not have the unanimous support of all the higher officials in that Ministry. Eventually the paper was surfaced at a Ministerial Level Committee meeting where in effect it was the Minister for Finance making the final decision. Fortunately, the then Minister, Richard Hu, understood the strategic importance of the synchrotron and put his support behind the project, even though not everyone in his own ministry had the same understanding of its importance.[12]

In a letter to me dated 21 April 1997, Lee Kheng Cheok, the NSTB's Deputy Director of its Planning Division, informed me that the Ministerial Committee had approved the three-year budget of S$31.2 million or 80% of the total project budget of S$39 million.[13] The remaining 20% would be sought from the Ministry of Education's Academic Research Fund. Three conditions were imposed on the project:

- 100% recovery of operating costs from Year 6 onwards.
- Aim to fully recover the original investment by Year 13.
- The following deliverables: development of a synchrotron-based research laboratory, utilisation of 10 beam lines by 2000, 12 higher degree students per year from 2002, 30 RSEs trained by 2000 and at least 10 industry R & D projects by 2002.

The funds were in fact disbursed to the University as though the project was a single research project, and I was named as the Principal Investigator for the whole project! It was gratifying that the funding approval for the synchrotron had come in April 1997, for my term as Dean of the Faculty of Science was coming to an end at the end of June 1997. Hence I had fulfilled my objective of getting the project approved before I stepped down as Dean of Science (to be succeeded by Lee Soo Ying). Lim Pin agreed that I should continue to be responsible for the synchrotron project as Chairman of the NUS Synchrotron Radiation Steering Committee, even after I had stepped down as Dean to become Head of Physics.

Contract with Oxford Instruments

The most immediate task which now lay ahead was to get a contract between NUS and Oxford Instruments for the purchase and delivery signed as soon as possible. The other key task was to plan for the design and construction of the building which was to house the synchrotron storage ring. We had envisaged that the machine be commissioned in the new building during 1999, which gave us just over a year to achieve our objectives.

For the acquisition of a major piece of equipment such as Helios-2, the normal procedure was to send out a set of specifications for the equipment and invite manufacturers to submit tenders and quotations which met the specifications. In our case, we had in effect already done an extensive evaluation of Helios-2 and its major competitors, and hence in view of the very tight time schedule we had for the commissioning of the synchrotron, we sought a waiver of tender from NUS.

Antony Bourdillon and I thus wrote to the NUS Tenders Board on 7 May 1997 to request a waiver of tender with the required justification and comparison between Helios-2 and its nearest competitor machines.[14] NUS Senior Assistant Bursar Jennifer Phang replied to me on 23 May 1997 that the NUS Tenders Board had approved our request for a waiver of competition for Helios-2 at a cost of S$25 million from Oxford Instruments.[15] Subsequently, the formal contract between Oxford Instruments and NUS was signed within a few days, and Alistair Smith was able to write to Jennifer Phang on 2 June 1997 to refer to the "recently signed contract" and to request NUS to transfer the initial payment of S$5 million as defined in the contract to Oxford Instrument's bank account with Barclays Bank.[16]

The Building and the Groundbreaking Ceremony

With the formal contract signed, we turned our attention to the building which was to house the synchrotron storage ring. We had already started discussions on the building with the NUS Estate Office and with INDECO Consultants (the government-owned building consultancy which had undertaken many projects for NUS), and a letter from Antony and me dated 16 April 1997 to Robin Wong of INDECO Consultants refers to a meeting on 14 April between us (Antony and me) and Robin Wong and Andrew Meier of INDECO and Tan Lien Seng, Head of NUS Campus Planning.[17] The letter to Robin Wong included a draft proposal for the building and a plan of the synchrotron storage ring within its concrete shielding in relation to the proposed site next to IMRE.

On 22 July 1997, I wrote to CC Hang to brief him on the progress of the synchrotron project.[18] I emphasised that we had started the planning of the building with the Estate Office and INDECO even before formal approval from NSTB. I also mentioned the proposed groundbreaking ceremony for the building, which was then envisaged by NSTB to be the first public announcement of the project. NSTB wanted the groundbreaking to be no later than September

1997, but I stated that under the present building schedule it would be difficult to hold it before November 1997. Indeed, NSTB had suggested that to expedite matters, we switch from INDECO to another consultant, but this had been strongly advised against by Tan Lien Seng as we would then have had to start the building design process all over again.

By then we had anticipated that the cost of the building might overrun the budget, in view of the rising building costs at that time. Indeed on that same day (22 July), Antony had written to Andrew Meier, the INDECO project architect, about the final design changes and the need to freeze the building design.[19] This was imperative since we needed to seek URA planning approval as soon as possible, as we did not know how long it would take for URA approval to be granted.

Tan Lien Seng then wrote to me on 28 July 1997 to confirm that the building design had been finalised and that INDECO would make a submission to URA for planning approval by the end of July.[20] I then arranged for the planning submission fee of S$3,090 to be paid directly from NUS to URA. In the meantime, we were still negotiating with Oxford Instruments on the contract options, and in a letter to Alistair Smith of 29 July 1997, in my capacity as Chairman of the Steering Committee, I listed the contract options to be taken up and to be declined.[21]

Lim Pin was understandably keen to announce the synchrotron project and wanted to do so in his convocation speech, but a memo dated 27 August 1997 to me from Koh How Eng of NSTB stated that they preferred a joint announcement at a later date, as the synchrotron was a very significant investment.[22] Indeed, the planned groundbreaking ceremony was planned to be the date on which the joint announcement would have been made, but the date of this ceremony became a major issue. The ceremony could only be held after the building contract had been awarded, and in a letter to Antony dated 9 September, Robin Wong of INDECO stated that the contract would be awarded by 15 December. With the impending wet season, site preparation for the ceremony could only be begun at the earliest in January 1998.[23]

I immediately sent a memo to Robin Wong declaring that January 1998 was far too late for the groundbreaking ceremony and that the wet season should not be a reason for the delay.[24] He then replied to me on 17 October 1997 that the tender would be called on 3 November 1997 as originally scheduled.[25] (Presumably the URA approval had by then been received in good time.) If we wanted the groundbreaking ceremony to be earlier than January, he would have to hire a separate contractor to prepare the site for the ceremony which would include costs of over S$60,000. This was obviously not an option, so on 14 November 1997, I wrote to Lionel Cheng of NSTB to request that the ceremony be held on 12 January 1998.[26]

The Building Tenders and Budget Deficit

The tenders for the building were revealed on 28 November 1997, with nine contractors bidding for the project.[27] The lowest tender was S$8.7 million, which was substantially above our budget figure for the building. As I was at that time away in Manila on official business, Antony wrote to Gong Wee Lik on 3 December 1997 to give details of the deficit.[28] The budgeted building cost was S$4.8 million, and together with the mechanical and electrical works, the total project budget was S$6.42 million. Though the lowest tender was S$8.7 million, the high tension cabling costs of S$0.83 million would be funded by the university, leaving a net building works cost of S$7.87 million, resulting in a funding shortfall of S$1.45 million. When I returned from Manila, I would discuss the shortfall with the architect to find ways of bridging the gap.

On 7 January 1998, I wrote to CC Hang to inform him about the shortfall.[29] We had reduced the mechanical and electrical requirements for the building by nearly S$600,000 (as described in a letter from Robin Wong dated 29 December 1997 to the contractor, SG Industrial) thus reducing the shortfall.[30] Furthermore, the Estate Office had actually budgeted S$1 million for the high tension power cables which would cost S$0.83 million, so we asked permission for the Estate Office to use the balance of the S$1 million to cover the costs of the optical fibres and telephone manhole for the building, both of which cost nearly S$300,000. In addition, we could use some of the contingency funds which had been set aside for the building costs, thus reducing the shortfall further.

Lee Siew Ling of the NUS Bursar's Office wrote to me on 9 January to confirm that the shortfall was indeed S$1.45 million.[31] However, on 16 January 1998, Jennifer Phang wrote to me to inform me that CC Hang had approved, as a special case, the provision of S$1.1 million from the Campus Upgrading Phase I (Electrical Upgrading) for the cost of the optical fibres, the telephone manhole and the installation of the power cables.[32] In my note to CC Hang of the same day, I thanked him for his kind assistance which I greatly appreciated.[33] Indeed, this was probably the main factor which allowed the building project to proceed in spite of the funding shortfall (mainly due to the unforeseen rise in building costs). However, this was not to be the end of our funding problems as will be seen later.

On 31 January 1998, I received official notification from NUS Bursar's Office through Lee Siew Ling that my application for the funds for the synchrotron had been approved by the Ministerial Level Research Committee.[34] This notification was in the form of a standard letter to NUS faculty who had applied for external research funding, stating all the procedures I had to go through in reporting the progress of the research. In the approval document, I was clearly listed as the Principal Investigator for the research project, as the synchrotron funds had been applied for and approved as a single research project, possibly the largest such project at that time.

The Steering Committee and the Press Release

On 18 February 1998, I formally asked Lim Pin for approval to set up a Steering Committee for Synchrotron Radiation, and in addition to myself as Chairman, to invite the following as members: the Deans of Science and Engineering, the Director of IMRE, the Executive Director of NSTB and the President of Chartered Semiconductor.[35] Lim Pin gave his approval on 23 February.[36] The Steering Committee was to operate until it became a fully-fledged Management Board. I then relinquished the Headship of Physics and became Director of a new NUS office, the Office of Student Affairs. On 2 March, I asked Lim Pin for his approval to retain the Chairmanship of the Committee after relinquishing the Headship,[37] and he gave his approval on 4 March.[38]

I now turned my attention to the proposed groundbreaking ceremony, which was going to be the occasion for the first public announcement of the synchrotron project. On 19 January 1998, Antony wrote to Koh How Eng of NSTB requesting approval to hold a seminar which would coincide with the groundbreaking, with invited speakers from overseas to speak on topics like MEMS, LIGA and other X-ray related research topics.[39] However, Lionel Cheng of NSTB wrote to me on 11 February 1998 to tell me that the date which had been set for the groundbreaking ceremony, 16 March, had to be changed to 2 April as the NSTB Chairman could not attend on 16 March.[40]

Indeed, as stated by Lionel, NSTB was still unsure of the appropriateness of the groundbreaking ceremony, which would be discussed at a meeting on 18 February. The upshot of this meeting was that instead of a groundbreaking ceremony, NSTB decided to hold a news conference on 17 March to announce the synchrotron radiation project. I wrote to Lim Pin on 9 March to inform him of this and to request a quote from him for the press release.[41] In the same letter, I informed him that NSTB had requested that NTU be represented on the Steering Committee and sought his approval, which he agreed to on

10 March.[42] NTU was represented by their Director of Research, Tan Hong Siang, and this gave the synchrotron the character of a national facility which was in fact what had been intended by us from the very beginning.

On 17 March 1998, the news conference was held and the press release stated that NSTB and NUS "jointly announce the establishment of a S$40 million synchrotron facility for research and for advanced industrial applications."[43] Researchers from NUS, NTU, IMRE, the Institute of Molecular and Cell Biology (IMCB) and the Institute of Molecular Agrobiology (IMA) would be working jointly with industry on the applications of the synchrotron light source. The Straits Times reported on the press conference and the news release the following day, stating that Singapore's most advanced facility for research and industrial applications would open early in 1999, catering to the wafer fab, pharmaceuticals, chemicals and biotechnology industries here and in the region, and that work on the S$40 million facility had begun at NUS.

Lee Soo Ying had, since his appointment as Dean, retained his interest in the project and had on 5 January 1998 written to me and Antony to arrange for a meeting on the progress of the synchrotron radiation facility.[44] An email from him to me on 23 February further proposed that a Centre for Synchrotron Radiation Science and Technology or CSRST be set up with myself as Director, and a Management Board with himself as Chairman.[45]

On 21 March 1998, I wrote to CC Hang to inform him that the new facility had been given the name of Singapore Synchrotron Light Source or SSLS at the news conference on 17 March, and requested permission for SSLS to be designated as an NUS Research Centre.[46] As Chairman of the Steering Committee and acting in an executive manner, I was effectively the Director of SSLS and would designate Antony as Technical Director of the Centre. The request was approved on the same day,[47] but on 25 March CC Hang informed me that NSTB had asked NUS to form a Management Committee for the new facility (confirmed in a letter to him from Koh How Eng on 27 March), and this would likely supersede the Steering Committee.[48]

The Pro-tem Management Committee

Hence on 28 March, I wrote to my successor as Dean of Science, Lee Soo Ying, congratulating him on his appointment as Chairman of the new Management Committee.[49] I informed him that due to my new duties as Director of the Office of Student Affairs, I would be drastically reducing my involvement in the SSLS. However, I strongly recommended that Antony be appointed Director of SSLS.

On 30 July 1998, Lionel Cheng sent me a draft paper written by him on the progress of SSLS for my comments, as I was still the Principal Investigator of the whole project.[50] The paper was for the RIMC (Research Institutes Management Council) meeting of 5 August 1998, and stated that a pro-tem management committee would be set up for SSLS to provide guidance during the developmental phase, and that a full management board would be formed once the facility was operational. Lee Soo Ying would chair the pro-tem management committee which would have representatives from NTU, Chartered Semiconductor, IBM and NSTB. I would be a member as Principal Investigator.

On 3 August 1998, I received an official invitation from Teo Ming Kian, Chairman NSTB, and Lim Pin to be a member of the pro-tem management committee,[95] which I accepted on 12 August. The first meeting of the pro-tem committee was on 25 August and as reported in its minutes,[52] a proposal to request for two general purpose beam lines for Helios-2 was approved by the committee, but the funds for these beam lines would probably have to be requested from the Academic Research Fund of the MOE through normal application procedures requiring external peer review.

At the same meeting, I reported that SAL Lithography, a leading company in the US developing X-ray lithography for semiconductor fabrication, had invited me to a Business Summit on X-Ray Lithography in Vermont which they were organising in September 1998. Antony and I jointly presented a paper at Vermont which enabled us to tell the lithography community about SSLS. We had also made contact with Franco Cerrina of the Centre for X-Ray Lithography at the University of Wisconsin. We started making efforts to get him attached to NUS as a Temasek Professor, but eventually did not succeed in doing so.

We had in the meantime been also occupied in building up the technical and support staff for SSLS. Our first employee was Tan Kah Bee who was already in place by January 1998. Of the four research engineers appointed in April 1998, Li Zhiwang is still with SSLS today as Senior Research Fellow. In November 1998, Yang Ping was appointed as Research Fellow of the SSLS and remains with the SSLS to this day as Principal Research Fellow. Both Zhiwang and Yang Ping have been crucial to the smooth running of the SSLS. In March 1999, we appointed Yuli Vladimirsky and his wife Olga as Senior Research Fellow and Research Fellow of SSLS respectively.

The Columns of the SSLS Building

The construction of the SSLS building had already commenced with the appointment of the contractor, and all seemed to be going well until we discovered a serious flaw in the design and construction of the floor and foundations, as reported in an email from Antony to CC Hang on 14 December 1998.[53] It had been clearly specified to the architect that the central section of the floor on which the machine would rest had to be vibration-free, and hence it was to be isolated from the rest of the building. In October 1998, on inspecting the floor, we discovered a serious design flaw — the architect had put the supporting columns for the roof within the central vibration-free area of the floor!

This meant that this area would not really be vibration-free, especially if external vibrations or loud sounds such as from an overhead aircraft could vibrate the roof and hence the columns and vibration-free floor. On 12 November 1998, we met with Robin Wong of INDECO to register our dismay at this development, and on 23 November he wrote to me to assure me that the design of the floor and columns would be changed to rectify this serious fault.[54] At the second meeting of the pro-tem committee on 14 December, I reported this serious defect to the committee, and they agreed that INDECO and the architect should be held responsible for the cost overrun due to this defect, which was eventually rectified at no extra cost to our budget.[55]

My Position as P.I. of the Project

A search for a full-time Director for SSLS had been initiated by the Chairman of the pro-tem management committee, as Kwa Siew Hwa, then Science Faculty General Manager and Assistant to the Dean, informed me on 28 December 1998 when she sent me a list of milestones for SSLS.[56] According to this list, in September 1998 the tender for the cryogenic equipment had been advertised, and the wiggler design study had been finalised. The wiggler was to be an additional component inserted into the storage ring which would extend its range of usefulness (but was never actually implemented). The building was scheduled for completion by February 1999, and the commissioning of Helios-2 would begin in April 1999. An advertisement for the Director had been placed in various scientific publications.

At this point, I felt that much of the control of the progress and development of SSLS had passed from my hands, since I was just a member of the pro-tem management committee. But I was still the Principal Investigator (P.I.) of the entire project from NUS's and NSTB's point of view, and hence I was fully accountable for the funds committed to the project. Therefore on 14 December 1998, I wrote to CC Hang to request that I relinquish my position as P.I. and that Lee Soo Ying be appointed P.I. instead.[57] On 28 December he replied to say that the Chairman of the pro-tem committee could not be P.I. simultaneously.[58]

On 30 December 1998, I nevertheless declared to CC Hang that I would like to relinquish my P.I. position, and that perhaps Antony or Andrew Wee of the Physics Department be appointed P.I. in my place.[59] I also wrote the same day to Lim Pin[60], Teo Ming Kian[61] and Su Guaning (Deputy Chairman of the NSTB)[62] that I wanted to relinquish my position as P.I. I then tried to reduce my involvement with the project, but still kept myself as fully informed about its progress as possible. Indeed, on 26 January 1999 CC Hang emailed Antony to say that I had agreed to regain a prominent role in SSLS, and that I had suggested that Antony be given a title like Technical Director as he was effectively serving such a role.[63] CC Hang then informed me on 28 January that he had asked for the Vice-Chancellor's agreement to appoint Antony as Director (Technical Operations) of SSLS.[64]

At a meeting on 27 January 1999 chaired by CC Hang in his capacity as Deputy Chairman of NSTB and attended by Lee Soo Ying, Antony and Kwa Siew Hwa from NUS, and Fong Yew Chan, Tay Meng Huat and Mike Loh from NSTB (I was invited but did not attend), CC Hang declared that I would still be responsible to drive the SSLS programme and that my role as the champion for the project would be unambiguous.[65] There were concerns about the reduced participation of Chartered Semiconductor due to a change in management, and Fong Yew Chan raised concerns that in the absence of a Director for SSLS, someone to champion the project had yet to be identified. In response, CC Hang stated that the pro-tem committee was no longer satisfactory to drive the SSLS, and that a full time management team including a Director should be formed, overseen by an Advisory/Management Board. He reiterated that I as the project P.I. (still the case at that point!) should be involved to champion the programme.

On 22 February 1999 Herbert Moser, then Head of the ANKA synchrotron radiation facility at Karlsruhe, indicated to Lee Soo Ying that he was interested in the position at SSLS, and the process of getting him to apply officially and interviewing him was set into motion.[66] Lee Soo Ying then informed CC Hang on 24 February that he had identified Herbert Moser as the best candidate for the Directorship.[67] At the same time, Antony was appointed as a Visiting Professor in the SSLS to supersede his appointment at the Materials Science Department, which enabled him to continue to remain heavily engaged in the setting up of SSLS.

With Lee Soo Ying as Chairman of the pro-tem committee and Antony acting in his capacity as a Visiting Professor at SSLS, this meant that SSLS was in effect without an executive head. I was acting by default in that capacity, since I was NSTB's Principal Investigator for the entire project which meant that I had sole accountability for the project funds. On 15 March 1999, I wrote to Chow Shui Nee, NUS Deputy Director of Research, to state that since I had originally been given authority in March 1998 by Lim Pin to be in charge of the synchrotron radiation project, I would be happy, in my current capacity as a member of the pro-tem committee, to advise him and the pro-tem committee chairman to enable them to authorise purchase orders, reports or personnel appointments relating to SSLS, thus enabling SSLS to go forward.[68]

Beam Line Development and the Temasek Professorship

At this stage, efforts were being made to get interested parties to invest in beam lines for Helios-2, since in our original proposal, no beam lines had been included. In an email to Lee Soo Ying dated 23 January 1999, Antony noted that Andrew Wee was interested in a surface science beam line.[69] Indeed, on 18 December 1998, Andrew Wee had emailed Antony to inform him that the Dean of Science had asked him to chair an SSLS R & D committee.[70]

One of the most important consequences of Andrew Wee's involvement was the eventual setting up of a dedicated surface science beam line, which is still in operation and which has proven to be one of the most productive and successful beam lines in SSLS, running under Andrew's supervision. IMRE had faced budget cuts and thus was unable to commit funds for a beam line, though several of their researchers were interested in working with SSLS. One beam line would be dedicated to MEMS work using X-ray lithography techniques for nano and microfabrication in clean room conditions, with funding from an NSTB-approved project applied for by Oxford Instruments.

Antony was still trying to get Franco Cerrina as a Temasek Professor to work on EUV lithography. Consequently, a seminar/workshop on next generation lithography was organised with Franco Cerrina as well as several other key leaders in next-generation lithography such as Henry Smith of MIT (who could not make it at the last minute but sent a message to be read), as indicated in an email from me to CC Hang dated 15 February 1999.[71] At about the same time, equipment for SSLS was beginning to arrive, such as the cryogenics as indicated in an email to me from Antony dated 24 February 1999.[72]

In spite of some disagreements about business class air fares for the speakers as indicated by emails between Antony and the Dean's Office, the seminar was successfully held on 8 March 1999 at the Institute of Molecular Agrobiology. I gave the welcome address to the participants, and in addition to Franco Cerrina, we had speakers from the major optical stepper manufacturers such as

Mitsubishi and Canon. This event was important in giving SSLS a degree of visibility in the international lithography community.

GST and Other Problems

At this stage, there appeared to be a lack of unanimity on the progress of the SSLS project and the building in particular, as emails and letters between SSLS, the Dean of Science's Office and the pro-tem committee show. This was indeed a complex project the likes of which had never been undertaken in NUS before, and its complexities not surprisingly sometimes resulted in a lack of complete agreement on its status, how it should proceed, and on its funding position. I gave my inputs to Antony, the Chairman of the pro-tem committee, the NUS Bursar's and Estate Offices, and to NSTB whenever necessary to smoothen matters out. The cooperation of my colleagues in all these agencies was extremely helpful in resolving most of the issues which did arise.

A letter dated 23 April 1999 from Antony to Francis Yeoh, Executive Director of NSTB, in requesting for a third budget variation for the project, detailed the complexities of the project and its funding.[73] One major budget item which we now had to deal with and which had not been taken into account in the original budget, was the Goods and Services Tax or GST which was to be levied on the NSTB grant, since the decision to impose GST came only on 8 May 1998, well after the award of the funds by NSTB. The GST sum was in the region of S$1.1 million, and in a note from me to Antony dated 30 April 1999, we had reduced the shortfall to about S$760,000 with S$430,000 for additional safety measures (Daresbury UK was advising us on the safety and shielding necessary for SSLS), but would need another S$450,000 for GST from NSTB.[74]

Appointment of Management Board for SSLS

On 13 May 1999, I was informed by NUS Registrar Joanna Wong that a Management Board for SSLS had been formed as of 2 May 1999, and I was invited to serve as a member of the Board which would presumably replace the pro-tem management committee.[75] The Chairman of this Board would be CC Hang, and the NUS representatives were to be Lai Choy Heng, Vice-Dean (Research) of the Faculty of Science, Daniel Chan who was Head of Electrical Engineering and myself. NTU was represented by Lee Sing who was Head of Physics, and the other representatives were from NSTB, Chartered Semiconductor, IMRE and IME. (A letter to me on 26 May 1999 from Teo

Ming Kian and Lim Pin announced the end of the pro-tem management committee.[51])

At the first meeting of the Management Board on 14 May 1999, CC Hang announced that he would chair the Board in his capacity as NUS Director of Research.[76] An executive committee was appointed consisting of the Deputy Director of Research (Chow Shui Nee) or his representative, the Dean of Science or his representative, Antony and myself. Lim Khiang Wee of NSTB informed the meeting that the position of Director would be offered to Herbert Moser. The meeting looked at the current status of the setting up of SSLS as well as its research plans, including the appointment of a Temasek Professor (which was intended to be Franco Cerrina). The budget shortfall was largely due to the GST and safety access issues, and on 24 May, Antony wrote to CC Hang to clarify that this shortfall amounted to S$781,000. NSTB which had provided 80% of the original funding made up its share of the shortfall, and later CC Hang negotiated with MOE for their 20% share.

By May 1999, the building had largely been completed and Helios-2 had arrived in Singapore and been unloaded and transported to the SSLS building, where the lengthy process for the commissioning of the machine had begun. By 8 July 1999, the first meeting of the SSLS executive committee or exco could be held in the new building, chaired by Chow Shui Nee and attended by Lai Choy Heng, Vice-Dean of Science representing the Dean, Antony and myself.

At the second meeting of the exco on 21 July 1999, Lai Choy Heng informed the meeting that NUS Personnel was negotiating with Herbert Moser on his terms and conditions.[77] He also mentioned that Andrew Wee had submitted a proposal for a surface science beam line to the Faculty Research Committee which was undergoing external review. Antony reported that the beam line for the Oxford Instruments MEMS project would be on beam line #14, and that a low-cost line was being designed by William Smith (formerly Head of Engineering at Daresbury). It was also reported that modifications had to be made to Daresbury's safety shield design enclosing Helios-2 due to a design fault.

Application for Licence from Radiation Protection Institute

Another key issue was that of radiation protection and safety. At the 3rd and 4th meetings of the exco, it was clear that there was a dispute between Oxford Instruments and NUS on who should apply for the L5 licence required for the operation of Helios-2. The Singapore Radiation Protection Institute or RPI had declared that as the manufacturer and vendor responsible for the installation and commissioning of Helios-2, Oxford Instruments should hold an L5 licence and be responsible for the safety of the machine, while NUS was responsible for the safety of the building. This was disputed by Oxford Instruments, leading to a delay in the commissioning of the machine. A letter from CC Hang to Alistair Smith on 1 December 1999 appears to have broken the log jam, with Oxford Instruments agreeing to apply for the L5 licence and thus allowing the commissioning to proceed.[78]

I wrote to Alistair Smith on 6 January 2000 to thank him for this and for resolving the safety issues which had been raised by RPI in regard to the radiation generated by Helios-2, with the responsibilities of Oxford Instruments and NUS for various aspects of safety now being clearly defined.[79] As the first large machine of its kind in Singapore, questions had naturally been raised as to whether there would be sufficient protection for the machine's operators, as well as for members of the public, from the radiation which the machine would produce while in operation. All pertinent issues relating to safety had now been settled satisfactorily, and the machine could be licenced for operation in Singapore.

Appointment of SSLS Director

On 22 March 2000, the NUS Research Professorial Review Board met and agreed to recommend that Herbert Moser be appointed to a Research Professorship in SSLS. SSLS also appointed Vince Kempson of Oxford Instruments as its Senior Facility Manager on 24 June 2000, as he had been closely associated with Helios-2 at Oxford Instruments. The commissioning of Helios-2 was proceeding smoothly, as I mentioned in my letter of thanks dated 22 June 2000 to Peter Williams for his recommendation of Vince Kempson[80]. (Peter Williams had by now left his position as Chairman of Oxford Instruments to become Master of St Catharine's College, Oxford.)

At this juncture, and with the appointment of Herbert Moser, Antony Bourdillon, who had done so much for the whole synchrotron radiation project since its inception, and who had performed the major part of the technical spadework as well as the technical process of grant application and the installation of the machine, regrettably decided that he did not see a future for himself at SSLS and decided to move on, resigning from NUS and leaving Singapore. In a letter to Antony dated 10 May 2000, I expressed my deep appreciation to him for all he had done for SSLS and wished him all the best for the future (in my capacity as Acting Director of SSLS which I had been since the setting up of the Management Board).[81] With Antony's departure, I wrote to Yang Ping on 23 May 2000 authorising him to act on my behalf on all technical matters relating to Helios-2.[82]

Final Acceptance and Commissioning of Helios-2

On 30 September 2000, Oxford Instruments completed the final set of acceptance tests for Helios-2, which were accepted by SSLS on behalf of NUS.[83] I wrote to Andrew Mackintosh, CEO of Oxford Instruments, on 9 October 2000 to seek clarification on Oxford Instruments' commitment to the maintenance of Helios-2 in view of recent manpower changes in Oxford Instruments.[84] He assured me in his reply of 10 October that Oxford Instruments would provide service, spares and support for five years.[85] I therefore wrote on 17 October 2000 to Chow Shui Nee, NUS Dean of Research, to inform him of this and to recommend the final payment of S$5 million to Oxford Instruments.[86]

Pamela Lee of NUS Human Resources wrote to me on 10 October 2000 to inform me that Herbert Moser would be arriving to take up his position on 20 October 2000.[87] After he had arrived, I wrote to CC Hang on 23 October 2000 to say that I was relinquishing my position as Acting Director of SSLS.[88] This was to make way for Herbert Moser's appointment as Director of SSLS.

The New Management Board

On 8 January 2001, Chow Shui Nee called a meeting with Herbert Moser (now Director of SSLS) and me to discuss the future direction of SSLS. The most urgent need was the appointment of the Chairman and members of the SSLS Management Board which was to be reconstituted following CC Hang's decision to step down as Chairman. Ho Ching, President and CEO of Singapore Technologies, very graciously accepted our invitation to be the new Chairman, and on 20 February 2001, I wrote to her to thank her for accepting this additional burden to her already onerous responsibilities.[89] I acknowledged her desire that SSLS be used to upgrade the level of research at the NUS Departments of Physics and Chemistry, and for the promotion of an interest in science among students and young people.

In a letter to Lim Pin on 14 March 2001, I informed him of Ho Ching's appointment as Chairman.[90] The other members of the new Management Board were to be CC Hang representing NSTB, Wee Heng Tin of MOE, Chow Shui Nee of NUS, Fong Hock Sun and Lee Sing Kong of NTU and myself. With

Ho Ching's guidance and her especially keen interest in ensuring that SSLS be a powerful stimulus for research in the physical sciences and for creating interest in science and technology amongst students and young people, SSLS was off to an excellent start.

At the first meeting of the Board on 29 June 2001, the Chairman reiterated that SSLS should not only provide the infrastructure for university research, but also bring in the schools to generate students' interest in this area of work.[91] These twin objectives of her vision for SSLS have provided the impetus for its progress and development, and in addition to providing new tools for research and industry, SSLS has been always mindful of its role in stimulating and encouraging the next generation of researchers in science and engineering. When Ho Ching stepped down as Chairman of the Management Board in June 2002 and was succeeded by me[96], she had firmly laid down the foundations for the development of SSLS, for which I was extremely grateful.

Beam Line Development at SSLS

The first two beam lines which became operational in SSLS were the LiMiNT and PCI beam lines. The LiMiNT (standing for LIthography for MIcro/NanoTechnology) beam line was the Oxford Instruments MEMS beam line which incorporated a clean room in which extremely delicate work on very small structures at the nanometer scale could be fabricated by X-ray lithography. The PCI or Phase Contrast Imaging beam line was a much simpler and cheaper line which enabled images of very small objects, including biological tissue and organisms such as insects, to be captured.

Herbert Moser became responsible for the further development of SSLS, and in this the guidance of Barry Halliwell (NUS Deputy-President for Research and Technology) and Lim Hock (Director, Research Governance and Enablement) was especially helpful and critical. In his Report on SSLS to the Provost dated December 2005, Herbert Moser could state that in addition to the LiMiNT and PCIT (as the PCI beam line was now known, PCIT standing for Phase Contrast Imaging and Tomography) beam lines, another three beam lines were operational and productive.[92]

These additional beam lines were: XDD (X-ray Development and Demonstration), ISMI (Infrared Spectro/MIcroscopy) and SINS (Surface, Interface and Nanostructure Science). The SINS beam line was the fruition of Andrew Wee's efforts to set up a surface science beam line, and it was extremely productive, supporting several techniques such as XPS (X-ray photoemission spectroscopy), XAS (X-ray absorption spectroscopy) and XMCD (X-ray magnetic circular dichroism).

Mark Breese Takes Over as Director

When Herbert Moser retired from the Directorship of SSLS in June 2010, he was succeeded by Mark Breese of the NUS Physics Department. Mark has very ably and successfully led the SSLS in the rational and well-considered expansion of its capabilities as well as bringing in many talented and capable researchers to work at SSLS. As of February 2016, SSLS has six fully operational beam lines and another under construction. The additional operational beam line is the XAFCA (X-ray Absorption Fine structure spectroscopy for CAtalysis) beam line and the beam line under construction is the RSXS beam line for electron and photon beam diagnostics.[93] The XAFCA line has particular significance as the first beam line set up by an A*STAR institute, as A*STAR is the successor to NSTB, whose funds enabled SSLS to be set up. This institute and now major user of SSLS is ICES, the Institute of Chemical and Engineering Sciences, whose Director Keith Carpenter's support was key to XAFCA's successful implementation.

Another beam line is being planned: an EUV line which will be an offshoot of the SINS line. This EUV line will be associated with Applied Materials, one of the leading international companies producing equipment and infrastructure for semiconductor wafer fabrication. This beam line brings SSLS into the area of wafer fabrication with which the Helios machines had been so closely associated.

As SSLS looks to the future, a degree of satisfaction can be gleaned from its achievements in providing a powerful research tool for physics, chemistry and materials science as well as for unique industrial applications. It has attracted research talents from within and outside Singapore, and also served as a means of inspiring and attracting young people with a passion for science and engineering to follow their dreams. It is very much hoped that SSLS will also inspire other similar initiatives to blaze new trials in science and technology for Singapore.

References

[1] Philip Willmott, *An Intrduction to Synchrotron Radiation*. Chichester, U.K.: Wiley, 2011.

[2] N.M. Wilson *et al*, "The Helios 1 compact superconducting storage ring X-ray source," *IBM Journal of Research and Development*, vol. 37, no. 3, pp. 351–371, May 1993.

[3] Chas Archie, "Performance of the IBM synchrotron X-ray source for lithography," *IBM Journal of Research and Development*, vol. 37, no. 3, pp. 373–384, May 1993.

4. Wikipedia. (2016, January) List of synchrotron radiation facilities. [Online]. https://en.wikipedia.org/wiki/List_of_synchrotron_radiation_facilities

5. Antony Bourdillon, Memo to members of Advanced Materials Committee, January 16, 1995.

6. Antony Bourdillon, Memo to Bernard Tan, February 27, 1995.

7. Advanced Materials Committee, NUS, "Advanced Materials — an Eye for the Future," 1995.

8. Advanced Materials Committee, NUS, Programme for NUS Seminar on X-Ray Lithography, September 22, 1995.

9. Antony Bourdillon, Memo to Bernard Tan and Lee Soo Ying, October 6, 1995.

10. Yeo You Huan, Synchrotron Radiation Facility for Economically Relevant and Applied Research, March 13, 1997.

11. Bernard Tan, Letter to Gong Wee Lik, March 31, 1997.

12. Yeo You Huan, Private Communication.

13. Lee Kheng Cheok, Letter to Bernard Tan, April 21, 1997.

14. Antony Bourdillon and Bernard Tan, Letter to NUS Tenders Board, May 7, 1997.

15. Jennifer Phang, Letter to Bernard Tan, May 23, 1997.

16. Alistair Smith, Letter to Jennifer Phang, June 2, 1997.

17. Antony Bourdillon and Bernard Tan, Letter to Robin Wong, April 16, 1997.

18. Bernard Tan, Letter to CC Hang, July 22, 1997.

19. Antony Bourdillon, Letter to Andrew Meier, July 22, 1997.

20. Tan Lien Seng, Letter to Bernard Tan, July 28, 1997.

21. Bernard Tan, Letter to Alistair Smith, July 29, 1997.

22. Koh How Eng, Memo to Bernard Tan, August 27, 1997.

23. Robin Wong, Letter to Antony Bourdillon, September 9, 1997.

24. Bernard Tan, Memo to Robin Wong, September 9, 1997.

25. Robin Wong, Letter to Bernard Tan, October 17, 1997.

26. Bernard Tan, Letter to Lionel Cheng, November 14, 1997.

27. NUS Tenders Opening Committee, Tender for Proposed Synchrotron Light Source at National University of Singapore, November 28, 1997.

28. Antony Bourdillon, Letter to Gong Wee Lik, December 3, 1997.

29. Bernard Tan, Letter to CC Hang, January 7, 1998.

30. Robin Wong, Letter to SG Industrial, December 29, 1997.

31. Lee Siew Ling, Letter to Bernard Tan, January 9, 1998.

32. Jennifer Phang, Letter to Bernard Tan, January 16, 1997.

33. Bernard Tan, Note to CC Hang, January 16, 1998.

[34] Lee Siew Ling, Letter to Bernard Tan, January 31, 1998.

[35] Bernard Tan, Letter to Lim Pin, February 18, 1998.

[36] Lim Pin, Memo to Bernard Tan, February 23, 1998.

[37] Bernard Tan, Letter to Lim Pin, March 2, 1998.

[38] Lim Pin, Memo to Bernard Tan, March 4, 1998.

[39] Antony Bourdillon, Letter to Koh How Eng, January 19, 1998.

[40] Lionel Cheng, Letter to Bernard Tan, February 11, 1998.

[41] Bernard Tan, Letter to Lim Pin, March 9, 1998.

[42] Lim Pin, Memo to Bernard Tan, March 10, 1998.

[43] NSTB/NUS, Press Release on Synchrotron Radiation facility, March 17, 1998.

[44] Lee Soo Ying, Letter to Antony Bourdillon and Bernard Tan, January 5, 1998.

[45] Lee Soo Ying, Email to Antony Bourdillon and Bernard Tan, February 23, 1998.

[46] Bernard Tan, Letter to CC Hang, March 21, 1998.

[47] CC Hang, Memo to Bernard Tan, March 21, 1998.

[48] CC Hang, Letter to Bernard Tan, March 25, 1998.

[49] Bernard Tan, Letter to Lee Soo Ying, March 28, 1998.

[50] Lionel Cheng, Letter to Bernard Tan, July 30, 1998.

[51] Teo Ming Kian and Lim Pin, Letter to Bernard Tan, May 26, 1999.

[52] NSTB, Minutes of SSLS pro-tem management commiittee meeting on 25 August 1998, August 25, 1998.

[53] Antony Bourdillon, Email to CC Hang, December 14, 1998.

[54] Robin Wong, Letter to Bernard Tan, November 23, 1998.

[55] NSTB, Minutes of SSLS pro-tem management commiittee meeting on 14 December 1998, December 14, 1998.

[56] Kwa Siew Hwa, Letter to Bernard Tan, December 28, 1998.

[57] Bernard Tan, Letter to CC Hang, December 14, 1998.

[58] CC Hang, Memo to Bernard Tan, December 28, 1998.

[59] Bernard Tan, Letter to CC Hang, December 30, 1998.

[60] Bernard Tan, Letter to Lim Pin, December 30, 1998.

[61] Bernard Tan, Letter to Teo Ming Kian, December 30, 1998.

[62] Bernard Tan, Letter to Su Guaning, December 30, 1998.

[63] CC Hang, Email to Antony Bourdillon, January 26, 1999.

[64] CC Hang, Letter to Bernard Tan, January 28, 1999.

[65] NSTB, Minutes of meeting with NUS on the Singapore Synchrotron Light Source chaired by CC Hang on 27 January 1999, January 27, 1999.

66 Herbert Moser, Letter to Lee Soo Ying, February 22, 1999.

67 Lee Soo Ying, Letter to CC Hang, February 24, 1999.

68 Bernard Tan, Letter to Chow Shui Nee, March 15, 1999.

69 Antony Bourdillon, Email to Lee Soo Ying, January 23, 1999.

70 Andrew Wee, Email to Antony Bourdillon, December 18, 1999.

71 Bernard Tan, Email to CC Hang, February 15, 1999.

72 Antony Bourdillon, Email to Bernard Tan, February 24, 1999.

73 Antony Bourdillon, Letter to Francis Yeoh, April 23, 1999.

74 Bernard Tan, Note to Antony Bourdillon, April 30, 1999.

75 Joanna Wong, Letter to Bernard Tan, May 13, 1999.

76 NSTB, Minutes of SSLS Management Board on 14 May 1999, May 14, 1999.

77 NSTB, Minutes of SSLS Exco meeting on 21 July 1999, July 21, 1999.

78 CC Hang, Letter to Alistair Smith, December 1, 1999.

79 Bernard Tan, Letter to Alistair Smith, January 6, 2000.

80 Bernard Tan, Letter to Peter Williams, June 22, 2000.

81 Bernard Tan, Letter to Antony Bourdillon, May 10, 2000.

82 Bernard Tan, Letter to Yang Ping, May 23, 2000.

83 Oxford Instruments, Final Acceptance Tests for Helios-2, September 30, 2000.

84 Bernard Tan, Letter to Andrew Mackintosh, October 9, 2000.

85 Andrew Mackintosh, Letter to Bernard Tan, October 10, 2000.

86 Bernard Tan, Letter to Chow Shui Nee, October 17, 2000.

87 Pamela Lee, Letter to Bernard Tan, October 10, 2000.

88 Bernard Tan, Letter to CC Hang, October 23, 2000.

89 Bernard Tan, Letter to Ho Ching, February 20, 2001.

90 Bernard Tan, Letter to Lim Pin, March 14, 2001.

91 NSTB, Minutes of SSLS Management Board meeting, June 29, 2001.

92 SSLS, Report to the Provost, 2005.

93 SSLS. Singapore Synchrotron Light Source. [Online]. http://ssls.nus.edu.sg

94 Bernard Tan, Letter to Gong Wee Lik, December 3, 1997.

95 Teo Ming Kian and Lim Pin, Letter to Bernard Tan, August 3, 1998.

96 Herbert Moser, Letter to Pamela Lee, November 18, 2002.

Chapter 19

CRISP — the Centre for Remote Imaging, Sensing and Processing

B T G Tan

The Centre for Remote Imaging, Sensing and Processing, better known as CRISP, is a university-level research centre of the National University of Singapore (NUS) which was established in the early 1990s as a centre for the reception and processing of images received from earth observation satellites. It is well-known in the region as the source of many of the satellite images which have revealed details of both natural and man-made disasters, such as the Indian Ocean tsunami of 2004 and the fires in Indonesia which have caused widespread hazy conditions in much of South-East Asia.

The Laboratory for Image Processing (LISP)

Sometime in the mid-1980s, MINDEF's Chief Defence Scientist Lui Pao Chuen, who was a classmate of mine in the NUS Physics Honours class of 1964/65 and with whom I had kept in close touch, discussed with me the possibility of establishing a joint laboratory between MINDEF and NUS. Collaboration between research universities and defence scientists was already widely accepted as a means by which academia could contribute to the national security capabilities of a country. In the United States, for example, the Lincoln Laboratory of the Massachusetts Institute of Technology is, as its webpage states, "a federally funded research and development centre that applies advanced technology to problems of national security."[1]

Pao Chuen's desire to establish a similar kind of laboratory between MINDEF and NUS was a visionary one which sought to leverage on the research capabilities of NUS to support our national security capabilities. As I was Dean of the NUS Faculty of Science at that time, it was appropriate that Pao Chuen discuss this with me, and we cast around for a suitable area of research in which this important initial collaboration could be fruitful.

At first it was thought that underwater acoustics would be a good area for joint work. We looked for an academic researcher who had an interest in this area, but failed to identify such a person within NUS. We therefore had to look for another area of research which was relevant and in which we could find academics with related expertise.

We eventually came to the topic of signal processing, which was certainly of relevance to national security, and there would certainly be academics whose research work had close links to signal processing.

One faculty member of the NUS Physics Department whom we greatly admired as a researcher and academic was Lim Hock, who had joined the Department relatively recently. He had done his PhD in meteorology at the University of Reading and had been one of my students in the early part of my academic career at the Department. Lim Hock was a brilliant young physicist who could bring his mind to bear on physics and engineering issues from first principles. As a meteorologist, he was well-versed in the use of complex techniques to model the weather and climate, and was certainly conversant with many signal processing techniques.

We thus decided, probably sometime in 1987, that the first joint laboratory between NUS and MINDEF should be focused on signal processing and its applications. The name of this new laboratory was the Laboratory for Image and Signal Processing or LISP, which made a good acronym. LISP was sited on the top floor of block S13 of the Physics Department at the Kent Ridge campus. This location was ideal as much of LISP's work would be in collaboration with researchers from the Faculty of Science, and in particular from the Physics, Mathematics and Information and Computer Science Departments.

Apart from LISP's founding Director Lim Hock, the staff strength of the new laboratory would be very small, due to financial constraints. Most of the funds available for the setting up of LISP would be for computing equipment, as signal processing would require computational power for many of the complex simulations and modelling problems which it would be tackling. By 5 July 1988, we were in the midst of recruiting our first (and only) Research Officer, Tan Ah Kaw, a recent Physics Honours graduate who had done his Honours project under my supervision.[2]

By the end of 1988, LISP was fully operational, and on 12 January 1989 we were able to receive a visit from Mark Richmond, Vice-Chancellor of the University of Manchester, who was visiting NUS.[3] Indeed, LISP was one of the two laboratories in Physics chosen to receive Mark Richmond and LISP and the other laboratory chosen — the equally new Surface Science Laboratory — had been designated as multi-disciplinary laboratories in the Faculty of Science serving more than one Department.

Though LISP was officially a signal processing laboratory, it was interested in any interesting problems which involved computer modelling and simulation, as it had computer and graphics workstations ideal for such projects. Hence one of the major projects embarked on by LISP was a computer simulation of galaxy evolution such as the change in the shape of spiral galaxies over time. We contacted the Singapore Science Centre to see if they were interested in a joint project on such a topic, and they did indeed express great interest. The Science Centre had quite recently opened their OmniMax Cinema Theatre, which was able to show spectacular movies on huge screens enveloping the audience.

LISP discussed with the Science Centre the possibility of developing movie sequences on galaxy evolution for the OmniMax theatre's huge screen. The OmniMax used a special large film format, and we spent some time trying to understand how the OmniMax system worked so that we could transfer the graphical output of the simulations to the OmniMax film. As it turned out, we never actually got to that stage, but working with our colleagues in the Defence Science Organisation (DSO), we were able to develop some video sequences which were the graphical output of our galaxy evolution modelling and simulations, and which could be presented for public viewing on computer screens.

One of the important image processing challenges which LISP undertook was the deblurring of images. For example, photographs may be blurred due to either the lens of the camera being out of focus or to motion blur i.e. a movement of the camera during the taking of the photograph. DSO was also interested in this topic, and one of their researchers, Tan Kah-Chye, who was an NUS Physics Honours graduate, was doing a PhD at NUS under the joint supervision of Lim Hock and myself on this topic.

This work took a different approach to deblurring from the conventional one, which worked with the signal to be deblurred in the frequency domain i.e. it was treated as a bunch of frequencies which were then processed accordingly. Our approach processed the signal in the time domain, i.e. as a phenomena proceeding in time, and gave results which were quite impressive. For example, a photograph of a newspaper article which was completely unreadable due to motion blurring could be easily read after going through our deblurring process.

Much interest locally and even internationally was aroused by our deblurring technique. For example the well-known UK television programme, *Tomorrow's World*, featured our technique in one of their episodes. Kah-Chye's groundbreaking work on image deblurring, in addition to earning him a well-deserved PhD, also eventually won him a Defence Technology Prize from DSO in 1993. We published four papers under the banner of LISP on various aspects of our deblurring technology in a leading international journal.[4-7]

LISP also undertook joint work with the NUS Department of Geography on image processing of images from the SPOT and LANDSAT satellites and with the Singapore Meteorological Service on the processing of meteorological satellite data. Other projects undertaken by LISP included neural network models for pattern recognition and the digitisation of terrain data from topological maps. These topics served as excellent preparation for the image processing work which was to be undertaken by CRISP later.

The Idea of a Remote Sensing Ground Station in Singapore

By 1991, we were very much heartened and encouraged by the progress of LISP and the work it had done, which convinced us that a more ambitious project involving signal processing was feasible. Pao Chuen had long believed that the emerging and rapidly progressing technology of satellite earth imagery would be strategically significant for Singapore. Satellite technology had already made a worldwide impact on telecommunications, with communications satellites such as Telstar and Syncom demonstrating that radio, telephone and television signals could be relayed efficiently and accurately across the globe.

Companies like the French company SPOT Image were launching satellites into earth orbit expressly for the purpose of acquiring images of the earth's surface using specialised digital cameras which could detect not just visible light images but also images outside the visible spectrum, such as infra-red radiation.[8] Such images could give information about vegetation, for example, which visible light could not detect. These satellites had great potential for mapping and analysing the earth's natural resources and hence greatly assist in their sustainable use. The monitoring of natural and man-made disasters such as earthquakes was another key potential application of these satellites.

Indeed, the entire field of the science and technology devoted to the processing and analysis of satellite images to recover as much information as possible was now known as remote sensing. We wanted to look into the setting up of a centre which would not just passively receive satellite images, but which would also have advanced research capabilities in the field of remote sensing. Such a centre would be Singapore's first step into space-related science and technology and would eventually bring the nation firmly into the space age.

As SPOT Image was one of the leading imaging satellite manufacturers, a visit was made to SPOT Image in France in June/July 1992 to better acquaint ourselves with the relevant technologies for the setting up of a receiving station for remote sensing. Together with Lim Hock and Quek Gim Pew, an expert in image processing, we first visited MATRA MS2i, a leading French technology company in satellite imagery in Paris, and then SPOT Image in Toulouse, the hub of the French Aerospace industry.

The Proposal for the National Science and Technology Board

When we returned from our trip, work started on preparing a proposal for the establishment of a remote sensing research centre in NUS which would receive satellite images from orbiting image satellites and process these images for various applications.

A formal proposal was prepared, with Lim Hock writing most of the proposal and its technical specifications. This proposal was then submitted to the main national research funding agency, the National Science and Technology Board (NSTB), which was the successor to the pioneering Science Council of Singapore, and which itself was succeeded by the current Agency for Science, Technology and Research (A*STAR).

The proposal to NSTB was entitled *Proposal for the Establishment of an Institute for Remote Sensing.*[9] The first of the proposal's two sections was entitled *Local Interest and Activities in Remote Sensing Technology* and gave the background to remote sensing technology, pointing out that "remote sensing technology has in the last two decades played a crucial role in the monitoring of the earth". It also listed the various parties in NUS and the government who might be potential users of the Institute. These included the NUS Departments of Geography and Zoology which had each already established a small-scale PC-based remote sensing data processing system.

For the Faculty of Science, the expertise built up by LISP and the recent establishment in 1991 of the new Computational Science course were very relevant to the proposed Institute. The Faculty of Engineering had recently established a Radar and Signal Processing Laboratory working on synthetic aperture radar or SAR, a technology which was to become an important component of CRISP's capabilities.

The main Government agencies which would be interested in remote sensing were the Meteorological Department which already operated weather satellite reception stations and the National Parks Board which had established a workstation-based remote sensing data processing system for monitoring the national parks and nature reserves. The Ministry of Defence, which had a Mapping Unit responsible for producing maps for the nation, would be another potential user.

In the second section entitled *Proposal for the Establishment of an Institute for Remote Sensing,* the rationale for the setting up of the Institute in relation to developments in the region was laid out. A current UN proposal for a Centre for Space Science and Technology Education or CSSTE was being competed for by countries such as Thailand and Indonesia, both of which already had satellite image receiving and processing infrastructure. Our proposed Remote Sensing Institute would enable Singapore to be a credible host for the CSSTE.

It was proposed that the new Institute be set up within NUS, and that it be equipped with a satellite station "to receive imagery from the orbiting satellites in real time". LISP would serve as the data processing laboratory for the Institute. The local research community would benefit in the following ways:

- Raw satellite data would be provided in real time, and data processed with minimum delay.
- Remote sensing research would be provided with the necessary human resources and related technology.
- Technical backup such as data processing services and consultation services would be provided to researchers.
- Postgraduate training in remote sensing would be provided to attract a steady stream of postgraduate students.

The Institute would be headed by a Director, under the guidance of a Management Board appointed by NUS. The Institute would have two sub-units: a Ground Receiving Station to be staffed by about 10 personnel, and an Image Analysis/Processing Centre, which would have about two full-time research scientists, 15 principal investigators or co-investigators who would be mainly NUS faculty members or researchers in government departments, five post-docs, 10 research assistants and 15 undergraduate or postgraduate students.

The proposal presented a four-year budget from 1992 to 1995 of S$31,648,000. Of this, S$19,000,000 was set aside for the satellite receiving equipment and image processing facilities, and S$1,000,000 for the building. It was envisaged then that this building budget would be mainly for the satellite station and that the staff would be located in existing Faculty of Science space. The remainder of the budget was made up of recurrent manpower expenditure, running and maintenance costs, and satellite licensing fees.

As the proposal was fully supported by NUS, and in particular by Vice-Chancellor Lim Pin, it was quickly approved by the NSTB in August 1992. At this stage, the name for the Institute was much discussed, and it was felt that "Institute for Remote Sensing" was a rather unimaginative name. Pao Chuen then came up with a brilliant suggestion — the Centre for Remote Imaging, Sensing and Processing, which made the very appropriate acronym CRISP (which suggested that the images received would be crisp and sharp!).

The Pro-tem Board of Directors

In order to guide the development of CRISP during these early days, a pro-tem board of directors was set up in December 1992. The members of the pro-tem board were Lui Pao Chuen, Quek Gim Pew, Lim Hock and Kwoh Leong Keong (even before his official appointment as Assistant Director of CRISP) as Secretary, and myself as Chairman. The first meeting of the pro-tem board was on 23 December 1992, at which Lim Hock briefed the Board on the possible vendors for the ground station as well as the satellite operators whose satellites we were likely to be using.[10] One concern was the higher than expected costs of licensing for the reception of images from the satellites.

Lim Hock was able to report that "CRISP had basically started functioning", since a cost centre had been established with the NUS Bursar's Office, enabling funds to be transferred from NSTB. He also reported that he had received his official letter of appointment as Director of CRISP from the NUS Director of Personnel. However, he had not yet accepted the offer formally as he needed endorsement from the Board meeting that CRISP would proceed in spite of the high satellite licensing costs. The meeting did confirm that CRISP would indeed proceed as planned, and proposed that Lim Hock should accept the appointment. He informed the meeting that Leong Keong, who was the second employee to be recruited by CRISP and was to be its Assistant Director, was awaiting processing of his appointment by NUS.

At the next meeting of the pro-tem board of directors on 28 April 1993, the conditions of the tender and the terms of the contract with the successful vendor were discussed.[11] The sites for the ground station antenna as well as the image analysis/processing centre, now known as the data processing facility or DPF, were also discussed. A site for the antenna had indeed been identified on Kent Ridge, but NUS was currently prohibited by the Urban Redevelopment Authority (URA) from building any structures on the Ridge which protruded over the treetops (as the antenna surely would). It was thus suggested that we should use the good offices of NSTB to resolve this with URA (the matter was of course eventually settled). I was also tasked to confirm the designation of the site for the CRISP antenna with the NUS Sites and Buildings Committee, and the site was secured without any problems.

Another concern was electromagnetic (EM) interference at the antenna site which could affect the reception of the satellite signals. It was thus proposed that we engage Yeo Tat Soon of the NUS Electrical Engineering Department to conduct an EM interference survey of the proposed site. This survey later confirmed that EM interference would not be a serious problem for the siting of the antenna.

However, the site for the DPF was still to be found. We had already decided to reserve the premises of the Faculty of Science's Centre for Industrial Collaboration or CIC for the temporary staff offices for CRISP. (CIC had recently merged with the Engineering Faculty's Innovation Centre to form the NUS Industry Liaison Office or ILO.) However, it was clear that the CIC premises were not going to be large enough to house the entire DPF equipment and technical staff, so we would make an estimate of the space required for the DPF and look for a suitable location.

The SPOT Satellites

Typically, an imaging satellite for remote sensing of the earth's surface, such as a SPOT satellite, orbits the earth in a low earth orbit. For example, the SPOT 2 satellite which was launched in 1990 and which was the first SPOT satellite to send images to CRISP, has an altitude of 832 km and orbits the earth in a circular polar orbit, i.e. it circumnavigates the globe by passing directly over the north and south poles, looking downwards at a "slice" of the earth's surface on each orbit.

Each successive orbit shifts in longitude and hence covers a different slice, and the satellite comes back to the same spot on the earth's surface within a period of a few days. In contrast, telecommunications satellites which are in a geostationary orbit circumnavigate the earth in the plane of the equator at the same rate as the earth rotates. This causes their position with respect to the earth's surface to appear stationary so that they seem to hover over the same geographical location. Geostationary satellites are much further from the earth compared to low earth orbit satellites, typically about 42,000 km. Remote sensing satellites need to be much lower in order to capture images of the earth's surface in useful detail.

The name "SPOT" is actually a French acronym, "*Satellite Pour l'Observation de la Terre*" which means "Satellite for observation of Earth". The SPOT series of satellites were developed by the French Space Agency CNES (*Centre national d'études spatiales* — meaning "national centre for the study of space") in collaboration with Belgium and Sweden. The first in the series, SPOT 1, was launched on 22 February 1986, and was followed by SPOT 2 on 22 January 1990 and SPOT 3 on 26 September 1993. All three satellites were launched by the European Space Agency's Ariane rockets.

The SPOT 1, 2 and 3 satellites all have the same pair of digital cameras. One camera is monochromatic i.e. sensitive to just a single wavelength of light (much like black and white film in a film camera) and is capable of resolving details on the ground down to 10 m in size. The second camera is multispectral, i.e. sensitive to three different wavelengths of light (much like a colour film). Unlike a conventional film which is normally sensitive to blue, green and red light, this camera is sensitive to green, red and near infra-red or NIR (which is invisible to human eyes) and is able to resolve objects 20 m in size.

The Tender for the Satellite Receiving Station

With NSTB's approval, we had to work hard to get CRISP up and running as soon as possible. LISP Director Lim Hock was of course the most logical choice as Director of CRISP, and he set to work immediately to prepare the complex tender paper specifying the technical details of the satellite ground receiving station which was to be the major part of CRISP's hardware installation. Remarkably, Lim Hock embarked on this challenging task with absolutely no prior experience in remote sensing, but he did a superb job. He not only drove the setting up of CRISP, but also gained international renown for the technical excellence of the new Centre very soon after it became operational.

Lim Hock thus taught himself the nuts and bolts of remote sensing concepts and technicalities in order to produce a tender document with precise specifications for the satellite ground receiving station hardware. This hardware was divided into two parts: the data acquisition facility or DAF, which was mainly comprised of the tracking antenna system, and the data processing facility or DPF, which made up the satellite data processing system. The tender specifications were completed in less than a year, and the tender was called on 15 May 1993. On 8 July 1993, the Business Times carried a report entitled *Two Research Centres for $41m* which stated that the NSTB had set up two new research centres at a total cost of S$41 million — the National Supercomputing Research Centre (NSRC) and CRISP.

By the close of the tender on 15 August 1993, six tenders had been received. Companies could tender for either the DAF, the DPF, or for both together as a complete turnkey satellite ground receiving station.[12] The evaluation team for the tender consisted of four people: myself as Chairman of the CRISP Management Board which had been set up earlier, Lui Pao Chuen who was a key member of the Management Board, Lim Hock as Director of CRISP, and Kwoh Leong Keong as Assistant Director of CRISP.

Four tenders were received for the complete turnkey system from MacDonald Dettwiler Associates (MDA), Electronic Space Systems Corporation (ESSCO), DI²S and Israel Aircraft Industries Ltd (IAI). Scientific Atlanta Inc (SA) tendered only for the DAF and MATRA CAP Systemmes (MCS) tendered only for the DPF, but these two companies later teamed up to offer a complete turnkey system together, making five complete systems to be evaluated.

The DAF was made up of the following components: (1) the tracking antenna system, (2) the data record and playback system, (3) the data switch unit and (4) the ground station controller. The tracking antenna accounted for 75% of the cost of the DAF and was its largest and most important component. The signals from the remote sensing satellite could only be received when there was line-of-sight visibility between the satellite and the antenna. As the satellite approached the location of the ground station, from the point of view of the station it would emerge from below the horizon, rise in the sky and traverse a path across the sky till it was overhead, and then move towards and drop below the opposite point on the horizon.

The tracking antenna would be a large parabolic dish pointed at the satellite as it traversed the sky. The dish needed to be able to track the satellite's path accurately in order to receive the satellite's signals optimally and in real time while the satellite was acquiring images of the earth's surface. This meant that the ground station could acquire real-time images only from an area of the earth defined by a circle centred on the location of the station. For the SPOT satellites, this circle had a radius of about 3,000 km, which meant that images of most of South East Asia and Sri Lanka, as well as parts of China, India, and Australia, could be acquired.

The DPF consisted of the following software modules: Record and Playback Subsystem (RPS), Production Control Module (PCM), Geometric Correction Module (GCM), Radar Processing Module (RPM) and Administrative Support Subsystem. A specialised piece of equipment called the Frame Formatter/Decommutator worked together with the RPS. Generally speaking, the role of the DPF was to ensure that the final processed images represented the actual surface being imaged as clearly and accurately as possible.

Evaluation of the Tender

In the tender evaluation document published on 17 March 1994,[12] CRISP's missions were defined as:

- To establish a remote sensing ground station.
- To implement a programme for the reception of satellite remote sensing data.
- To develop research programmes in remote sensing.

- To develop value-added remote sensing products.
- To provide training programmes in remote sensing technology.
- To explore commercial exploitation of remote sensing technology.

CRISP would receive signals from satellites with imaging capabilities in the optical region, defined as the visible and infra-red parts of the electromagnetic spectrum, as well as from satellites using the synthetic aperture radar (SAR) technique operating in the microwave region. The ground station would have a tracking antenna which could be locked onto the satellite to receive its signals, and the images beamed down to the station would be decoded. The station also needed to have the computational capability of processing the received images to give them the required radiometric and geometrical accuracy.

The four most important satellite programmes relevant to the region were identified as:

- Landsat — a series of optical remote sensing satellites operated by NASA, the National Aeronautics and Space Administration of the US.
- SPOT — the series of satellites operated by SPOT Image.
- ERS — a series of SAR remote sensing satellites operated by the European Space Agency.
- Radarsat — A series of Canadian SAR remote sensing satellites to be launched in 1995.

The evaluation of the DAF tender showed that DI²S's system was superior to those of the other tenderers in technical terms, with its proposed tracking antenna system outperforming the others by a significant margin. For the DPF tender evaluation, DI²S's system was also judged to be superior to those of the other tenderers. As DI²S's tender sum for both systems together was the lowest, it was recommended that the turnkey ground station tender comprising both the DAF and DPF be awarded to DI²S.

However, this was recommended only after a thorough check of the two companies which were the partners who had set up DI²S, i.e. Datron/Transco Inc (DTI), a subsidiary of Datron Systems Inc, and International Imaging Systems Inc (I²S). Our legal advisors, Tan Rajah and Cheah, advised us that both DTI and I²S should explicitly give their commitment to ensuring that they would be fully responsible for DI²S discharging their contractual obligations and liabilities for the project, before we awarded the tender to DI²S.[13]

The total cost of the turnkey remote sensing ground station was S\$13,197,448, which did not include the computer hardware on which the DPF was to run. This hardware was to be purchased from Silicon Graphics Pte Ltd in Singapore at a cost of S$368,237. Together with the cost of the turnkey ground station, the total cost was well within the S$19,000,000 specified in the approved budget. On 28 April 1994, an NUS purchase order amounting to US$7,143,888, signed by Lim Hock, was sent to Erik Elmar of DI²S for the bulk of the ground station equipment, including the antenna dish.[14]

The Siting and Installation of the Ground Station

In order to receive a signal from remote sensing satellites which generally transmit radio signals within the frequency range of 8–8.45 GHz, it was expected that the size of the tracking antenna's parabolic reflector dish should be at least 10 m in diameter. The antenna dish proposed by DI²S was in fact 13 m in diameter. The dish would have to be able to point at a satellite emerging from the horizon and traversing overhead, until the satellite dropped below the opposite point on the horizon.

Hence the tracking antenna had to be sited at a location large enough to accommodate a 13 m diameter parabolic dish as it swung from one direction to the other to track the path of the satellite. Furthermore, it was imperative that this location should have as unobstructed a view as possible down to the horizon, without any buildings or other obstacles obstructing this field of view in any direction. In urban Singapore, we knew this was not going to be easy, but it was obvious that the higher in elevation the site we could obtain, the better the field of view would be.

A survey had been conducted in May/June 1993 to look for such a site within the boundaries of the NUS campus at Kent Ridge, with an unobstructed view in every direction above 3 degrees elevation above the horizon. The best site was an elevated site on a small hill off Kent Ridge Road which wound along the top of Kent Ridge. We had already obtained provisional planning approval for the siting of the antenna and a small adjacent single storey building housing the essential control and microwave equipment.

The main data processing equipment and facilities i.e. the DPF would be housed some distance away in a laboratory within the NUS Faculty of Science, and fibre optic cables would link the antenna to this laboratory. Technical officers at the laboratory would be able to control the antenna remotely from the laboratory, and the signals received by the antenna would be sent to the laboratory for processing and storage. We had to look for a suitable location within the Faculty of Science for the DPF, and it became clear that we would have to do some major construction and renovation work at an existing covered area near the Science Library to convert it into the temporary laboratories and office space for the DPF.

Funds would be required for these renovation works, so we requested NSTB to allow us to transfer funds from the balance of the sum allocated for the ground station to pay for the construction and renovation costs. On 6 July 1994, Koh How Eng of NSTB wrote to Lim Hock to confirm that CRISP would be allowed to transfer funds from the sums allocated for the ground station and building renovation amounting to S$2,757,000.[15] This would enable the construction of a new floor near the Science Library as the temporary premises for CRISP, as well as the renovation of one floor of the new S17 building as the permanent offices of CRISP.

Lim Hock wrote to D R Rondeau, President of DI²S, on 29 April 1994 to inform him of his company's successful bid for the tender.[16] Both sides then moved quickly to proceed with the preparation of the site and the installation of the antenna. On 3 May 1994, Erik Elmar of DI²S sent a preliminary site preparation plan to Lim Hock, to enable us to begin getting the antenna site ready for the antenna installation.[17]

On completion of the Preliminary Design Review by DI²S, NUS was able to make the first advance payment to DI²S for the ground station on 26 July 1994, which was a sum of US$714,388.80.[18] The stage was then set for a formal contract signing ceremony for the ground station.[19] The ceremony was held on 30 September 1994 at the Multipurpose Room on level 3 of the NUS Faculty of Science block S16 at 9.30 am and was attended by key NUS and NSTB personnel. It was a simple ceremony with four speeches: first by Lim Hock as Director of CRISP, then by me as Chairman of CRISP, then by a representative of DI²S and finally by Vijay Mehta, NSTB's Executive Director.

One matter which had to be settled before CRISP could start operations or even perform any tests of the antenna was the licensing of CRISP's use of the radio spectrum in communicating with the remote sensing satellites. Lim Hock therefore wrote to Lim Choon Sai, Director of Engineering of the Telecommunications Authority of Singapore (TAS), to seek a waiver of frequency management fees from TAS.[20] This was on the grounds that while CRISP would be receiving radio signals from such satellites, it would not be transmitting any radio signals to these satellites. The waiver was of course granted without any problems.

The plans for the ground station and its equipment building had to be checked by competent civil engineers for their structural soundness. A memo from Karen Wong of CRISP to Raymond Chua of the NUS Estate Office dated 8 March 1995 referred to the payment of invoices to the Engineering

Consultancy of Dr Lee Chiaw Meng and Associates. This was for the professional checking of the plans for the tracking and site equipment shelter which was to house the control and communications equipment for the antenna.[21]

The CRISP Management Board

The pro-tem board of directors appears to have undergone a name change to become the CRISP Management Board, as evidenced in a letter to me from Leong Keong on 3 August 1996 to which the minutes of the Board meeting on 26 July 1994 were attached.[22] Lim Hock's progress report on CRISP presented at that meeting reported the award of the ground station tender to DI²S, as well as his meetings with SPOT Image and other satellite operators.[23] A new staff member, Tay Geok Kee, had also assumed his appointment as System Officer of CRISP. The minutes of the meeting also record that the Board agreed that CRISP should recruit an administrative assistant.[24]

The official appointment of the new Management Board as sanctioned by NSTB was confirmed in a letter dated 22 December 1994 from John Tan of NSTB to Huan Tzu Hong, Registrar of NUS, informing NUS that the Ministry of Trade and Industry had approved the appointment of six members of the CRISP Management Board i.e. myself as Chairman, Lui Pao Chuen, Lim Hock, Tay Siew Choon who was Executive Vice-President of Singapore Technologies Industrial Corporation or STIC, and James Leo in his capacity as Executive Director of the Port of Singapore Authority.[25] They were to be appointed for a two-year term as from 1 December 1994 to 30 November 1996. A representative from NSTB, Koh How Eng, joined the Management Board on 2 December 1995, and K K Phua, Chairman of World Scientific Publishing, became a member of the Board in October 1997.

The first meeting of this new CRISP Management Board was held on 21 February 1995 in the Conference Room of the Dean's Office, Faculty of Science.[37] The progress report for this meeting reported on the recruitment of four laboratory technologists for CRISP as well as one administrative officer.[26] The four technologists were Chng Ngai Kun, Chee Kin Mun, Edwin Chow Sui Sen and Yew Chiat Keng. The administrative assistant was Karen Wong May Yin who became a familiar figure to visitors to CRISP and who is still in CRISP today.

The meeting was also told of the Factory Acceptance Tests (FAT) for the DAF and the DPF.[27] The FAT for the DAF would be on 20 March 1995 and the FAT for the DPF on 29 April 1995. It was proposed, and agreed to by the meeting, that the DAF FAT be witnessed by one CRISP staff member and the DPF

FAT be witnessed by two CRISP staff members. The DAF was to be shipped to Singapore on 29 March 1995, and would arrive at the site on 29 April 1995. However, there would be adjustments to the delivery date of the DPF as DI²S were not likely to complete the whole DPF in time.

The site preparation and construction at the antenna site was progressing well, as reported to the meeting. However, construction work on the floor housing the DPF and its offices had not yet commenced as the NUS Tenders Board's approval was being awaited.

Another key issue which board members enquired about was that of lightning protection for the antenna. Lim Hock stated that the antenna would be equipped with two lightning rods. To provide additional protection, a 3-tower lightning protection system would be built around and above the antenna by a local contractor. Board members also suggested that the advice of Liew Ah Choy of the NUS Electrical Engineering Department be sought on lightning protection.

The permanent site for CRISP in block S17 was also an issue at this meeting, as the layout for block S17 was then being planned. The board members felt that the current plan to put the CRISP offices on the 6th floor should be changed to put them on the ground floor instead. (Today, the CRISP main offices are in S17 ground and 2nd floors.)

The meeting then turned to the planned agreements on the reception and distribution of images from the SPOT and ERS satellites. It was planned that the data reception and distribution agreements for these two satellites be signed in the second quarter of 1995. Indeed, the board agreed that both agreements should be signed in time for CRISP's operation. Discussions were ongoing on a similar agreement for the Radarsat satellite.

Lim Hock also clarified that in general, the operators would not allow a ground station to market images received outside of its national territory. In order to get around this and to allow CRISP to market images of other national territories, it would be necessary for CRISP to enter into joint marketing arrangements with the satellite operators.

On possible partnerships with commercial and other agencies on remote sensing applications, three such possibilities were raised during the meeting. One was with MapIndo, an Indonesian consultant company, on a pilot project in forestry applications. CRISP was also discussing possible oceanographic applications with the Singapore Meteorological Service. A third possible application was brought up by Tay Siew Choon, who informed the meeting that an Indonesian pineapple plantation owner was interested in monitoring pineapple plantations in the region.

Acceptance Tests and Installation of Ground Station

As is turned out, the factory acceptance tests (FAT) for the DAF and DPF were conducted slightly later than the schedules given at the Management Board meeting. At the next Management Board meeting on 2 August 1995, it was reported that the FAT of the DAF took place at Datron/Transco Inc at Sun Valley, California from 24 to 28 April 1995, witnessed by Lim Hock. The FAT for the DPF took a longer period, due to its complexity, from 31 May to 10 June 1995, witnessed by Kwoh Leong Keong and Tay Geok Kee.[28] The DAF passed its FAT easily except for two units of High Density Digital Recorders or HDDRs, whose problems would be rectified by the time of the on-site acceptance tests. The DPF also passed its FAT except for the ERS data processing module, which would be modified to ensure compliance.

The DAF was then shipped from California and arrived in Singapore on 13 June 1995. DI²S's engineers arrived a few days later and immediately started the installation and calibration of the DAF. At the time of the progress report, the system was already able to track and receive signals from the SPOT and ERS satellites. The DPF was sent by air to Singapore in three batches, the first of which arrived on 28 June 1995 and the last on 24 July 1995. It was expected that major installation of the DPF and the integration of the DAF and DPF would begin on 27 July 1995.

Liew Ah Choy had already started an in-depth study of the lightning protection requirements for the DAF and CRISP had also contacted Singapore Telecoms for advice concerning lightning protection of antenna systems.

Unfortunately, the preparation of the site for the antenna had run into some problems, as the construction work which had been already done, and the antenna foundation in particular, had been found to be defective, and had to be removed completely. A completely new foundation was then constructed, with DI²S airfreighting to Singapore a new set of bolts and sending over a foundation engineer at its own expense to work with the local contractor, resulting in a very professionally constructed antenna foundation.

The antenna construction team then, due to other works delays, was forced to work in rather unsatisfactory conditions such as lack of telephones (these were the days before mobile telephones), proper on-site toilets, and messy site conditions. Our building consultant INDECO and the NUS Estate Office thus had to take action against the local contractor and had to engage other contractors to remedy the defects and jobs left uncompleted, such as unreliable power supply and leaking roof and windows, as well as improvement of site security.

By then, construction and renovation work on the CRISP temporary premises had been completed in mid-June 1995. These premises were constructed in a covered but unused floor of block S6 next to the Science Library, right next to the earth slope which was part of Kent Ridge. DI²S engineers were already installing the DPF and part of the DAF systems in these temporary premises. As for the permanent location of CRISP offices and DPF in block S17, discussions had already been held with the INDECO architect to move the offices from level 6 as originally planned to the lower floors, as proposed by the Management Board.

In the meantime, CRISP had signed a formal agreement on 2 May 1995 with SPOT Image for the reception and distribution of data from the SPOT satellites. To enable CRISP to distribute and market images of regions outside Singapore territory, CRISP also signed an agreement with SPOT Asia for the distribution of such images through SPOT Asia. Discussions were continuing with the respective satellite operators on the reception and distribution of images from the ERS and Radarsat satellites. CRISP had also initiated negotiations with the operators of high resolution (i.e. of the order of 1–2 m or so) satellites such as Eyeglass, CRSS and Worldview.

The Ground Station Becomes Operational

By the time of the next Management Board meeting on 3 January 1996, Lim Hock was able to report to the board that operational acquisition and processing of images from the SPOT 2 and ERS-1 satellites had begun in September 1995. This was accomplished in spite of some ongoing problems with the tape recorders for data recording, and some parts of the DPF. Kwoh Leong Keong distinctly remembers to this day that the first SPOT image acquired was one of Singapore island with some cloud cover over the island.

The visit by Education Minister Lee Yock Suan to NUS on 8 November 1995 was able to include the now operational CRISP. This was followed by a visit by Deputy Prime Minister Lee Hsien Loong on 26 January 1996. CRISP quickly became an important part of the visit schedule for important visitors to NUS, and particular for those who were interested in the University's scientific and technological development. As the satellite passes over CRISP were generally in the late morning, such visits would usually be scheduled at a time when they could see live images coming from a satellite making an overhead pass.

Once CRISP became operational and data reception and distribution agreements with the major satellite operators had been signed, Lim Hock very quickly built up the Centre's technical expertise, starting from scratch, to a level which quickly established CRISP as one of the leading remote sensing satellite stations in the world. Lim Hock and Leong Keong were soon joined by Liew Soo Chin from the NUS Physics Department who helped to establish and who still heads CRISP's research programme, and Mak Choong Weng who became the manager of the ground station. Both of them are still with CRISP today.

In 1999, the permanent CRISP premises at block S17 were completed and CRISP staff moved from the temporary premises next to the Science Library (which had been christened "Siberia" by the CRISP staff stationed there for its frigid temperatures) to S17. The main CRISP administrative offices and the DPF remain at S17 today, and many seminars and workshops organised by CRISP are held in lecture theatre 33 conveniently located just adjacent to the CRISP offices in S17.

The Song San Oil Spill

The acquisition of the ERS-1 SAR images enabled CRISP to detect oil floating on the surface of the sea very easily, as this showed up on the SAR images as a much darker patch of sea compared to the unaffected sea surface. In August 1996, a major oil pollution event occurred in the Singapore Straits, when oil discharged illegally from a vessel drifted to the beaches of the East Coast and Southern Islands of Singapore, causing major pollution damage which cost S$1 million to be cleaned up.

It so happened that CRISP had been acquiring SAR images of the Singapore Straits from ERS-1 at that time, and a check of the images clearly showed a 5 km oil plume originating from a vessel anchored in the Straits — the Song San. This enabled the MPA to make investigations of the vessel, which conclusively proved that it was indeed the culprit responsible for the oil discharge. The vessel's owners were subsequently charged in court for the offence and convicted successfully.

This was the first case in the world of earth observation imagery being accepted in a court of law as evidence for illustrating the extent of oil pollution.[29] The owners, agents and master were fined S$400,000, the largest ever fine then in Singapore maritime history, and the master was sentenced to a three month prison term for the oil spill charge. Subsequently, a Memorandum of Understanding between the Maritime and Ports Authority or MPA and CRISP was signed on 25 June 1997 to develop remote sensing applications for the maritime environment.

ERS-1 SAR data could be used for many maritime applications, such as ship detection and monitoring. In August 1997, CRISP hosted the prestigious International Geoscience and Remote Sensing Symposium (IGARSS) at Suntec City Convention Centre, and the many papers presented by CRISP researchers, which included maritime applications such as ship wake detection using SAR, showed that the range and quality of CRISP's research had advanced very quickly since its recent establishment.

The Straits Times of 26 November 1998 highlighted CRISP's first map of oil slick pollution comprising 3,500 SAR images which showed the extent of oil pollution in the seas around Singapore.

CRISP started receiving images from the other key SAR satellite, Radarsat, on 25 January 1996. SAR imagery has enormous potential for our region which is perpetually covered by clouds, as SAR techniques can acquire land and sea images through such cloud cover.

The Haze of 1997–98

Singapore had been very fortunate in enjoying generally unpolluted atmospheric conditions most of the time, this due in no small measure to stringent anti-pollution regulations imposed by the government on industry and commerce, as well as on the general public, who greatly appreciated the resultant clean air. However, during the periods May-November 1997 and February-April 1998, a virtually unprecedented pollution event occurred, when thick smoke haze covered much of the region including Singapore, which greatly impacted on air quality and on health.

The approximate locations of the fires causing the haze were first identified using images supplied by the Meteorological Service of Singapore (MSS) from the US National Oceanic and Atmospheric Administration (NOAA) AVHRR or Advanced Very High Resolution Radiometer sensors on the NOAA satellites. CRISP then used the SPOT 2 satellite images to zoom in on the identified "hot spots" to accurately determine the location of smoke plumes which showed exactly where the fires were located. Working virtually around the clock, CRISP personnel daily supplied such images to the Ministry of Environment who could then take up the necessary action with the authorities responsible.

The haze episode of 1997/1998 (which was worsened by the "El Nino" phenomenon) indeed brought CRISP into the public eye rather dramatically. Images of smoke plumes billowing from forest fires became familiar in the pages of the newspapers, and effectively illustrated the power of remote sensing.

The Straits Times of 10 October 1997 highlighted CRISP's work in identifying the forest fires in Indonesia which caused the haze pollution and on 23 November 1998 reported that CRISP had identified fires in Borneo which had razed 43 million hectares of forests. On 17 February 1998, CRISP and the Ministry of Environment signed a Memorandum of Understanding under which CRISP would use satellite images to monitor hot spots in Sumatra and Kalimantan.[30] This has become a long-standing arrangement which is still in operation today and has been effective through subsequent haze crises.

CRISP becomes NUS RIC and Lim Hock Steps Down

The setting up of CRISP had been funded by a grant from the NSTB and hence the status of CRISP was that of a research centre under the umbrella of NSTB. In the year 2000, this status changed and CRISP was transferred to NUS to become an NUS RIC (research institute/centre) completely under the umbrella of NUS.[31] From then onwards, CRISP became totally accountable to NUS, reporting to the Provost who was then responsible for the RICs under NUS, and later to the NUS Deputy President (Research and Technology).

CRISP was still funded by a annual grant from NSTB (and subsequently from A*STAR, the Agency for Science, Technology and Research), but this grant was to be requested through NUS, which now became institutionally accountable to NSTB for the proper use of these funds. By then, CRISP had been able to generate a modest revenue through its sales of satellite images, as for example through its partnership with SPOT Image's Asian subsidiary SPOT Asia to market images outside of Singapore's territorial boundaries. Researchers and other users who did not need the full resolution of these images could have access to reduced resolution images on the CRISP website free of charge.

In October 2000, after building up CRISP from scratch and swiftly bringing CRISP to international prominence for the excellence of its research and technical capabilities, Lim Hock stepped down as Director to take on a new position as Director of the NUS Temasek Laboratories. Kwoh Leong Keong was the logical successor to Lim Hock, having been at CRISP from the very start and being completely conversant with every aspect of CRISP. Leong Keong has very ably built on the solid foundations laid by Lim Hock and maintained and enhanced CRISP's international reputation, expanding the range and scope of CRISP with access to new satellites and new hardware.

High Resolution Satellite Imagery

One of the more exciting developments in satellite imagery after CRISP had been set up was the advent of high resolution satellites which were capable of acquiring images of the earth's surface with resolutions superior to those of existing satellites such as the SPOT satellites, which were only capable of discerning features of 10 m size or larger. These high resolution satellites were the outcome of years of technological development by countries such as the US, and could see ground features much smaller than satellites like SPOT 2 could observe.

The declassification of some of this technology thus led to the commercial deployment of high-resolution satellites capable of imaging ground features of 1 m size or even smaller. For example, the IKONOS satellite, launched on 24 September 1999 by Space Imaging (later acquired by DigitalGlobe), was capable of 0.82 m resolution in panchromatic mode.[32] As reported by the Straits Times of 31 August 1997, CRISP opened the IKONOS Data Reception and Processing Facility in August 1997. This facility would receive and process high resolution images of the earth's surface from the IKONOS satellite, and CRISP had in fact already received its first IKONOS images in the same month.

The astonishing detail which was captured by the IKONOS images demonstrated that a new era in commercial satellite remote sensing imagery had indeed been opened up. Perhaps the most dramatic images showed cityscapes which looked like they had been captured by overhead aircraft cameras, showing building details, streets and individual vehicles clearly. The resolution was not quite good enough to show individual persons, but one memorable image clearly showed a long queue of people waiting to enter the Mausoleum of the late Chairman Mao Zedong in Beijing.

In more recent years, CRISP has started receiving images from the even more advanced high resolution satellites, GeoEye-1 and Worldview, on 21 June 2009 and 9 November 2009 respectively. GeoEye-1 is capable of a resolution twice as good as that of IKONOS, i.e. 0.41 m, and Worldview has a similar resolution capability.

The Indian Ocean Tsunami

Many natural and man-made disasters and phenomena were captured by satellites such as the SPOT satellites and IKONOS which CRISP was licensed to access. However, the most dramatic and far-reaching event involving CRISP satellite imagery was probably the Indian Ocean earthquake and resultant tsunami of December 2004.

On 26 December 2004, i.e. Boxing Day or the day after Christmas Day, I returned home after lunch with my family, switched on my television set and learnt that a disaster of major proportions was occurring at that very moment. A huge undersea earthquake had occurred in the vicinity of the Andaman Islands in the Indian Ocean, causing a devastating tsunami which was most acutely felt in North Sumatra and Aceh in particular, and which had severely affected countries as far afield as India and Sri Lanka.

I immediately phoned Leong Keong, and he promptly alerted key CRISP ground station staff who proceeded to work around the clock acquiring images of the devastated areas, and of Sumatra and Aceh in particular. Many of these images were to be critical in helping the Singapore relief mission to Aceh.[33] The available maps were not of much use to the mission as the effects of the tsunami had rendered the physical features of the area, and the coastal features in particular, to be virtually unrecognisable using existing maps.

Hence the CRISP images enabled the relief mission to navigate their way to the devastated areas to provide much needed relief for the stricken areas. CRISP had already acquired many images of these areas before the tsunami, and a comparison of the images before and just after the tsunami showed how severe the devastation was. Many of CRISP's images, which were made available to governments and relief agencies freely, were utilised not just for relief efforts but also for the evaluation of the damage caused, which was essential for future redevelopment efforts.

The Sea-to Space Building

When the 13 m antenna was erected on Kent Ridge as part of the CRISP ground station in 1995, the area around Kent Ridge was relatively unobstructed. After more than 10 years at that location, the construction of many tall buildings around the Kent Ridge campus had rendered the antenna's view from horizon to horizon somewhat less unobstructed. After the 13 m antenna was installed, CRISP had installed a 6 m diameter antenna on top of the S17 building in 2000, (which was itself replaced by a 6.1 m antenna in 2014). Another CRISP antenna of 6.1 m diameter was installed on the roof of the new Temasek Laboratories building in Kent Ridge in 2013.

Technology had advanced sufficiently in the meantime to allow a smaller antenna of about 8.5 m diameter to match or even outperform the 13 m antenna. CRISP thus commenced the search for a suitable site for an 8.5 m diameter antenna to supplement the 13 m antenna, and it was determined that a site at the other end of Kent Ridge Road would be suitable if the base of the antenna could be elevated by a couple of storeys.

CRISP and the Tropical Marine Science Institute (TMSI) then joined forces to construct a new building which would house both CRISP and TMSI offices and laboratories, and on top of which would sit the 8.5 m antenna. Lui Pao Chuen was very much the driver behind this new building, which he christened the S2S or Sea-to-Space building, reflecting its tenants' contrasting interests. After overcoming some construction cost problems, the S2S building was officially opened on 19 June 2009 and with its distinctive blue-green façade stands proudly on Kent Ridge as a testament to NUS's research horizon spanning from Sea to Space!

Other Milestones and the Future

Due to space constraints, we have been able to list only the major milestones and events in the short history of CRISP. Apart from the SPOT 1, 2, 4, 5 satellites, the ERS 1 and 2 satellites, the Radarsat satellite, the IKONOS, the GeoEye 1 and the Worldview I and II satellites, CRISP also receives signals from the SeaWIFS satellite, the TERRA (MODIS) and the AQUA (MODIS) satellites as well as the Singapore designed XSAT satellite.

On 16 December 2015, Singapore's space history took an ambitious new turn when an Indian rocket launched six Singapore-designed satellites into equatorial orbit. These included the TELEOS-1, a 400 kg earth imaging satellite produced by ST Electronics (Satellite Systems) Pte Ltd, and the Kent Ridge 1, a 77 kg hyperspecral imaging satellite designed and built in NUS. For TELEOS-1, the most ambitious Singapore satellite launched so far, CRISP is providing ground station facilities as well as value-added image processing.[34] Together with the other members of the TELEOS-1 team, CRISP was given the President's Technology Award in October 2016,[35] adding to the Excellence for Singapore Award which CRISP won in 1999 for its work in the haze crisis.[36]

CRISP has now been operational for more than 20 years, and it is renowned in the remote sensing and space community for the excellence of its operational standards and its research and value-added products. Much of the credit for this is due to its founding Director, Lim Hock, who built its reputation from ground up. Its current Director, Kwoh Leong Keong, has maintained the international reputation of CRISP and expanded CRISP's operations into new areas.

I must also pay tribute to the vision and drive of Lui Pao Chuen who was responsible for the original idea of a remote sensing ground station for Singapore and who saw the project through to its fruition. I am sure CRISP's next 20 years will be even more successful!

References

[1] MIT Lincoln Laboratory. Lincoln Laboratory, Massachusetts Institute of Technology. [Online]. https://www.ll.mit.edu/about/History/history.html

[2] Bernard Tan, Email to Ong Jin Soo, July 5, 1988.

[3] Bernard Tan, "Report on Sir Mark Richmond's Visit to the Faculty of Science," Faculty of Science, National University of Singapore, 1989.

[4] Hock Lim, Kah-Chye Tan, and B. T. G. Tan, "Edge errors in inverse and Wener filter restorations of motion-blurred images and their windowing treatment," *Computer Vision Graphics and Image Processing*, vol. 53, no. 2, pp. 186–195, March 1991.

[5] Hock Lim, Kah-Chye Tan, and B.T.G. Tan, "New methods for restoring motion-blurred images derived from edge error considerations," *Computer Vision Graphics and Image Processing*, vol. 53, pp. 479–490, September 1991.

[6] Kah-Chye Tan, Hock Lim, and B.T.G. Tan, "Restoration of real-world motion-blurred images," *Computer Vision Graphics and Image Processing*, vol. 53, pp. 291–299, May 1991.

[7] Kah-Chye Tan, Hock Lim, and B.T.G. Tan, "Windowing techniques for image restoration," *Computer Vision Graphics and Image Processing*, vol. 53, pp. 491–500, September 1991.

[8] Wikipedia, the free encyclopedia. [Online]. https://en.wikipedia.org/wiki/SPOT_(satellite)

[9] Lim Hock, "Proposal for the Establishment of an Institute for Remote Sensing," National University of Singapore, 1992.

[10] Kwoh Leong Keong, Minutes of first pro-tem board of directors of CRISP meeting on 23 December 1992 at 1400 hrs, December 23, 1992.

[11] Kwoh Leong Keong, Minutes of the pro-tem board of directors of CRISP meeting on 28 April 1993 at 1130 hrs, April 28, 1993.

[12] Bernard Tan, Lui Pao Chuen, Lim Hock and Kwoh Leong Keong, "Supply, Delivery, Installation and Commissioning of a tracking antenna system and a satellite data processing system for CRISP," 1994.

[13] Lim Hock, Letter to Bernard Tan and Lui Pao Chuen, January 18, 1994.

[14] Lim Hock, Invoice to Erik Elmar of DI²S , April 28, 1994.

[15] Koh How Eng, Letter to Lim Hock, July 6, 1994.

[16] Lim Hock, Letter to D.R. Rondeau, April 29, 1994.

17 Erik Elmar, Fax to Lim Hock, May 3, 1994.

18 Suzannah Tan, Letter to DI²S, July 26, 1994.

19 Lim Hock, Letter to Vijaykumar Mehta, August 16, 1994.

20 Lim Hock, Letter to Lim Choon Sai, Director Engineering, Telecommunications Authority of Singapore, January 21, 1995.

21 Karen Wong, Memo to Raymond Chua, March 8, 1995.

22 Kwoh Leong Keong, Letter to Bernard Tan, August 3, 1994.

23 Lim Hock, "Progress Report , Centre for Remote Imaging, Sensing and Processing, 25 July 1994," 1994.

24 Kwoh Leong Keong, Minutes of CRISP management board meeting on 26 july 1994, July 26, 1994.

25 John Tan, Letter to Huan Tzu Hong, December 22, 1994.

26 Lim Hock, Progress Report , Centre for Remote Imaging, Sensing and Processing, 1 August 94–15 February 95, February 15, 1995.

27 Kwoh Leong Keong, Minutes of CRISP management board meeting on 21 February 95, February 21, 1995.

28 Lim Hock, Progress Report, Centre for Remote Imaging, Sensing and Processing, 16 February - 26 July 1995, August 1, 1995.

29 Geraldine Goh Escolar, "The Use of EO Data as Evidence in the Courts," in *Evidence from Earth Observation Satellites: Emerging Legal Issues*. Leiden, Netherlands: Martinus Nijhoff, 2013, pp. 93–112.

30 "Satellite images to monitor hot spots," *The Straits Times*, February 1998.

31 Bernard Tan, "A proposal for the Repositioning of Centre for Remote Sensing, Imaging and Processing, National University of Singapore," National University of Singapore, 2000.

32 Wikipedia. [Online].https://en.wikipedia.org/wiki/Ikonos

33 L.K. Kwoh, S.C. Liew, P. Chen, and A.S. Chia, "Satellite remote sensing imagery for tsunami relief operation and damage assessment," in *Asian Association on Remote Sensing — 26th Asian Conference on Remote Sensing and 2nd Asian Space Conference, ACRS 2005*, 2005, pp. 92–96.

34 eoPortal. eoPortal Directory. [Online]. https://directory.eoportal.org/web/eoportal/satellite-missions/t/teleos-1

35 "Lauded for taking Singapore to space", The Straits Times, October 19 2016.

36 CRISP Milestones. CRISP, National University of Singapore. [Online]. https://crisp.nus.edu.sg/Milestones/index.html

37 Kwoh Leong Keong, Letter to Chairman and Members, CRISP Management Board, February 21, 1995.

Chapter 20

The NUS Nanoscience and Nanotechnology Initiative (NUSNNI)

Seeram Ramakrishna, T Venkatesan and Andrew T S Wee

Introduction

The prefix *nano-* is a unit prefix meaning one billionth, and is derived from the Greek word for dwarf. Hence *nanoscience*, the commonly used term for nanoscale science, deals with the study of atoms, molecules and nanoscale particles, entities that are typically measured in *nanometres* (billionths of a metre or 10^{-9} m). The modern origins of nanotechnology are commonly attributed to physics Nobel Laureate Richard Feynman, who delivered his talk "There's Plenty of Room at the Bottom" on 29 December 1959 at the annual meeting of the American Physical Society at Caltech.[1] In his talk, he predicted that the principles of physics should allow the possibility of manoeuvring things atom by atom. He also highlighted a number of interesting problems that arise due to miniaturisation since "all things do not simply scale down in proportion". Atoms do not behave like classical objects, for they must satisfy the laws of quantum mechanics. Furthermore, nanoscale materials tend to stick together by van der Waals attractions.

A major breakthrough occurred in 1981 when the Scanning Tunnelling Microscope (STM), was invented by a group at IBM Zurich Research Laboratory.[2] The STM uses a sharp tip that scans very close to a surface thereby mapping its atomic topography, creating for the first time a real space atomic scale image. This landmark achievement earned Gerd Binnig and Heinrich Rohrer the 1986 Nobel Prize in Physics "for their design of the scanning tunneling microscope".

Around the same time in 1985, Robert Curl, Harold Kroto and Richard Smalley discovered that carbon can exist in the form of very stable spheres, which they named fullerenes, and for which they won the 1996 Nobel Prize in Chemistry.[3] In 1991, Iijima *et al.* reported Transmission Electron Microscopy (TEM) images of hollow graphitic tubes or carbon nanotubes, which form another member of the fullerene structural family.[4] The strength and flexibility of carbon nanotubes makes them potentially useful in many nanotechnology applications such as in composite fibres, field emitters, energy storage materials, and so on. Thus began the era of modern nanoscience and nanotechnology research, which continues to this day.

Genesis of NUSNNI

The United States led the world in funding nanotechnology research with the establishment in 2000 of the National Nanotechnology Initiative (NNI), a US Government research and development initiative involving the nanotechnology-related activities of 20 departments and independent agencies.[5] The Singapore research community took notice of this major US initiative, and in August 2001, the then NUS President Professor Shih Choon Fong set up a campus-wide Nanoscience and Nanotechnology 'task force' with Professor Seeram Ramakrishna from the Faculty of Engineering and Professor Andrew Wee from the Faculty of Science as co-chairs. In September 2001, the first Nanoscience Initiative Panel meeting of Singapore institutions was held. A report on Nanoscience and Nanotechnology research worldwide and at NUS was prepared in November 2001. NUS acknowledged the global assessment that nanotechnology is expected to be a key technology underlying a wide range of applications in the fields of information and communication, biotechnology and medicine and engineering sciences. As nanotechnology is multifaceted and encompasses very broad areas, the NUS Nanoscience and Nanotechnology Initiative (NUSNNI) was formed to bring together research groups within the university as well as from local and overseas organisations.

NUSNNI is a joint effort between the Faculties of Engineering and Science spearheaded by its two founding Co-Directors, Professors Seeram Ramakrishna and Andrew Wee. NUSNNI was launched on 14 January 2002 (Figure 1) with the following objectives:

- To galvanise and coordinate multidisciplinary research effort (across departments, faculties and with the RIs) in nanoscience and nanotechnology.
- To help set research priorities and directions for high impact nanoscience and nanotechnology research.
- To develop research human capital and long-term research capabilities in the strategic field of nanoscience and nanotechnology.

NUS allocated a two-million-dollar equipment grant to NUSNNI to help establish the necessary scientific infrastructure. In February 2004, the Economic Development Board (EDB) awarded a total grant of $720,000 to

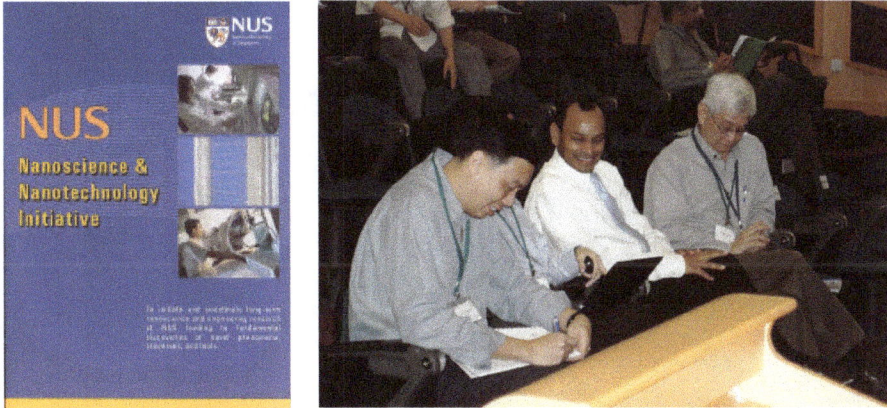

Fig. 1. NUSNNI brochure (left); and 2002 photograph of NUSNNI Co-Directors (from left) Professor Andrew Wee, Professor Seeram and Professor Ong Choon Nam.

NUSNNI to support 30 postgraduate MEng (Nanoengineering) scholarships under the joint industry postgraduate program (Figure 2). Areas of impact envisioned then included the following:

- Sensors with faster and finer sensing abilities
- More efficient solar cells for energy harvesting
- High density and high capacity energy storage devices
- Lighter and stronger structural materials
- Impact resistant materials
- Rapid and miniaturised diagnostic tools
- Scratch-resistant and self-cleaning coatings
- Anti-corrosion and anti-fouling coatings
- Sustainable food packaging
- Membranes for efficient water filtration
- Improved air filters
- Regenerative medicine and tissue engineering
- Nanomedicine and targeted drug delivery systems
- Faster electronic chips and processors
- High density memory storage devices and systems

Then acting Minister for Education Mr Tharman Shanmugaratnam officially launched NUSNNI on 9 July 2004 (Figure 3). He said, *"The 21st Century belongs to cities and nations which bring people together to catch each new wave of ideas, and which spur enterprise to create new value added around these ideas. Singapore aspires to be at the forefront of nanoscience. We have some competitive*

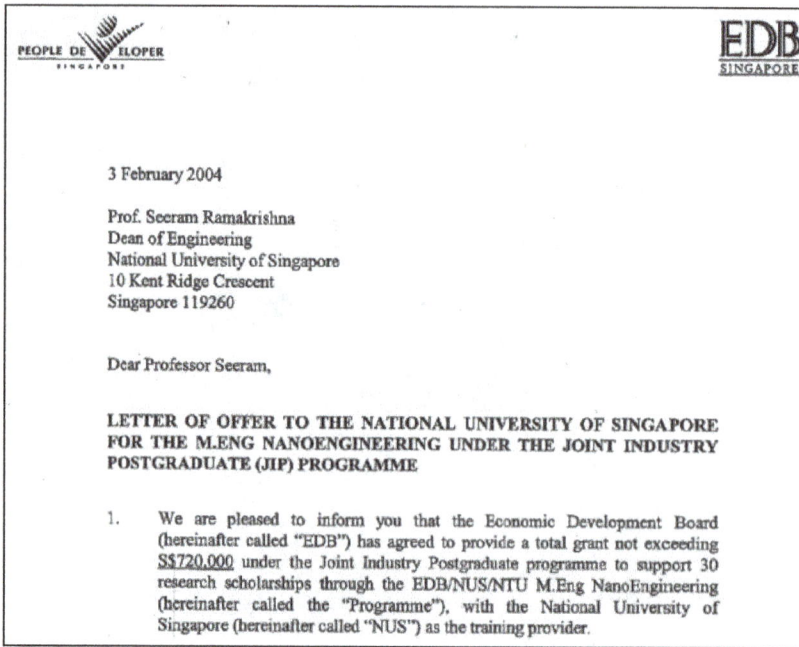

Fig. 2. EDB letter of offer for MEng (Nanoengineering) research scholarships.

Fig. 3. Photographs of the official launch of NUSNNI on 9 July 2004.

advantages. Despite being a resource-scarce nation of just four million people, we are a centre for high-value manufacturing and research. We also rank highly in intellectual property protection, transparency, and economic and physical security. And at the most basic level, standards of scholastic achievement in mathematics and science remain high in Singapore schools — by most international assessments among the highest in the world. Investment in nanotechnology is therefore an extension of the solid human capital and technological infrastructure we have built up over the years".

Also speaking at the Launch was NUS President Shih Choon Fong, who said, *"Nanoscience and nanotechnology are areas of strategic importance to NUS because we believe they hold great promise. Nanotechnology is making great impact in a wide range of areas and applications, including electronics, defence, transportation, medicine and health care, and agriculture. NUS aspires to position itself at the leading edge of exciting developments in nanoscience and nanotechnology. As such, NUS is committed to strengthening its capabilities and collaborations in nanotechnology research and development."*

NUSNNI Activities

During its early years, NUSNNI organised numerous activities to facilitate research collaborations within NUS as well as internationally. Some of the conferences and workshops co-organised by NUSNNI are shown in Figure 4.

Between 2002–2010, NUSNNI-affiliated laboratories formed an extensive network across the Faculty of Engineering (FOE) and Faculty of Science (FOS), as shown in Figure 5. These included the four Nanotech Corridors in FOE, and the newly set up common NUSNNI Labs I and II in FOS as well as other PI labs listed. NUSNNI postgraduate research scholars had co-supervisors in FOE and FOS, which encouraged researchers to work across several laboratories. Major equipment and facilities were open for use to all NUSNNI researchers, thus maximising the resources available and encouraging cross-fertilisation of ideas.

To facilitate multidisciplinary collaborations in strategic nano-related research areas, six NUSNNI focus areas were formed (Figure 6):

- Nanomagnetics and Spintronics
- Nanophotonics
- Nano/Micro Fabrication
- Health and Environmental Effects of Nanomaterials
- Nanobiotechnology
- Nanofiber Science and Engineering

Fig. 4. (Top) 2006 Slide showing some NUSNNI co-organised conferences during 2003–2005; (Bottom) Joint CNSI-NUSNNI-IMRE workshop at UC Santa Barbara.

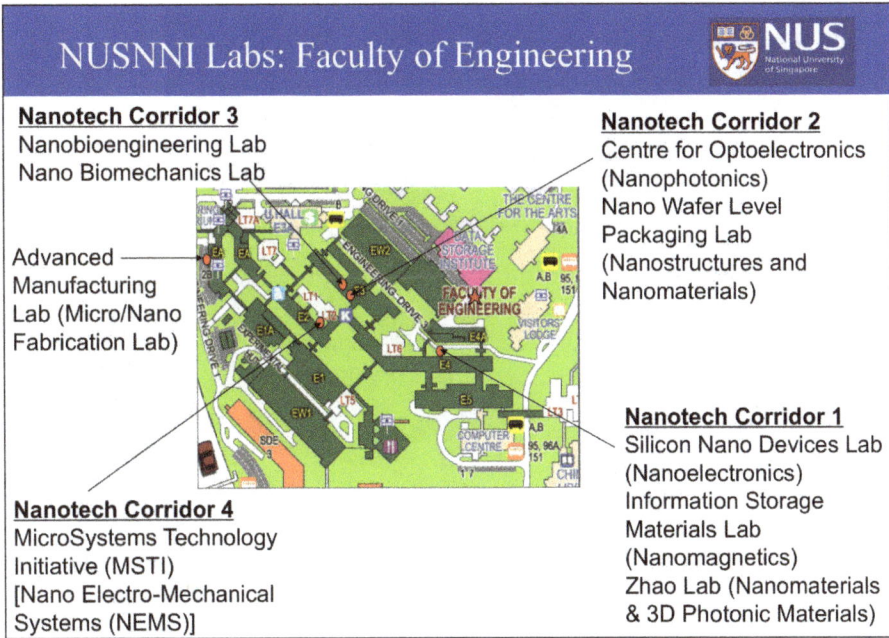

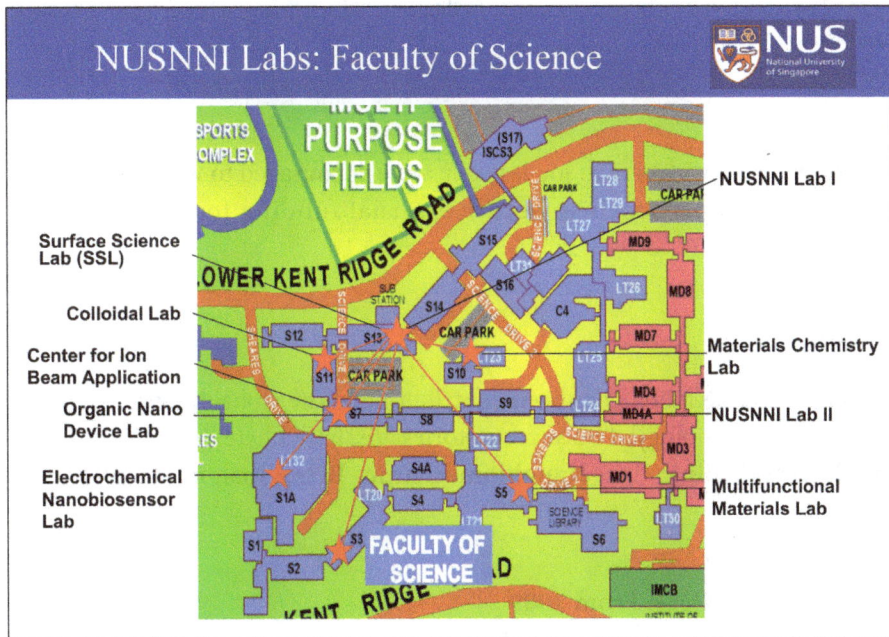

Fig. 5. Laboratories affiliated with NUSNNI in the Faculties of Engineering and Science.

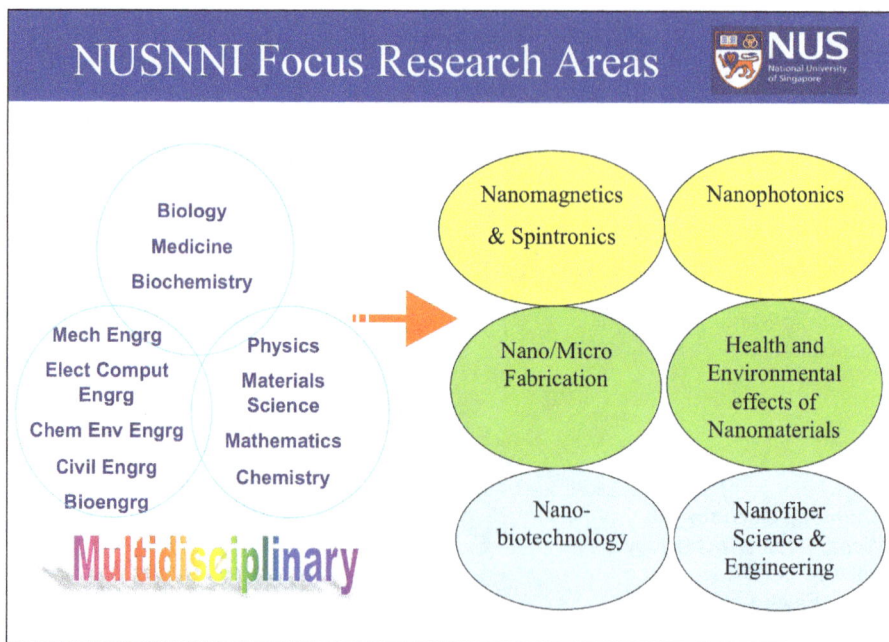

Fig. 6. The six NUSNNI focus areas formed to facilitate multidisciplinary collaborations in strategic nano-related research areas.

Researchers within each focus group were encouraged to write grand challenge research proposals to secure external funding for their research programmes.

Success Stories

Through NUSNNI, NUS has been contributing to the global knowledge reservoir of nanotechnology. Two success stories are shared here. (Box: **Nanofibers by Electrospinning**; Box: **Nanobiomechanics Bring Insights into Human Disease**).

NUSNNI-NanoCore: From 2009

Prof T Venky Venkatesan joined the ranks of the NUS faculty to start a Nano Institute here with centralised laboratory equipment. This facility was called NanoCore and was fully established by the year 2009 in the new TLab building

Nanofibers by Electrospinning
Seeram Ramakrishna

In 2001, NUS granted $150,000 to the NUSNNI Principal Investigator Professor Seeram Ramakrishna to kick start research on *Processing & Characterisation of Nanometer Scale Fibres* (research project number RP 3012698; WBS number R-398-000-002-112). This seed funding led to nurturing a world leading research

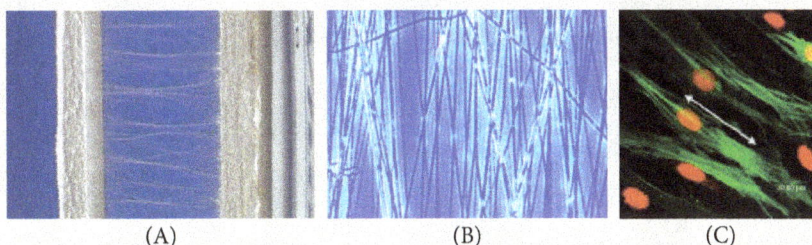

(A)   (B)   (C)

Fig. 7. (A) Electrospun polyethylene oxide nanofibers, (B) higher magnification photo, and (C) aligned smooth muscle cells on nanofiber scaffold.

in nanofiber science and engineering in Singapore. His team conceived innovative methods of producing three-dimensional as well as aligned nanofibers

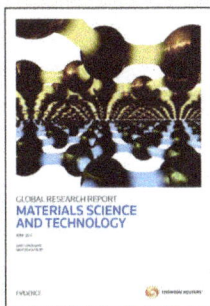

(Figure 7), and demonstrated their usefulness in air filtration, water treatment, wound dressing, regenerative medicine, tissue engineering, energy storage batteries and super capacitors, noise mitigation, thermal protection, lighter and tougher materials, and food packaging. Several hundred machines inspired by his advances in electrospinning are now operating in various companies and university research labs around the world. Thomson Reuters's Global Research Report dated 2011 on Materials Science and Technology ranked the National University of Singapore as number one in this field (Figure 8). This work also placed Professor Seeram Ramakrishna among the Highly Cited Researchers (www.highlycited.com). According to European Commission FP7 project led by Leiden University aimed at understanding research trends, he is ranked fourth among several scientists in Singapore. Thomson Reuters identified him among the World's Most Influential Scientists.

(Continued)

(Continued)

Ranking of countries and institutions by their output of research papers on electrospun nanofi-brous scaffolds for tissue engineering, 2000 – May 2011.

Papers	Country	Rank	Institution	Papers
657	USA	1	National University of Singapore	144
448	China	2	Songhua University	120
438	S. Korea	3	SUNY Stony Brook	58
161	Singapore	4	Virginia Commonwealth University	56
92	UK	5	Seoul National University	53
80	Italy	6	Chinese Academy of Sciences	42
70	Germany	7	Hungnam National University	35
66	Japan	8	Chulalongkorn University	34
49	Australia	9	Ohio State University	28
39	Thailand	10	University of Pennsylvania	27

Fig. 8. Thomson Reuters's Global Research Report dated 2011 on Materials Science and Technology for the field of Electrospun Nanofibrous Scaffolds for Tissue Engineering.

Nanobiomechanics Brings Insights into Human Disease
Sow Chorng Haur and Lim Chwee Teck

Professors Sow Chorng Haur (SCH) and Lim Chwee Teck (LCT) were amongst the founding members of NUSNNI. LCT is from the Department of Mechanical Engineering and Department of Biomedical Engineering. SCH comes from the Department of Physics. During one of the many workshops organised by NUSNNI in 2002, LCT and SCH met and started talking about potential collaboration. With LCT's interest in the biomechanics of biological cells and SCH's interest in the bioapplication of optical tweezers, the two researchers readily got into serious discussion to apply the technique of optical tweezers on biological samples. Under NUSNNI, LCT and SCH initiated a joint research using optical tweezers to stretch human red blood cells and DNA molecules. In the early phase of the project, the experiments were conducted using the optical setup in SCH's lab (Figure 9). This was the first instrument SCH developed after he returned to Singapore. It is a simple but functional setup that made use of an infra-red laser beam as the optical tweezers in the laser stretching of biological samples. The first experiment was carried out by a final year project student, Keng Thiam, who used it to stretch healthy human red blood cells and determine its elasticity. In the meantime, another MEng student, Gregory Lee, used the optical tweezers to stretch a single strand of l-phage DNA in

(Continued)

(*Continued*)

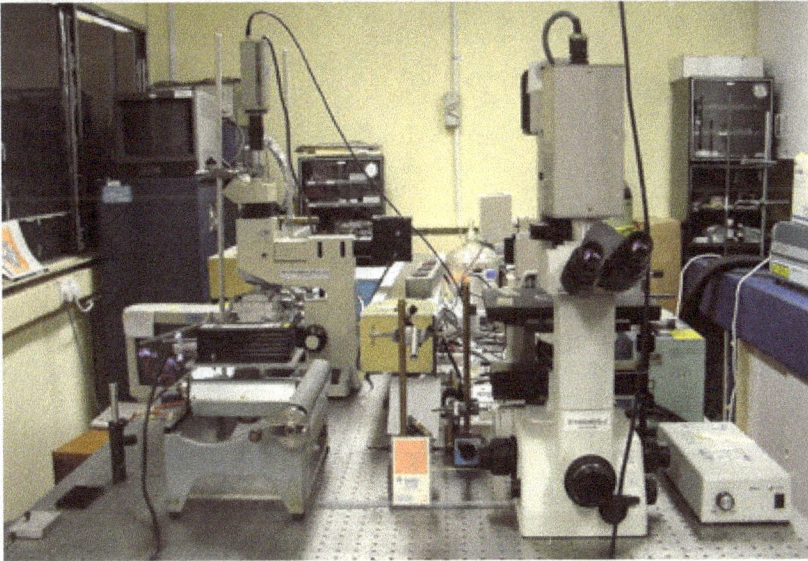

Fig. 9. Optical Tweezers used in the earlier experiment for the stretching of red blood cell and DNA biomolecules.

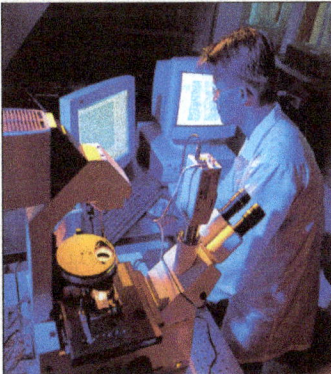

Fig. 10. Optical Tweezers setup at CT's lab to study the nanobiomechanics of human diseased cells and biomolecules.

order to understand the unfolding mechanism of the DNA strand. Arising from interest in this research, LCT also acquired a dedicated optical tweezers for his lab (Figure 10).

The stretching of human red blood cells caught the interest of Prof Subra Suresh, who was then from MIT and visiting the Faculty of Engineering as one of its International Scientific Committee members. He quickly got involved in this research collaboration and contributed in terms of performing computational modelling of the cell stretching experiments. With these results, both LCT and Prof Suresh published their first bioengineering related journal paper. They then extended their studies to include malaria infected red blood cells in collaboration with Prof Kevin Tan from the Department of Microbiology. The collaboration between NUS and MIT was fruitful and went on to produce several landmark papers and showed how mechanics can bring important insights into the pathophysiology of human diseases.

Fig. 11. TLab Opening Ceremony in September 2010: (from left) Prof T Venky Venkatesan, Director of NUSNNI; Prof Tan Chorh Chuan, NUS President; Mr Teo Chee Hean, Deputy Prime Minister and Minister for Defence, Singapore; Dr Ng Eng Hen, Minister for Education and Second Minister for Defence, Singapore.

(Figure 11). Even though NanoCore had several slots for hire after the initial hiring of Barbaros Oezyilmaz, Hyunsoo Yang, Andrivo Rusydi, Ariando and Qing Wang, the focus went in to integrating established scientists within NUS into the NUSNNI program. As of today, over 25 faculty members from the University participate in a variety of NanoCore related programs with close to 10 faculty members having a facility within NanoCore. NanoCore was a heavy user of facilities such as SSLS, CIBA and NUSNNI on the Campus and the NUS administration decided to integrate all of these facilities under a single roof and it was decided to name the new organisation NUSNNI with the "I" representing Institute. The new Directors of the combined facility were T Venkatesan and Mark Breese.

The integration of all of these facilities have enabled better participation of researchers on the campus on various programs enabling the building of more optimised research teams to tackle various problems. The number of NRF CRP programs generated by the associated faculty is a tribute to the quality of the researchers associated with NUSNNI.

In addition, the transformed NUSNNI continued to build up the infrastructure available for researchers. The following are some of the latest initiatives that were completed:

Ultra-fast Laser Facility: Two sophisticated femto-second laser system were installed with the following capabilities:

- Optical pump probe measurements — 2 beamlines with capability to do spatially scanned measurements as a function of temperature

- Coherent Raman anti-Stokes measurement system
- A high energy nanosecond laser surface interaction laboratory to study the dynamics of laser surface modification
- An advanced Raman and photoluminescence measurement system with variable temperature capability.

Resonant Soft X-ray Beamline: This will probably become the most important beamline on the SSLS as the energy range of the synchrotron fits perfectly with such a system which is best tailored to the study of oxides where one can bridge the oxygen s and p levels and the cationic d levels very nicely. This sophisticated, unique beamline will have various capabilities such as high energy optical reflection, angle-resolved photoemission, and *in situ* film preparation, so that a surface can be prepared *in situ* and studied without degradation of the surface due to atmospheric exposure. This beamline is expected to come online by the summer of 2016.

The Zeiss Microscopy Lab at NUS: As a joint collaboration with Zeiss, an advanced microscopy facility has been set up in the Engineering building with this Lab housing some of the latest electron and ion microscopy tools (Figure 12).

Fig. 12. The Zeiss Microscopy Lab: (from left) Prof T Venky Venkatesan, Director of NUSNNI; Dr Daniel Pickard, Head of the Plasmonics and Advanced Imaging Technology Laboratory; Mr Manfred Hanke, Managing Director of Carl Zeiss SMT Pte Ltd Singapore; Prof Barry Halliwell, NUS Deputy President (Research and Technology); Dr Nick Economou, President of Carl Zeiss SMT Inc, USA.

The Energy Lab at CREATE: A new energy lab is being set up at the NRF CREATE Building for studying photon to electrons and photons to fuels. This Lab is a joint collaboration involving NTU, NUS and UC Berkeley. The NUS directorship of this program will rest with Venkatesan and NUSNNI will be a strong player in this program.

Global Collaboration: A number of Laboratories around the world are collaborating closely with NUSNNI including UC Berkeley, MIT, University of Twente, University of Hamburg, Cornell, Maryland, Weizmann Institute, Trinity College, IIT Madras, IIT Delhi, IISER (Pune), IISc (Bangalore), Rutgers, CLRI (Chennai), University of Bremen, Japan Atomic Energy Commission, University of Tokyo. We have over 30 international collaborators and MOUs with many leading institutions. Numerous international conferences have also been organised.

The overall research focus of the reconstituted NUSNNI includes:
- Spintronics
- Topological insulators
- Atomically Controlled Oxide Heterostructures
- Graphene Electronics
- Oxide Electronics
- High Density Memories
- Nano-Drug Delivery
- Nano Toxicology
- Sustainable Energy Materials & Systems
- Nano-Scale Imaging & Patterning
- Charge transport in Mesoscopic Systems
- Active Plasmonics
- Nanoparticles & their Applications
- Composites and Nanocomposites
- Functional Polymers and Self-assembly
- Materials Nanowires
- Bio/Inorganic Interfaces

Innovative programs have been introduced such as:

Nanospark: This initiative helps promote entrepreneurship among the researchers and helps them with Proof-of-Concept grant proposal preparation, as well as organising occasional Meet-the-Entrepreneur programs.

PhD-MBA Program: This is new program initiated to recruit to NUS outstanding entrepreneurial students.

At NUSNNI, we believe that the combined strengths of a collaborative team are much bigger than the sum of the individuals. In a multidisciplinary field such as nanotechnology, this is all the more true. We see nanotechnology solutions in some of the most critical areas of importance to humanity, energy, water purification, environment (CO_2 sequestration), electronics-sensors and health. At NUSNNI, we hope to train our young people to prepare them for academia, industry and the wonderful world of entrepreneurship. We envision a research institute where the students can interact with any of the faculty or researchers and pursue their dreams in a nurturing environment.

References

[1] Feynman, R. P. (1960). There's plenty of room at the bottom. *Eng. Sci.*, **23(5)**, 22–36.

[2] Binnig, G., Rohrer, H., Gerber, C., Weibel, E. (1982). Tunneling through a controllable vacuum gap. *App. Phys. Lett.*, **40**, 178–180.

[3] Kroto, H. W., Heath, J. R., O'Brien, S. C., Curl, R. E., Smalley, R. E. (1985). C60: Buckminsterfullerene. *Nature*, **318**, 162–163.

[4] Iijima, S. (1991). Helical microtubules of graphitic carbon. *Nature*, **354**, 56–58.

[5] http://www.nano.gov/

Chapter 21

Two Decades of Quantum Information in Singapore

Kuldip Singh, Kwek Leong Chuan, Artur Ekert, Chan Chui Theng,
Jenny Hogan and Evon Tan

Today Singapore is recognised as a world-player in the fast-moving field of quantum technologies. This is thanks to the country's investment in the Centre for Quantum Technologies, established in 2007 at the National University of Singapore as the first national Research Centre of Excellence.

It's a long way from the beginnings of quantum information science in Singapore. That can be traced back to a small journal club in 1998. Every Friday evening, a group of theoretical physicists would gather in a room on the third level of the Physics Department at the National University of Singapore (NUS) to talk about developments in their respective areas.

Discussion topics at these meetings ranged from particle physics to mathematical physics because the people who gathered mainly worked on high-energy physics. Regulars at the meetings included Wang Xiangbin, Lai Choy Heng, Oh Choo Hiap, Kuldip Singh, Belal Baaquie, and Kwek Leong Chuan. Even Tan Eng Chye, then Dean of Science, sometimes attended despite his heavy administrative duties.

The group's foray into quantum information was sparked by an exciting development in the area. A 1995 paper by American mathematician Peter Shor had created a buzz in the community. Shor's seminal paper spelt out a quantum algorithm that could reduce the computational difficulty of factorising large numbers. Although quantum theory was familiar ground for the participants in our journal club, the *use* of quantum mechanics for something applied was novel.

But why would an academic-sounding claim about factorisation kick up such a storm? The quick answer is that it meant trouble for information security. The difficulty of discovering the factors of large integers is the basis for encryption by common methods such as the RSA protocol (named after its inventors Rivest, Shamir and Adleman).

The idea behind the protocol is simple: multiplying two large prime numbers is very easy (at least for computers), and this gives a public key for locking information. But to decrypt the information you have to perform the reverse process, obtaining the factors, which is extremely difficult with any known classical algorithms. To give an idea, it would take roughly five calendar months (thirty 2.2 GHz Opteron-CPU years) using classical algorithms to factor a 640-bit number.

Shor showed that a quantum mechanical machine could reduce the difficulty of factoring numbers to a level roughly comparable to that of multiplication. Although this was only a theoretical possibility, its implication for national defence was quickly realised. The conversation around 'quantum-safe encryption' has only grown more urgent in the years since, as researchers in academia and industry have made progress towards building quantum computers.

Even before Shor, theoretical work on the possibility of such machines for the purpose of computation was known. In 1982, based in part on work on reversible engines by Charles H. Bennett,[1] Edward Fredkin, Rolf Landauer, Paul Benioff[2] and Tommaso Toffoli, Richard Feynman[3] mooted the idea of using a quantum mechanical machine for computation.

Independently, around the same time, David Deutsch introduced the idea of a quantum Turing machine. In 1985 Deutsch showed that there exists a universal quantum computer — and that it is possible to achieve better than classical performance.

A Turning Point

The big venture into quantum information science in Singapore began with a Millennium Conference on Frontiers in Science. Around 1999, KK Phua, Chairman of World Scientific Publishing Company, suggested a conference that would anticipate the big problems in the upcoming millennium.

With generous support from the Singapore government, the high-profile event in May 2000 drew distinguished scientists from various fields of physics: Neil Turok came, so did Benoit Mandelbrot, Boris Altshuler and Robert Laughlin. Artur Ekert was there too, and gave a fantastic talk on quantum computation based on the geometric phase. Little did the local organisers know that this was going to be a turning point.

Soon after, a workshop on quantum information[4] was organised with Ekert, Bennett, Isaac Chuang and Sandu Popescu as speakers. During the workshop, there was also a meeting between some of the people from the workshop and NUS — Bennett, Chuang, Oh and Kwek — and the team led by Lim Khiang Wee from the National Science and Technology Board (NSTB). The NSTB is somewhat the precursor to the Agency for Science, Research and Technology (A*STAR). The meeting was a fruitful one as it seeded the idea of a consolidated effort in Quantum Information Science in Singapore.

Following up on this, Ekert was persuaded to take up a Temasek Professorship[5] at NUS. Ekert was already well known for his pioneering work in quantum cryptography. In 1995, he had been awarded the 1995 Maxwell medal and Prize by the UK's Institute of Physics.

Box Story: Quantum Fundamentals

Quantum systems behave in ways that seem bizarre. Properties that characterise a system, such as a particle's position or momentum, are indeterminate before we attempt to measure them. We can only describe the probabilities of different outcomes. This is like saying that the colour of your shirt isn't fixed until someone has looked at it, and then it has a 50% chance of being red and a 50% chance of being blue.

Another unexpected property of quantum systems is 'entanglement'. When particles are entangled they exist in a shared quantum state: the relationships between the particles are fixed but the individual properties are not. For example, two photons can have correlated polarisation — an intrinsic property that is measured as pointing in a particular direction. You can't know beforehand which way either polarisation points, but as soon as you've measured one photon you can infer with certainty the polarisation of the other.

Einstein worried about this giving rise to "spooky action at a distance". Imagine two entangled particles are taken far apart and then measured too quickly for light signals to travel between them. The maths of quantum physics says that the correlation is still there. How can this be if there is no communication between them, and the results are probabilistic? He suggested there must be some "hidden variables" guiding the process, implying quantum physics was incomplete.

Since quantum physics described so well what scientists saw in experiments, such philosophical concerns were easily side-lined. But what is now called the EPR Paradox continued to rankle some researchers. One of those was John Bell, a physicist working at the European particle physics facility CERN. In the 1960s he described a way to test whether hidden variables exist.

Scientists soon performed such "Bell tests" and found that particles behaved just as quantum physics predicted, leaving no room for hidden variables. These experiments have continued, getting more precise and closing loopholes, with quantum physics always sailing through. Physicists are left to conclude that quantum physics really doesn't respect a 'local realistic' picture of the Universe: the theory is non-local. Remarkably, it turns out that we can make use of Bell tests — as in quantum cryptography.

Ekert had shown that quantum entanglement and non-locality, two unique properties of quantum systems (see box story, Quantum Fundamentals), can be used to distribute cryptographic keys with perfect security. His 1991 paper *Quantum Cryptography based on Bell's Theorem* has by now become the most cited paper in quantum cryptography.

Ekert's expertise had a particular appeal for the local group as it had all the elements that would make a good candidate for a grant proposal. While Shor's algorithm signalled a problem with secure communication, quantum cryptography offered a remedy. Quantum cryptography offers security for communication and computation that is rooted in physics rather than computational hardness, making it resilient to progress, be it technological or the development of new algorithms.

The first concerted effort to set up a quantum information group in Singapore began in 2002, when the group helmed by Ekert won an A*STAR Temasek grant of $5M for a proposal under the title *Quantum Information Technology (QIT)*.

The argument put forth was simple. Singapore, at that point, was already fast becoming one of Asia's leading countries in computer technology and with Europe and USA's increasing investments in quantum technologies, the move towards this was inevitable. It was clear that the long-term strategy in IT was to develop people with the right expertise.

At this time the group adopted the slogan *We do IT with qubits*. It also came to be known informally as 'quantum lah', borrowing the Singlish 'lah' (this expression lives on in the Centre's website, found at quantumlah. org).

Beyond the project's immediate goals, the local group had even bigger ambitions. They envisioned Singapore assuming a leadership position in areas such as quantum communication and cryptography.

The Rise of a Quantum Island

The investment began to pay off quickly. The project attracted more academics to Singapore: theorist Berthold-Georg (Berge) Englert and experimentalists Christian Kurtsiefer and Antia Lamas-Linares (see Figure 1).

On the theoretical front, the group made significant contributions in proposing novel quantum key distribution protocols. On the experimental front, the group demonstrated free space quantum key distribution with entangled photons and developed sources of polarisation entangled photons. More

NUS >> QIT Lab >> October 2004

Fig. 1.   Looking young! An early photo of members of the quantum information group in Singapore.

importantly, the project was crucial in establishing the infrastructure for experiments and technology for photonic quantum processing. The research generated over 70 papers in leading international journals.

By 2004, Singapore's research in quantum communication had even come to the attention of the media. *New Scientist* magazine hailed "the rise of a quantum island" on its 10 January cover.

Inside the magazine, the feature "We have seen the future…" looked at forthcoming developments in science — including plans in Singapore to develop quantum communication. It described a proposal to send entangled photons between the rooftops of buildings to "investigate how efficiently the photons can be detected through air turbulence, wind and rain". By 2009, the team had reported[6] "successful generation of an encryption key at a rate of a few hundred bits per second over several days and nights of variable weather."

From a broader perspective, the group's research was acknowledged by leading research centres and laboratories abroad. This led to close links with the National Institute of Standards and Technology (NIST, USA), the University of Cambridge and the University of Oxford (UK), the Ludwig-Maximillans University in Munich (Germany) and the Moscow State University (Russia).

Singapore's visibility as a new player in quantum technologies was also underscored by two popular international workshops: the 2nd Asia-Pacific Workshop on Quantum Information Science in December 2003 and the 3rd Asia-Pacific Workshop (in collaboration with the Institute of Mathematical Sciences) in January 2005.

In August 2006, at the conclusion of the QIT project, the group was given another boost with a second A*STAR project grant of $5M, under the leadership of Oh. The project on Quantum Information and Storage (QIS) cemented the group's visibility as a world player.

Birth of the RCEs

The year 2006 was also a critical year for the research landscape in Singapore. The Research, Innovation and Enterprise Council (RIEC),[7] chaired by Prime Minister Lee Hsien Loong, was set up to promote research, innovation and enterprise nationally, with a view of encouraging new initiatives in knowledge creation and catalysing new areas of long term economic growth.

The Prime Minister, in the inaugural meeting, noted that[8] *"Our priorities are firstly, to build up core R&D capabilities in selected strategic areas, and secondly, to attract and develop a significant concentration of talent to sustain a critical mass of advanced research activity into the long term."* This paved the way for the establishment of Research Centres of Excellence (RCEs) in the local universities.

Endorsed in 2007 by the RIEC, the RCEs were conceived as research centres that would focus on medium to long term world-class-investigator-led research.[9] Among other objectives, RCEs were primed to attract, retain and support world-class academics who would catalyse and transform the local universities into research-intensive universities. Through engendering interest in research among local students and encouraging them to pursue research careers, it was envisioned that this initiative would raise the international standing of the local universities. A key difference from previous initiatives was that the focus was primarily on investigator-led programmes as opposed to mission-oriented ones.

In tandem with the developments at the national level, NUS also started to ramp up its support for programmes that were of strategic interest. The Quantum Information group under Ekert's leadership was immediately identified as a potential candidate for the impending RCE proposals. With a $1.5m strategic budget allocation the group embarked on further consolidating its efforts by drawing another two academics to its ranks: Murray Barrett, an experimentalist who had extensive experience in working with trapped ions and Valerio Scarani, a theorist who had worked in quantum cryptography and foundations of quantum physics.

Shortly after, in September 2006, in response to a national call for the establishment of RCEs, the Quantum Information group submitted a white paper outlining the group's vision of conducting interdisciplinary theoretical and experimental research. Thereafter the group was invited to submit a full proposal. In a matter of a few months, at the recommendation of the Ministry of Education's (MOE) Academic Research Council and the National Research Foundation (NRF), the proposal was put up to the RIEC for their endorsement. This essentially paved the way for the establishment of Singapore's first RCE — the Research Centre of Excellence in Quantum Information Science and Technology (RCE in QIST). The name was subsequently changed to the Centre for Quantum Technologies (CQT). The Centre was allocated $158m in core funding from MOE and NRF, with a further award of $36.9m from NRF

Box Story: How it happened, as told by Artur Ekert

There is no narrative, simple or embellished, that can possibly capture the pioneering spirit of those days when Singapore took the first step towards becoming a quantum island. It is not even clear by who and how and when it was decided to give it a go. You might ask me how I joined this endeavour. There is no good answer to that question either. Or perhaps there is, but it is personal, incoherent and far too long to be presented here, so let us just say that it happened. It happened in a truly quantum fashion, in more than one way.

One narrative will lead us to the corridors of power. The year was 2000 and Tony Tan, at the time Deputy Prime Minister and Minister for Defence, took a personal interest in the Millennium Conference on Frontiers in Science (suggested by KK Phua) at which I gave an overview of quantum computation. My subsequent meeting with Tony Tan, and a long over-dinner conversation, left me confused. I was not used to politicians talking in an intelligent and persuasive way about big problems in the upcoming millennium that science can address. I was impressed.

Another narrative will take us to the hawker stalls and numerous Indian eateries that I frequented with Kwek Leong Chuan and Kuldip Singh, talking mostly about quantum physics. Their enthusiasm for this new emerging field was contagious. The journal club they used to run brought together faculty members, to mention only Lai Choy Heng, Oh Choo Hiap and Tan Eng Chye, who despite heavy admin duties, found time to discuss science. I was impressed.

(Continued)

Then I can tell a story about Chan Chui Theng, a brave woman who decided to risk her job security in order to offer temporary administrative support (now in its tenth year) to a wacky ang moh talking some quantum mumbo-jumbo, and also about her younger assistant, Evon Tan, who from day one defied causality (things got done before anyone asked for them). I was impressed.

Yet another narrative must include a wonderful duo from the Ministry of Education, Benny Lee and Perry Lim, who spent days drafting policies for our newly created centre, doing their best to protect us from red tape. I shall not tell you any anecdotes about them, of which there are many, for they are very serious people today (Perry is Chief of Defence Force of the Singapore Armed Forces and Benny is Principal Private Secretary to the President) but they stayed in touch, showing a genuine interest in our work. The two of them redefined for me the meaning of civil service. How could I possibly not be impressed.

Or perhaps I can tell you a story about one outreach event, during which Christian Kurtsiefer explained the magic of quantum interference to a Singaporean boy. The moment was nicely captured by Antia Lamas-Linares (see picture). The glow of thoughtful curiosity and a sense of wonder that emanated from his face made a big impact on me. You can almost feel his inquisitive mind in action. At this very moment I understood that no matter how many groundbreaking papers we publish, no matter how many patents we file, and no matter how many quantum contraptions we design, there would be another, less tangible but perhaps more important, legacy. We can shape how the next generation sees and

(*Continued*)

in 2014, to fund its first ten years of operation. Ekert became the Centre's founding Director. He shares his memories of this time in the box story "How it happened, as told by Artur Ekert".

The Heroic Months

In the early months of 2007, the group swung into full action. The mandate was to have a Centre up and running before the year was up — a challenge that was by no means trivial. Some members of the group fondly recall them as the "heroic months". The room that was designated to the group even came to be called the "War Room".

On the administrative front, the group had to work from ground zero. From the outset, the RCEs were planned to be units within a host university

understands the world, how they think and solve problems and how they will make a good use of quantum technologies. Singapore offered this opportunity on a silver platter. How could I possibly ignore it!

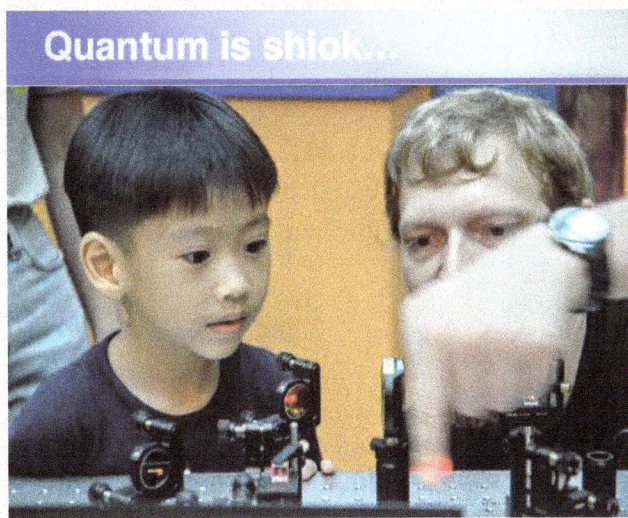

Quantum is shiok...

And I can tell many more stories, big and small, funny and sentimental, but none of them giving a truly comprehensive answer to the question of how it all happened. It just happened and I am so glad that it happened, for I truly believe we were given an opportunity to create something exceptional.

that would have significant autonomy in pursuing their research agenda. The rationale was simple. It was acknowledged that, given the competitive nature of research, the flexibility to strategize and innovate was a necessary prerequisite for such centres to compete on a global stage.

For the group, this meant drawing up policies, establishing committees and getting the physical infrastructure up. A first formal meeting in early March 2007 with all stakeholders marked the start of a process that would ultimately lead to the establishment of the centre. The RCE would have a governance framework different from the other centres and institutes in the university. This difference included the establishment of a Governing Board that would be empowered to approve policies that deviated from those of the host.

On the infrastructure front, the group was faced with an even a bigger challenge — that of getting labs, workshops, offices and seminar rooms functional in a short span of time. Right after the formal announcement of the centre, four floors of S15, a building occupied by the School of Computing (SoC) was identified as the Centre's physical space.

The specialised requirements of quantum laboratories made the challenge bigger. For one, air-conditioning designs had very stringent requirements in terms of temperature fluctuations. Shifts in temperature can nudge optical components out of alignment — an issue that means even today, some CQT researchers prefer to take data at night. Such specialised needs meant engaging contractors that could meet strict standards. The challenges were compounded by the constraint that renovation could only begin after September 2007. Fortunately, under very trying circumstance, the group was able to complete the renovations in a short span of three months.

CQT Launches

The flurry of activities before the Centre's official launch was punctuated with visits from important officials and formal meetings. The Chairman of NRF at the time, Dr Tony Tan (Figure 2) held a media briefing in May 2007, formally announcing the decision to set up the Centre. This was followed by a visit by the then President S.R. Nathan in July 2007 (Figure 3), who took a tour to see early

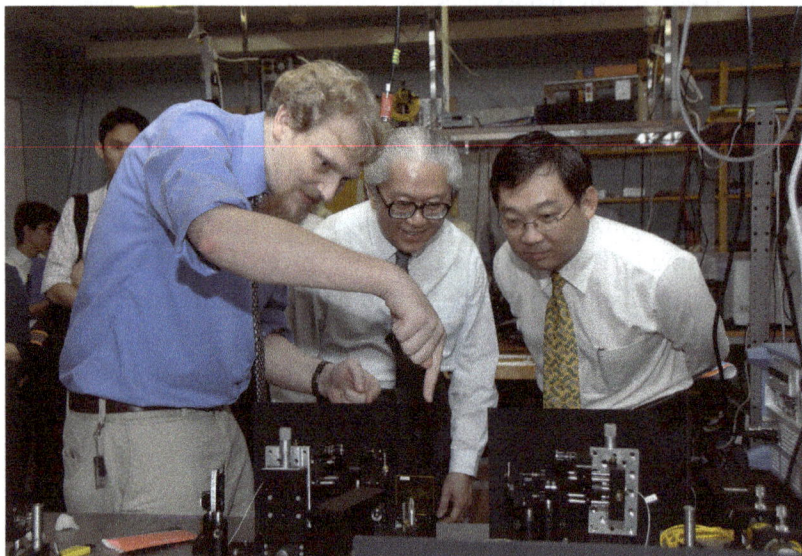

Fig. 2.   Tony Tan (centre), President of Singapore since 2011, is pictured touring the quantum optics laboratory at NUS in 2007 when the decision to set up CQT was announced. At that time, he was chairman of Singapore's National Research Foundation.

Fig. 3.   Then President S. R. Nathan (front, right) visited the quantum information group at NUS in 2007 while preparations for the launch of CQT were underway. Here he is being presented with a model qubit — a quantum bit existing in a superposition of the states 0 and 1 — by CQT's soon-to-be director, Artur Ekert (front, left).

setups of free-space quantum cryptography kit, and other photonic and atomic systems. Much later, the Centre would also host visits by Singapore's Deputy Prime Minister Teo Chee Hean, in 2012 (Figure 4), and the Prime Minister Lee Hsien Loong, in 2013 (Figure 5).

The Centre had its first Pro-tem Governing Board meeting in August 2007. Chaired by Mr Lam Chuan Leong, who was then Ambassador-at-Large with the Ministry of Foreign Affairs, the meeting formalised the governance principles and practices for the Centre. The Centre's Scientific Advisory Board (SAB) was also established. (see box story, How CQT is Structured).

The heroic months culminated with the Centre for Quantum Technologies' official launch on 7th December 2007 (Figure 6).

Science from Day One

Following the official launch, the Centre embarked on its quest to explore the quantum nature of reality and the limits of information processing. Led by an initial group of 12 Principal Investigators (PIs) and a small group of researchers (numbering 27), the science took off from day one.

Fig. 4.   In 2012, Singapore's Deputy Prime Minister Teo Chee Hean (pictured right), visited CQT in his capacity as Chairman of the board of the National Research Foundation.

Fig. 5.   In 2013 CQT received a visit from Singapore Prime Minister Lee Hsien Loong (pictured right, in pink shirt) with a delegation from the National Research Foundation. He later posted on his Facebook page: "…There are many potential uses for CQT's research, e.g. for online banking and secure internet transactions. I hope talented young people interested in physics and computing will learn more about these subjects, do research and make new breakthroughs one day!"

Box Story: How CQT is Structured

Being an autonomous research unit means the Centre has a self-contained organisational structure with a Governing Board (GB), Scientific Advisory Board (SAB) and Executive Committee comprising the Centre's Principal Investigators.

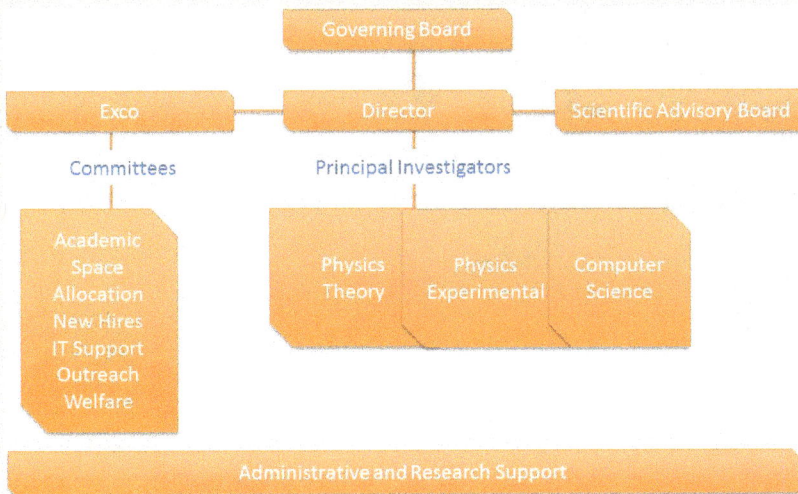

Governing Board

Exco — Director — Scientific Advisory Board

Committees   Principal Investigators

Academic
Space
Allocation
New Hires
IT Support
Outreach
Welfare

Physics Theory   Physics Experimental   Computer Science

Administrative and Research Support

The Governing Board has been chaired by Mr Lam Chuan Leong since its inception. Mr Lam is a Senior Fellow at the Lee Kuan Yew School of Public Policy, NUS and Director, ST Electronics (Info-Software) Systems Pte Ltd. He formerly held senior roles in Singapore's civil service. In an interview in 2011, he said: *"I think CQT set an excellent model for research that is not immediately to be applied. It is a significant milestone in that it demonstrates Singapore's greater appreciation of fundamental knowledge and not just the immediate practical returns. The results are good for humanity and not just for short run business benefits. This sort of value resonates in the hearts of all true scientists."*

The SAB comprises seven eminent scientists with research expertise overlapping CQT's activities. They make an annual visit to the Centre to discuss its scientific direction and management, offering advice and a formal report.

In the first year of operation, the Quantum Optics Group, headed by Kurtsiefer and Lamas-Linares, in collaboration with Scarani, demonstrated some notable results. They falsified a model of quantum mechanics proposed by Nobel-laureate Anthony Leggett. This non-local variable model would have allowed photons to have well-defined if unknown polarisation, in contrast to the orthodox quantum picture where this property is indeterminate until measured. This

Fig. 6. Counting down to CQT's official launch on 7 December 2007. The occasion was presided over by then NUS President Professor Shih Choon Fong.

experiment received significant attention in the international community and resulted in a National Science Award for the three PIs.

The year 2008 also saw the inception of the CQT PhD Programme. One of the Centre's principal missions includes the training of graduate students. The programme had its first intake of nine students in that year (see box story, CQT's PhD Programme).

There were other big milestones for the Centre in its early years too. In 2009, Barrett's group earned CQT the nickname of "the coolest place on the equator" by creating the Centre's first Bose-Einstein condensate in the lab (Figure 7). A condensate is a quantum form of matter in which particular kinds of atoms — bosons, those with integer spin — fall into a shared wavefunction and act in unison. It happens only when the atoms are cooled to within fractions of a degree of absolute zero ($-273°K$).

The feat wasn't new to science. Such condensates were predicted in the 1920s and first created in the mid-90s. Making them still wasn't easy, however, meaning that CQT's condensate marked the Centre's arrival in the international cold atoms research community. Barrett and his then PhD student Kyle Arnold found themselves featured in Singapore newspaper the *Straits Times*.

On the theoretical side, the Centre's researchers also made a splash. Particularly notable were a paper in the prestigious journal *Nature* in 2009 and a best paper award at the computer science conference known as STOC in 2010.

Box Story: CQT's PhD Programme

The Centre was established with a mission to train 80 students over its first ten years — which it's on target to achieve. CQT celebrated the first student to graduate from the CQT programme in 2012.

Students at CQT are trained to a high standard, developing not only expertise in the academic areas of quantum physics and computer science, but also general technical and problem-solving skills. They are encouraged to use their initiative and explore their topic areas.

Most CQT graduates have taken postdoctoral positions in academia as their next jobs but a few have moved into industry and government too, including positions at the Ministry of Defence and DBS Bank. Students graduate with degrees from the National University of Singapore.

Arun (right) was the first CQT PhD student to defend his thesis. He graduated in 2012 for work on "Hybrid Quantum Computation". He is pictured here with his supervisor, Berge Englert, who is also the head of CQT's graduate programme.

Some of CQT's graduates in 2013 (looking at camera, Han Rui and Kyle Arnold) pictured at the NUS commencement ceremony.

(A)

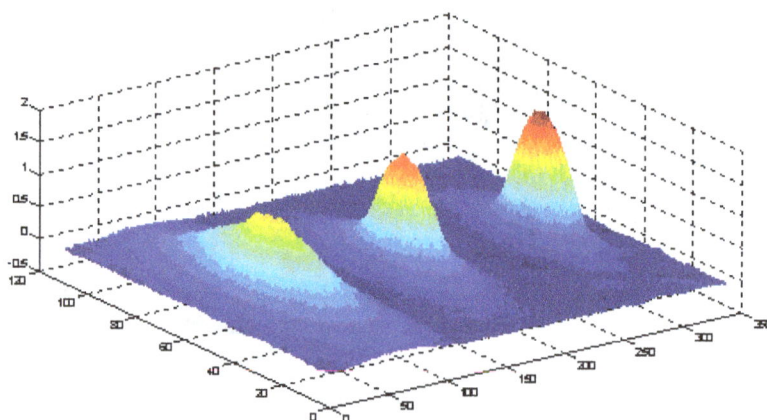

(B)

Fig. 7.   In 2009, Murray Barrett (top, left) and his then student Kyle Arnold (top, right) used the set-up pictured (A) to create a quantum form of matter known as a Bose Einstein condensate. The graph (B) shows absorption images of cold atom clouds. On the left the atoms are mostly in the thermal phase, and on the right in condensate phase.

The *Nature* paper, by CQT's Tomasz Paterek, Andreas Winter, Dagomir Kaszlikowski, Scarani and their collaborators, put forward a new physical principle the authors dubbed "information causality". In the same way that Einstein formulated special relativity from the idea that the speed of light is constant, even for a moving observer, some researchers in the quantum field hope to find simple principles underpinning the mathematics of quantum physics.

In this case information causality[10] fell short of defining quantum effects exactly, but did better than other principles that have been considered — such as simply requiring that information can never travel faster than light. With just the 'no-signalling principle', the correlations seen between entangled particles could be even stronger than those allowed by quantum physics. The backstory of this result exemplifies what the Centre hopes to achieve: a concentration of talent and collaboration that makes the sum of its research greater than its parts (see box story, The Story of Information Causality)

One of the defining features of CQT is that it has computer scientists among its staff. This interdisplinary mix is not found in all quantum centres. Here, computer scientists work in tandem with physicists to explore the potentials and limits of quantum computation, developing new algorithms and protocols for communication and computation that exploit the extraordinary features of quantum mechanics. Back in 2010, the STOC best paper award to CQT's Rahul Jain and his collaborators announced their presence.

STOC is the Annual ACM Symposium on Theory of Computing, considered one of the top meetings in the field. Jain and his colleagues had proved[11] the equivalence between two classes of computation problems, a landmark discovery resolving a decade-old problem in complexity theory. Essentially, they showed that quantum interactive proof systems provide no more computational power than classical ones (QIP=IP).

Fast Forward

Over the years, the Centre grew. It has hosted visitors from all over the world, and organised workshops and international conferences (see box story, Conferences). The Centre is also active in outreach, from supporting books (see box story, Quantum Reading), to hosting artists (Figure 8) and taking part in public and school events (Figure 9).

The science continued apace too. There have been new breakthroughs in the areas of strength evident in the Centre's early years — in quantum foundations, with new understanding of the quantum uncertainty principle, in quantum communication, with the successful launch of a satellite testing technology for a global quantum network and in computer science, overturning long-held beliefs about query complexity, including showing the

Box Story: The Story of Information Causality

What does quantum theory tell us about the nature of our underlying reality? Do its rules emerge from more intuitive principles? In 2009, CQT researchers contributed to a paper in *Nature* proposing that information causality could be a fundamental principle.

CQT's Dagomir Kaszlikowski recalls his first encounter with the idea of 'information causality' being over coffee with a visitor from Poland, Marcin Pawlowski. He had an intuition, but no mathematical proof, that the idea could be important. They worked together, roping in other researchers. "Days passed and we still couldn't find a complete solution. We knew intuitively that Marcin was right and we knew the way to prove it but we were missing various bits and pieces necessary to close all the loopholes in the reasoning. It was exasperating. Suddenly I realised we had one more chance: Andreas Winter," wrote Kaszlikowski in an account for CQT's annual report. "When I presented this problem to him he simply said 'Beautiful'. The next morning, he had the full proof."

To understand the concept, imagine someone is sending you a 10Mb file from a travel guide. They don't know what you're interested in so they pick the pages at random. Could they send the data in such a way, using quantum correlations, that you could choose any 10Mb of information from the book, or would you be stuck with the pages they'd picked? It turns out that requiring you're stuck with the chosen information comes close to matching what quantum physics allows. More

(Continued)

Box Story: Conferences

These are some of the major international conferences and schools organized or supported by the Centre for Quantum Technologies (not listing a host of smaller and local workshops):

- International Conference on Quantum Communication, Measurement and Computing, 4-8 July, 2016
- 22nd International Conference on Laser Spectroscopy, 28 June–3 July, 2015

(Continued)

Box Story (*Continued*)

technically, information causality can recover the Tsirelson bound in an intuitive way, without invoking Hilbert space formalism.

Researchers (pictured left-right) Dagomir Kaszlikowski, Valerio Scarani, Andreas Winter and others collaborated to come up with mathematical evidence for there being a principle of information causality.

Box Story (*Continued*)

- Theory of Quantum Computation, Communication and Cryptography, 21–23 May, 2014
- QCRYPT, 10–14 September, 2012
- Quantum Discord Conference, 9–13 Jan, 2012
- 5th Asia-Pacific Workshop on Quantum Information Science, 25–28 May 2011
- 14th Workshop on Quantum Information Processing (QIP), 8–14 Jan, 2011
- Les Houches School of Physics in Singapore, Ultracold Gases and Quantum Information, 29 June–24 July, 2009

Box Story: Quantum Reading

In 2009, CQT Principal Investigator Valerio Scarani is at a poster session for students of the NUS High School of Math and Science when he has an idea for a project. He invites two students to help him co-author a booklet on quantum physics for high school students. "Our task was to write the concepts at a level targeted at high school students," recalled Chua Lynn — one of those students.

Six Quantum Pieces: A first Course in Quantum Physics by Valerio Scarani, Chua Lynn and Liu Shi Yang was published by World Scientific in 2010. The book is now used in some quantum courses, including at CQT in programmes for pre-University students enthusiastic in physics and maths.

Targeting an even younger audience (and the young at heart) is the book *Sir Fong's Adventure in Science Book 5: The Quantum Bunny*. The result of a collaboration with Otto Fong, a Singaporean comic book author, this book retells the Chinese legend of the monkey king causing uproar in heaven with physicist characters and quantum plot elements. This book was launched at the Singapore Writer's Festival in 2015.

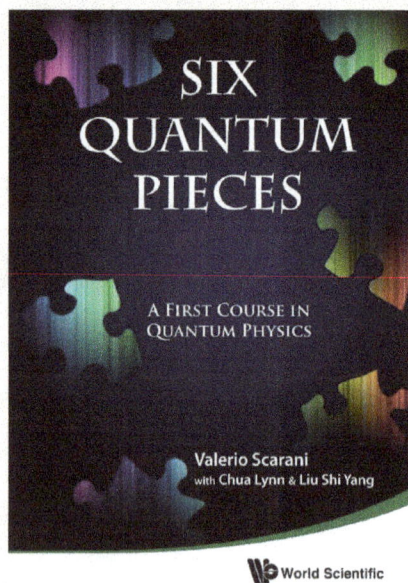

"The art is fantastic, the science is wild (but true), and the bunny is incredibly cute. I'm in awe of Otto Fong!"

Larry Gonick, creator of educational comics including *The Cartoon History of the Universe*

greatest quantum advantage yet known for a total function (see box story, More Research Highlights).

Meanwhile, CQT's world of quantum has expanded to new topics (Figure 10). On the experimental side, from its early focus in quantum

Fig. 8.   The Centre has offered "Quantum Immersion" residencies for artists and writers. This picture from 2011 is a prototype installation of Timensions by Linda Sim Solay and Dario Lombardi, the Centre's first artists in residence. The work was inspired by the many worlds interpretation of quantum theory and ideas from string theory.

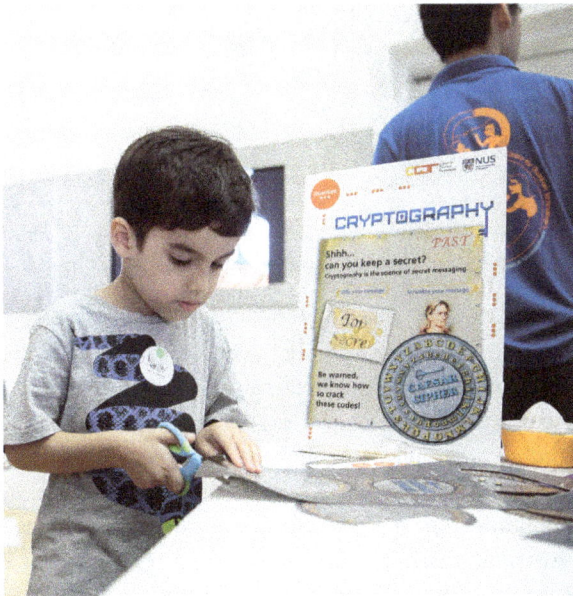

Fig. 9.   A young visitor learns about cryptography — classical to quantum — at a one-day exhibition organised by CQT at the ArtScience Museum with the NUS Science Demonstration Laboratory. The event was held in December 2015.

Box Story: Research Highlights

Probing Uncertainty: Although quantum physics has for 100 years successfully described the phenomena of Nature, its counterintuitive features continue to be a rich source of inspiration for fundamental research. Researcher Stephanie Wehner, then a Principal Investigator at CQT, contributed to unravelling a deep connection between the uncertainty principle and non-locality in quantum theory. Pursuing investigation of the uncertainty principle further, Wehner also showed the violation of the uncertainty principle implies a violation of the second law of thermodynamics and that wave-particle duality is equivalent to the uncertainty principle.

Work by Stephanie Wehner and her collaborator showed that the uncertainty principle puts a limit on the amount of non-locality allowed by quantum physics. The result was published in *Science*.[12] Cartoon by Frans Bartels, concept by Haw Jing Yan.

Aiming for orbit: From pioneering work in device-independent quantum information processing to the first implementation of bit commitment, the Centre is prominent in setting the agenda of and leading worldwide research efforts in the field of quantum cryptography. One high-profile project is the development of a compact and robust entanglement source to fit into a nanosatellite. Distributing entangled photons from space is one idea to extend the distance over which quantum key distribution can be performed. This project is led by Principal Investigator Alexander Ling, a former PhD student of the quantum information

(*Continued*)

Box Story (*Continued*)

group in the pre-CQT era, and has involved collaboration with the NUS Centre for Remote Imaging, Sensing and Processing and the Department of Electrical and Computer Engineering. The group's first device launched on an ill-fated rocket in October 2014 — the rocket exploded on the launchpad. On the second attempt, however, in December 2015, the device reached space and successfully returned data. This is the first test of a quantum entanglement setup in the challenging space environment.[13]

CQT researchers have launched a sandwich-sized source of entangled photon pairs into orbit on a nanosatellite, with the goal of developing technology for a global quantum communication network.

Query complexity: CQT computer scientists Troy Lee, Miklos Santha and their collaborators busted a hypothetical ceiling on quantum speed-ups,[14] the theoretical speediness of a quantum versus a classical program in solving a problem. Whereas the previous best quantum advantage for 'total functions' was the quadratic speed-up of Grover's algorithm — for searching unsorted databases — the new speed-up is that squared: it's quartic. The quadratic separation of Grover's algorithm had remained the best known separation for total functions since the algorithm was described in 1996, despite intensive search for something better. Given this situation, researchers thought Grover's might have hit a limit. The team's work uprooted this belief. The new 'super Grover' algorithm does not solve any particular real-world problem, but its existence raises the possibility that a super-quadratic quantum speed-up is possible for useful functions.

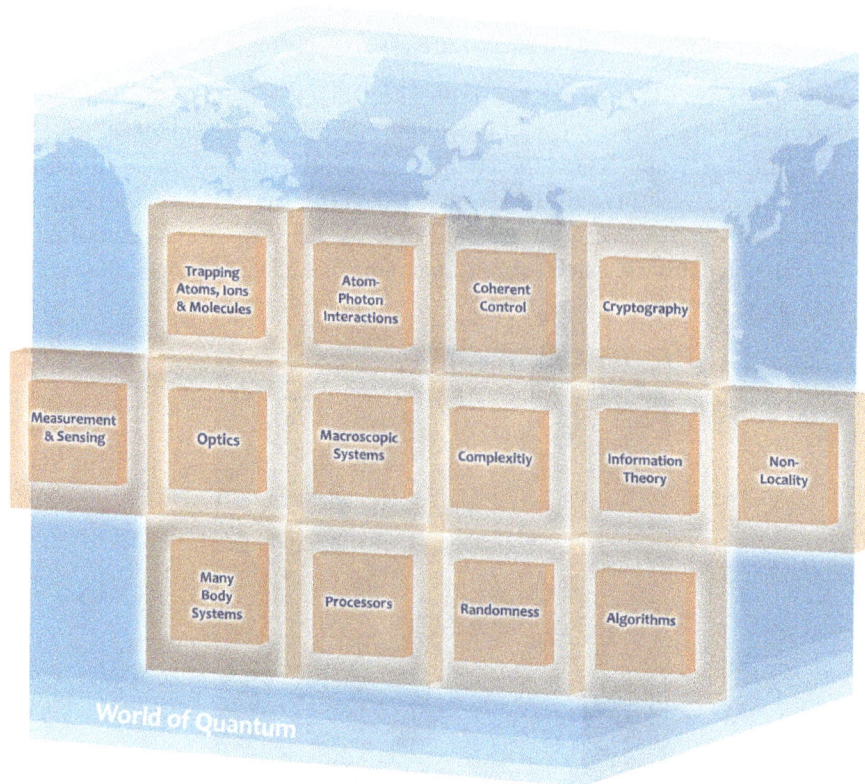

Fig. 10. This graphic illustrates the range of topics under research at CQT in 2016.

optics, the Centre's projects diversified, fuelled by the range of active research topics in the field and the many different routes for building devices that harness quantum phenomena. The exquisite control of photons was complemented by control over individual atoms, ions and molecules. CQT added ultra-cold gases, NV centres in diamonds and atom chips to its experimental expertise. CQT researchers also began working to build 'hybrid' quantum systems, with interfaces between light and matter. It takes years to build the kinds of setups required to do cutting-edge experimental work in quantum physics (Figure 11). As these experiments came online, research results followed.

By the time MOE convened the Centre's second international review in 2015, its researchers had contributed to over 1000 journal publications. The Centre was by then at a steady state size of some 170 staff and students, with research led by around 20 Principal Investigators and a small number of CQT Fellows.

Fig. 11.   This 360° panorama shows the density of equipment inside a quantum lab. This setup is to study quantum matter. The red laser on the left-hand table helps cool atoms close to absolute zero. The green laser on the right-hand table controls the atoms for experiments to help us understand phenomena such as superconductivity and magnetism. This project is run by CQT Principal Investigator Kai Dieckmann. Photo by Daniel K. L. Oi.

The Future

The Centre's 2015 international review came at a critical juncture, as the end-point of its initial ten years of funding approached. The continuation of the centre beyond the end of 2017 depended on a favourable assessment. Thick files of reports, statistics and papers were compiled for the visiting panel of distinguished scientists, who spent four days at CQT meeting with its scientists and stakeholders.

Thankfully, the panel reported that it was "impressed", summarising its general report as follows: "*The International Review Panel has been impressed by the achievements of the Centre for Quantum Technologies. In a short time span, it has attracted world-class researchers, both senior and junior, established a strong relationship with local Universities and built the requisite infrastructure to support the Centre. These elements culminated in making a strong scientific impact in the international community and built a reputation of excellence for Singapore.*"

Singapore's Academic Research Council, which advises on research spending, considered the panel's review as one input in deciding the Centre's future. It's thanks to this Council that the history does not end here: CQT received a commitment to a further five years of core funding from NRF and MOE and further support from NUS. The Nanyang Technological University (NTU) is also supporting some CQT staff and laboratories on its campus. CQT also holds competitive grants from both local and international funders, which support projects across the Centre.

Elsewhere there is new funding for quantum technologies too. Big corporations such as Google, Microsoft and Intel are investing in quantum. Proposals extend from the national to multinational: Europe in 2016 approved a 1-billion flagship initiative supporting quantum technologies research across the continent, on top of existing national programmes. Quantum centres are multiplying.

The hope is that research in this field will lead not just to quantum computing — that's probably still years away — but a whole new wave of technologies for communication, simulation, sensing and measurement.

What can we expect in the Lion City? Ekert, in his Director's letter for CQT's 2015 annual report, wrote: "While going through the review process I realised that the 2022 horizon, six years away, is almost another era in this rapidly moving field. So much progress has been made in the past six years (not necessarily in the directions we originally anticipated) that one can hardly speculate what will happen by then. Of course, I can stretch my imagination and suggest quantum random number generators in local casinos, island-wide quantum key distribution networks, quantum simulators which help to design new drugs, and super-precise atomic clocks leading to a super-accurate global positioning system. But the thing is, each year brings surprises.

...Even though I cannot tell you what exactly we will be working on six years from now, I do know that we will be doing something interesting. I can say this with some degree of certainty because I see genuine quality and potential in our team."

More information about the Centre for Quantum Technologies can be found at quantumlah.org. The Centre's activities since 2007 are documented in detail in annual reports available via the website.

References

[1] Logical reversibility of computation. *IBM Journal of Research and Development,* **17(6),** 525–532 (1973).

[2] Quantum mechanical hamiltonian models of turing machines. *Journal of Statistical Physics,* **29(3),** 515–546 (1982).

[3] Simulating physics with computers. *International Journal of Theoretical Physics,* **21(6/7),** 467–488 (1982).

[4] This is the precursor of the current Asia Pacific Conference and Workshop on Quantum Information Science.

[5] The Temasek Professorship Programme, an A*STAR initiative, is a platform which serves to draw renowned international R&D research leaders to lead strategic research projects identified as imperative to Singapore's scientific and economic development.

[6] Daylight operation of a free space, entanglement-based quantum key distribution system. *New J. Phys.* **11**, 045007 (2009).

[7] http://www.nrf.gov.sg/about-nrf/governance/research-innovation-and-enterprise-council-(riec)

[8] http://www.nas.gov.sg/archivesonline/data/pdfdoc/20060707990.pdf

[9] http://www.nas.gov.sg/archivesonline/data/pdfdoc/20070316997.pdf

[10] Information causality as a physical principle. *Nature* **461**, 1101 (2009).

[11] QIP=PSPACE. *Proc. 42nd ACM STOC* 573 (2010), also invited to *Journal of the ACM* doi:10.1145/2049697.2049704 (2011).

[12] The Uncertainty Principle Determines the Nonlocality of Quantum Mechanics. *Science* **330**, 1072 (2010).

[13] Generation and Analysis of Correlated Pairs of Photons aboard a Nanosatellite. *Physical Review Applied* **5**, 05402 (2016).

[14] Separations in Query Complexity Based on Pointer Functions. Preprint at http://eccc.hpi-web.de/report/2015/098

Chapter 22

2D Materials*

Andrew T S Wee, Kian Ping Loh and Antonio H Castro Neto

Beginnings of 2D Research in Singapore: NUS Surface Science Laboratory (1986–2005) [A Wee]

Surfaces, the two dimensional (2D) termination of a bulk solid, have been notoriously difficult to study. The Physics Nobel Laureate Wolfgang Pauli was quoted as saying:

"God made the bulk; surfaces were invented by the devil."[1]

Indeed the studies of solid surfaces and 2D materials have historically been difficult to study due to their sensitivity to the external environment at the interface, be it a gas, liquid or another solid. Nevertheless, surfaces are critically important in modern technology, such as in the fields of heterogeneous catalysis, optical and electronic devices, protective coatings, adhesion, sensors, energy storage and generation and so on.

The field of surface science encompasses the study of physical and chemical phenomena that occur at the interface of two phases, including solid–liquid interfaces, solid–gas interfaces, solid–vacuum interfaces, and liquid–gas interfaces. It started to gain prominence in the 1970s, when ultrahigh vacuum technology was maturing and it was possible to study surfaces reproducibly in well-controlled conditions at the solid-vacuum interface. Today the fields of surface chemistry and surface physics underpin the foundations of surface engineering applied to the many technologies as mentioned in the preceding paragraph.

The story of surface science and 2D research can be traced to the establishment of the Surface Science Laboratory (SSL) in 1986 at the Department of Physics, National University of Singapore.[2] The first ultrahigh vacuum (UHV) system delivered in 1987, possibly the first in Singapore, was a Vacuum

---

*Adapted from *50 Years of Materials Science in Singapore*. Published by World Scientific Publishing, Copyright © 2016 by World Scientific Publishing.

Generators (VG) ESCALAB Mk 2/SIMSLAB (Figure 1), acquired with a Science Council grant of just under S$1 million. This effort was initiated and driven by then NUS Deputy Vice-Chancellor, Prof Huang Hsing Hua, who is a distinguished chemist himself and who had a clear vision of the importance of surface science. The first Director, physics Professor Tan Kuang Lee, and his chemical engineering collaborators Professors Kang En-Tang and Neoh Koon Gee did pioneering research in the photoelectron spectroscopy of polymers on this instrument, and their achievements were recognised when they won the 1996 National Science Awards (Figure 2).

This author (A Wee) joined NUS in 1990, and participated in the development of the Surface Science Lab from a single UHV system to several state-of-the-art UHV systems today. As one of the earliest major research laboratories in NUS, the Surface Science Laboratory is recognised today as one of the world's leading surface science groups producing numerous international journal publications, patents and international collaborations. The lab also actively engages the local industry, and has organised surface and interface analysis workshops to raise Singapore companies' competency in state-of-the-art materials characterisation techniques (Figure 3).

The laboratory currently houses five ultrahigh vacuum (UHV) systems (as of mid-2015), including the ESCALAB, Omicron low-temperature scanning tunneling microscope (LT-STM), Omicron variable-temperature (VT)-STM, ultraviolet/x-ray photoelectron spectroscopy (UPS/XPS) system, and Cameca IMS 6f Secondary Ion Mass Spectrometry (SIMS) system, along with an array of thin film growth systems, and other characterisation equipment. In 2000, we received an NSTB[3] grant to set up the Surface-Interface-Nanostructure-Science (SINS) beamline and end-station (Figure 4) at the Singapore Synchrotron Light Source (SSLS).[4] The synchrotron source produces highly monochromatic and tunable x-rays and serves as a unique probe to study surfaces and interfaces.

Our journey into graphene and 2D materials research was serendipitous. In the early 2000s, PhD candidate Chen Wei, currently an Associate Professor in NUS Chemistry, was investigating the surface structure of the carbon nanomesh phase formed on top of an annealed silicon carbide surface, SiC(0001). He was first author of a 2005 paper published in the journal *Surface Science* that proposed the structure of the carbon nanomesh, "whereby isolated carbon islands one atomic layer thick assemble to form the nanomesh structure" (Figure 5).[5] These one atomic layer thick carbon islands were actually *graphene* islands, and we had not recognised then the significance of graphene, the reason being that the landmark 2004 paper by Physics Nobel Laureates Novoselov and Geim had only just been published.[6] As a result of the surge of interest in graphene in the years that followed, our paper eventually became a top cited article in *Surface Science* (2005–2010), serendipitous indeed!

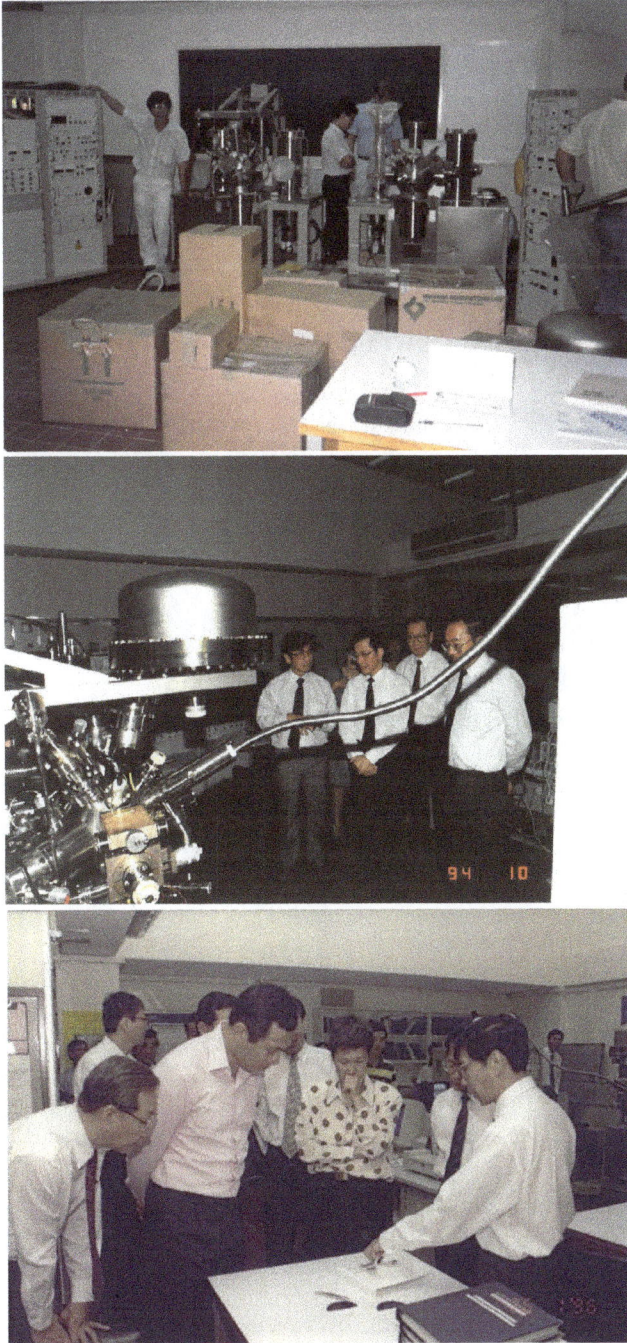

Fig. 1. [Top] Delivery and installation of the VG ESCALAB Mk 2/SIMSLAB at the NUS Surface Science Laboratory; Professor Tan Kuang Lee is standing on the left. [Middle] Prof Tan hosting a 1994 visit of the then Minister for Education, Mr Lee Yock Suan, accompanied by the then Vice Chancellor Professor Lim Pin. [Bottom] Prof Tan hosting a 1996 visit of PM (then DPM) Lee Hsien Loong.

Fig. 2. [Top photo, left to right] Professors Tan Kuang Lee, Kang En-Tang, Neoh Koon Gee posing with the VG ESCALAB Mk 2/SIMSLAB system; and [bottom] winning the 1996 National Science Award. (adapted from *lian he zao bao* article 3 Sep 1996).

RESEARCH TRENDS

Vol 4 No 2   A Publication of the Centre for Industrial Collaboration, Faculty of Science, NUS   July/Aug 1992

ISSN 0129-1890

Surface and Interface Analysis Workshop

The Surface and Interface Analysis Workshop was the first of its kind jointly organised by the Centre for Industrial Collaboration (CIC), the Department of Physics and the Institute of Physics. The workshop was designed to meet the growing needs of industry in materials and surface analysis, and give a comprehensive coverage of modern methods used to characterise surfaces and interfaces. It was held over two days in the Faculty of Science. A total of 18 participants from industry, NTU and SISIR attended the workshop including one participant from Malaysia.

At the opening ceremony, The Dean of the Science Faculty, A/P Bernard Tan, gave a warm welcome to the participants and stressed the importance of collaboration between the University and industry. He also made known to participants the role of CIC in the Faculty and encouraged greater interaction between the Faculty and industry in areas of common research interest.

The Deputy Vice-Chancellor,

Professor Huang Hsing Hua, in his opening address then gave an account of how the field of surface and interface analysis had progresses over the past years, as reflected by the scale of the biennial European Conference on Applications of Surface and Interface Analysis (ECASIA). He highlighted four important areas of activity: grain boundary embrittlement, microelectronics, adhesion and tribology and corrosion protection. He added that this workshop would provide opportunities for mutual learning and interaction amongst participants and the possibility of collaborative research could be explored.

During the morning sessions, participants were introduced to a wide range of surface and interface analysis techniques available in the Department of Physics, NUS. On the first morning, lectures were given on X-ray photoelectron spectroscopy, Rutherford backscattering spectroscopy/proton induced x-ray emission and scanning electron microscopy. The lectures on the second morning

covered scanning Auger microscopy, secondary ion mass spectrometry and surface analysis in thin films. The afternoon sessions were devoted to practical demonstrations where some samples brought by participants were analyses by the various techniques. Participants were thus made more aware of the capabilities as well as limitations of the techniques applied to their specific problems.

The workshop achieved its goal of bringing together academics and industrialists, providing a platform for discussion of possible collaborative work. The diverse backgrounds of the participants reflected the widespread interest in this important field. ■

Inside

Professor Huang Hsing Hua (Deputy Vice-Chancellor, NUS) delivering his opening address at the workshop.

Fig. 3.   The NUS Faculty of Science newsletter highlighting the Surface and Interface Analysis Workshop organized for industry participants in 1992.

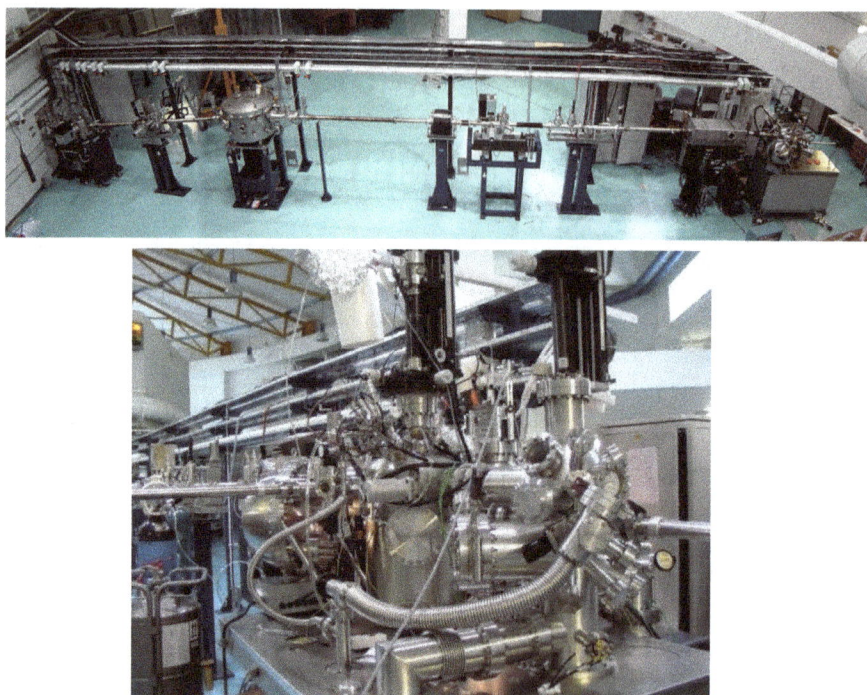

Fig. 4.   The Surface-Interface-Nanostructure-Science (SINS) beamline [top], and end-station [bottom] at the Singapore Synchrotron Light Source (SSLS).

Since then the Surface Science Laboratory has published numerous papers on graphene, 2D monolayers, and hybrid organic-inorganic heterostructures. Our LT-STM system has been particularly instrumental in allowing us to visualise atomically resolved images of graphene and adsorbed molecules. Figure 6 shows a photograph of our LT-STM with the growth chamber attached, as well as a typical STM image of epitaxial graphene, which was highlighted in *Nature News* on 25 March 2009.[7] As of 2015, we are currently engaged in STM studies of 2D transition metal dichalcogenide (TMD) monolayers. Unlike graphene, TMDs such as MoS_2 and WSe_2, are semiconductors with tunable direct bandgaps dependent on the number of atomic layers, and have potential electronic and optoelectronic applications.

From Diamond to Graphene (1998–2010) [KP Loh]

Loh KP started his research in diamond growth in 1999 in collaboration with A Wee. A home-built hot filament CVD system constructed by MSc student Lin

Available online at www.sciencedirect.com

SCIENCE ($) DIRECT·

ELSEVIER

Surface Science 596 (2005) 176–186

SURFACE SCIENCE

www.elsevier.com/locate/susc

Atomic structure of the 6H–SiC(0001) nanomesh

Wei Chen [a], Hai Xu [a], Lei Liu [a], Xingyu Gao [a], Dongchen Qi [a],
Guowen Peng [a], Swee Ching Tan [a], Yuanping Feng [a], Kian Ping Loh [b],
Andrew Thye Shen Wee [a,*]

[a] Department of Physics, National University of Singapore, 2 Science Drive 3, Singapore 117542, Singapore
[b] Department of Chemistry, National University of Singapore, 3 Science Drive 3, Singapore 117543, Singapore

Received 10 May 2005; accepted for publication 14 September 2005
Available online 6 October 2005

ELSEVIER

Surface Science
Top Cited Article 2005-2010

Awarded to:

*Chen, W., Xu, H., Liu, L., Gao, X., Qi, D., Peng, G., Tan, S.C., Feng, Y.,
Loh, K.P., Wee, A.T.S.*

For the paper entitled:

"Atomic structure of the 6H-SiC(0 0 0 1) nanomesh"

This paper was published in:
Surface Science, Volume 596, Issue 1-3, 2005

David Clark
Senior Vice President, Physical Sciences I
Amsterdam, The Netherlands

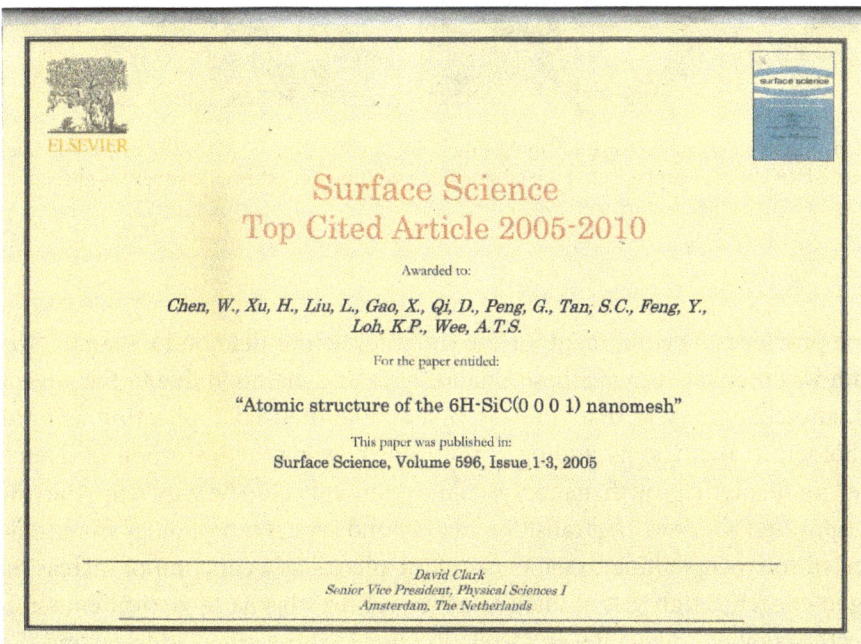

Fig. 5. [Top] Paper published in the journal *Surface Science* that proposed the structure of the carbon nanomesh "whereby isolated carbon islands one atomic layer thick assemble to form the nanomesh structure" (Chen *et al.*, *Surface Science* **596** (2005) 176); [bottom] Certificate in recognition that this paper was a top cited article in *Surface Science* (2005–2010).

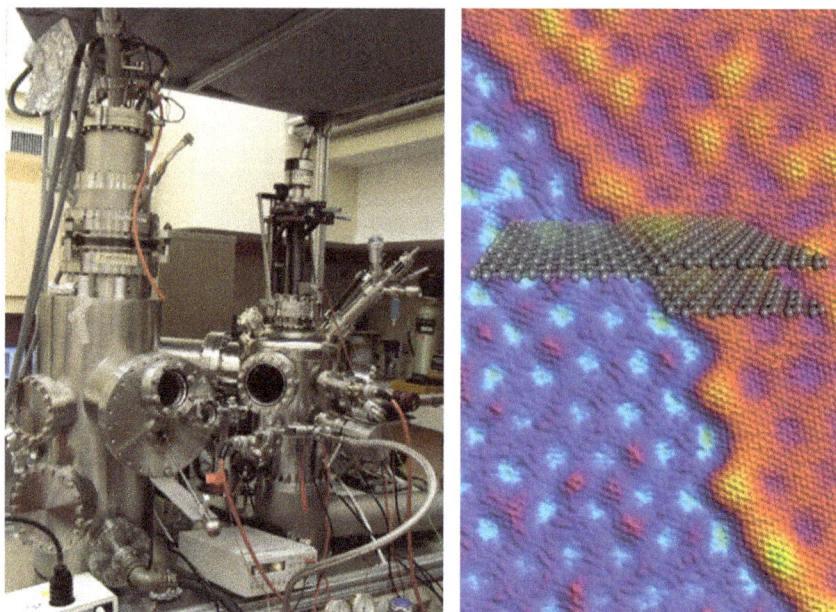

Fig. 6. [left] Low temperature scanning tunnelling microscope (STM) with growth chamber attached; [right] STM image of epitaxial graphene with model schematic superposed (Nature News 25 March 2009; Nature 458, 390–391 (2009)[8]).

Ting provided the equipment for their first foray into diamond research.[9] The team was motivated to synthesise nanocrystalline diamond due to the advent of nanotechnology at that time, as well as its potential application as hard tribological coatings. By adding argon into the usual hydrocarbon feedstock used for diamond growth, nanocrystalline diamond could be grown in a narrow composition window. The transition in diamond crystal morphology from well-faceted microcrystalline to nanocrystalline phases as a function of increasing argon concentration was studied. Analysis of the plasma by optical emission spectroscopy revealed a linear correlation between the argon addition and the occurrence of C_2 dimers in the plasma. This work had since received very good citations by the international diamond community.

Recognising that the research focus on diamond at that time was largely engineering driven, Loh KP decided to carve a niche in the surface chemistry of diamond, with a view towards their application as biosensors[10–12] (Figure 7). One perspective he introduced to the community was the notion that a hydrogenated diamond surface can be viewed as the organic analogue of solid hydrocarbon.[13,14] The C-H bonds on the surface of diamond can be activated

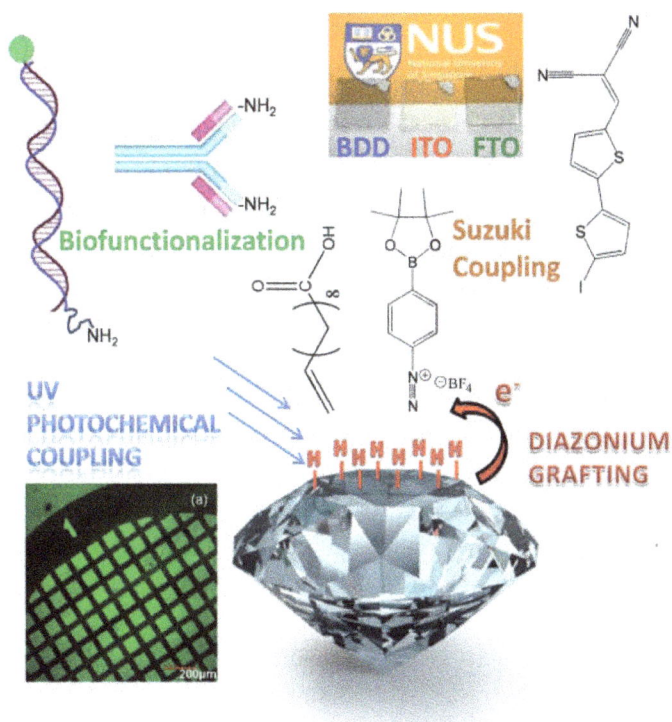

Fig. 7.   Surface chemistry of diamond.

for C-C coupling in analogy to the mainstream chemistry effort on C-H activation. Diamond surface chemistry research allowed him to establish a unique niche in diamond research, and Loh was invited to be the associate editor of *Diamond and Related Materials* (2009–2012). He was also an active member in the advisory panels and scientific boards of international diamond conferences such as the European Diamond Conference series, New Diamond and Nanocarbon and the Hasselt Diamond workshops.

Loh KP started researching the 6H-SiC surface in 2000. Dr Xie Xianning, now the assistant director of NUS Suzhou Research Institute, was his first PhD student and he worked on surface reconstructions on 6H-SiC (0001). In a paper published in *Diamond and Related Materials* in 2001 entitled "Atomic beam etching of carbon superstructures on 6H-SiC (0001) studied by reflection high energy electron diffraction (RHEED)", the various stages during the segregation of carbonaceous super-structures at high temperatures were studied by RHEED.[7,8,15,16] A smooth silicate-terminated root3 x root3 R30 surface could be obtained after a hydrogen-plasma beam treatment of 6H-SiC at 800 degrees C. Annealing the root3 x root3 R30 face to 900 degrees C readily resulted in the segregation of a few layered graphene on the surface, with the basis vectors of the graphite unit cell rotated 30 degrees with respect to the bulk SiC. The importance of this as a means to form single layer graphene had not been recognised at that time. Following Xie's work, Loh KP and A Wee co-supervised a PhD student Chen Wei, who continued to research on the carbon superstructure on 6H-SiC, and this led to further work on epitaxial graphene which paralleled the emergence of the graphene research field.[5]

Without appreciating the significance of 2D materials at that time, Loh KP initiated research on MoS_2 (molybdenum disulphide) and h-BN (hexagonal boron nitride) as early as 2004. One rather advanced notion at that time was the synthesis of edge-oriented MoS_2 nanosheets using a single source precursor based on Mo(IV)-tetrakis(diethylaminodithiocarbomato),[10] as opposed to the conventional methods using dual sources of Mo and S. Edge-oriented MoS_2 films exhibit a high density of nanowalls, which had been demonstrated then to exhibit excellent electrochemical charge storage properties.[17] Interestingly, such edge-oriented MoS_2 nanosheet-like films had been found also to exhibit weak magnetism (similar to 1–2 emu/g) and 2.5% magnetoresistance effects with a Curie temperature of 685 K.[18] The magnetisation is related to the presence of edge spins on the prismatic edges of the nanosheets according to spin-polarized calculations.

The growth of h-BN on 6H-SiC (0001) using plasma-excited borazine was studied in 2005 with PhD student Chen Wei. On 6H-SiC (0001), the growth of a pin-hole free, compact h-BN film was difficult due to poor wetting properties between h-BN and 6H-SiC.[19] The strained BN layer would release its elastic energy due to a morphological instability at the interface. This strain relief mechanism gave rise to a buckling of the film into longitudinal islands and round trenches between 500–700°C. The work was selected as the Editor's choice in *Physica Status Solidi*. The expertise developed in the growth of h-BN film at that time was useful to later research in 2D materials, especially in view of the role of h-BN as a dielectric and passivation layer in 2D research.

Loh KP's entry into graphene research started in 2007 when he was encouraged by A Wee to organise a National Research Foundation (NRF) Competitive Research Programme (CRP) project on graphene. His team won the inaugural CRP award in 2008 centred on the theme of Graphene and Related Materials. At the end of this project, more than 10 *Nature* series papers were published as a result of the efforts, ranging from the use of graphene as a broadband polariser to the face-to-face transfer growth of graphene.[20–29] Not forgetting his first love, Loh KP relished the opportunity to marry diamond with graphene, which he achieved by attaching a layer of graphene on diamond and fusing the interface by desorbing the surface hydrogen at a high temperature.[29] To his delight, it was discovered that residual water from the transfer process trapped in the bubbles formed by graphene on diamond became superheated at 600° C. This table-top hydrothermal anvil had allowed the dynamics of supercritical water entrapped between a graphene membrane and diamond to be studied using optical spectroscopy methods. Amazingly, super-heated water could be rendered highly corrosive and etches diamond. This work captured the imagination of the local and international media and was reported extensively in the *Sunday Times*, *Today* and *Lian He Zao Bao*. In 2010, Loh KP met AH Castro Neto and got involved in the founding of the Graphene Research Centre.

From Graphene to 2D Materials/GRC to CA2DM (2010-Present) [AH Castro Neto]

In November 2008, Prof Barbaros Oezyilmaz invited me (AH Castro Neto) to come to Singapore to give a talk at the Asia Nano conference that he was helping to organize. At that time graphene research was at its peak with many papers being written every day and, at least in the USA where I was a Professor, we had conferences with thousands of participants. It was obvious to some of us that graphene was going to lead to a Nobel Prize because of the overwhelming response from the whole scientific community. I remember that in 2005, soon after the publication of the famous Manchester Group Science paper on the "scotch tape" exfoliation of graphite, I met Prof Andre Geim personally at the March Meeting of the American Physical Society in Los Angeles. I told my two main collaborators at the time, Prof. Francisco Guinea from Spain and Prof. Nuno Peres from Portugal, that graphene would become the subject of a Nobel prize.

Singapore was already on the world map for graphene with the efforts of Prof Barbaros Oezyilmaz (NUS), Prof Loh Kian Ping (NUS), Prof Andrew Wee (NUS), Prof Yu Ting (NTU), Prof Shen Ze Xiang (NTU), whom I knew from papers. At the time, Prof Andrew Wee suggested we should apply to the National Research Foundation (NRF), under my leadership, for the creation of a S$ 150 million Research Centre of Excellence (RCE) on carbon based materials (fullerenes, carbon nanotubes, graphene, diamond, etc). My wife and I were really positively impressed with Singapore and decided we should try. I remember having great meetings with NUS President Prof Tan Chorh Chuan and NUS Provost Prof Tan Eng Chye who were enthusiastic supporters of the idea.

During the first semester of 2009 we organized several events in Singapore in order to bring together the group that would seed this new RCE. This group included Kostya Novoselov who, at the time, was still a Royal Society Research Fellow in Manchester and who I had convinced to move to Singapore. In the second semester of 2009 I became a Visiting Professor at NUS and worked on the proposal for the new RCE. Unfortunately that year the RCE funding was terminated. By the end of my visit in December 2009, on my way to become a Visiting Miller Professor in Berkeley, I said my goodbyes to Singapore and thanked everybody who gave support to the project. During my stay in Berkeley I received a phone call from Prof Tan Eng Chye who proposed the creation of a smaller research centre, with "only" S$ 40 million, focused on graphene. One has to keep in mind that this was before the Nobel Prize and I give a lot of credit to the NUS administration regarding their vision that graphene would play an important role in science and technology. That night I remember calling two key people to discuss Prof Tan Eng Chye's proposal, Prof Andre Geim and Prof Philip Kim (at the time at Columbia University), and both were very supportive of the idea. The support from the key graphene people was fundamental to my decision to move forward and take leave of absence from Boston University in the USA and move to Singapore in August 2010.

With the guidance of Prof Novoselov and Prof Oezyilmaz and the outstanding assistance of Dr Peter Blake (Manchester University) and Mr Ang Han Siong (Graphene Research Centre's facilities manager), together with the invaluable administrative support by Ms Lee Wei Fen, we took the steps towards the creation of the Graphene Research Centre (GRC) micro and nanofabrication facility which was envisioned to be the first of its kind in the world completely focused on graphene. The experience of creating this facility was daunting given the lack of expertise in Singapore in building high technology facilities of this nature. However, the support of the Faculty of Science (FoS), at the time under the leadership of Dean Andrew Wee was extraordinary. We were so lucky to work with people such as Mr Syam Kumar Prabhakaran and Ms Belinda Beh Hui Min, who not only facilitated our life tremendously but who were also directly involved in all the steps that would lead to a world class facility.

By September 2010 we had made all the arrangements for Prof Novoselov to move to Singapore. However, the rumours of a Nobel Prize for Geim, Novoselov and Kim made me worry a little bit but I tried to bring myself to think that it was too soon for a Nobel Prize, but that would mean Kostya would be awarded it as a Professor at NUS, which would have been a major coup for NUS. During an award ceremony in early October 2010 I recalled telling Prof Tan Eng Chye and Prof Andrew Wee that I was crossing my fingers that they would not get the Nobel Prize that year so that Kostya would join us at NUS. I was convinced that the United Kingdom would not let him leave if he had won the Nobel Prize while still at Manchester.

On October 6th 2010, Prof Vitor Pereira (NUS) and I were in my office at NUS on a teleconference with Kostya discussing a joint project when Kostya phone rang. The announcement had been made and Kostya together with Andre had been awarded the 2010 Nobel Prize in Physics. We were witnesses of the phone call that had brought Kostya the good news. For me it was a bitter sweet experience since I was happy for my friends but realised that to bring Kostya to Singapore was now essentially out of the question. My own consolation was that Kostya invited me and my wife for the Nobel Prize award ceremony in Stockholm in December 2010 which was an unforgettable experience (Figure 8).

Although Kostya did not move to NUS, he continued to play a big role in helping us build our facilities. It took almost one year of planning plus one year of actual construction and in June 2012 our nano-fabrication facility was ready. We had the honour with the visit of Deputy Prime Minister Teo Chee Hean who experienced, under the guidance of Prof Novoselov, the thrill of producing a graphene device inside our cleanroom (Figure 9). In June 12, 2012, Prof Tan Chorh Chuan inaugurated our world class facility (Figure 10).

Prof Andre Geim has also been a constant presence at GRC since he became Distinguished Visiting Professor at NUS. Since 2012 Prof. Geim has visited GRC one month per year (Figure 11). During his visits Prof Geim acted as a mentor to many of our assistant professors, their research fellows and students. Moreover, collaborations had been born out of these visits and a memorandum of understanding had been signed between Manchester University and NUS.

Since 2005 after the Manchester group published the first paper demonstrating that other 2D crystals besides graphene could be exfoliated by the "scotch tape" method, it was clear that graphene was not alone and that a very large class of layered materials could be exfoliated. Many of these materials, unlike graphene, have many-body electronic states that are not trivial. Transition metal dichalcogenides, for instance, can be charge density waves (CDW) and superconductors, layered cuprate oxides can be high temperature superconductors, layered manganites can be ferromagnets and anti-ferromagnets.

Fig. 8.   [Top] Prof Novoselov, Prof Castro Neto, and Prof Geim at the Nobel Ceremony, December 2010, Stockholm, Sweden. [Bottom] Prof Geim and Prof Novoselov receiving the 2010 Nobel Prize in Physics from the King of Sweden.

Furthermore, graphene is a semi-metal, that is, a metal with a very low density of states, and hence conductive. For that reason, graphene cannot be readily used in digital applications. It had been rather clear that for applications in modern electronics we needed 2D semiconductors and there is a large family of those with different lattice structures, different gap sizes, and different electronic mobilities. Already in 2005 we had known that progress in this area would be in the production and synthesis of new 2D materials with properties complementary to those of graphene. The expansion of GRC's research scope into new 2D

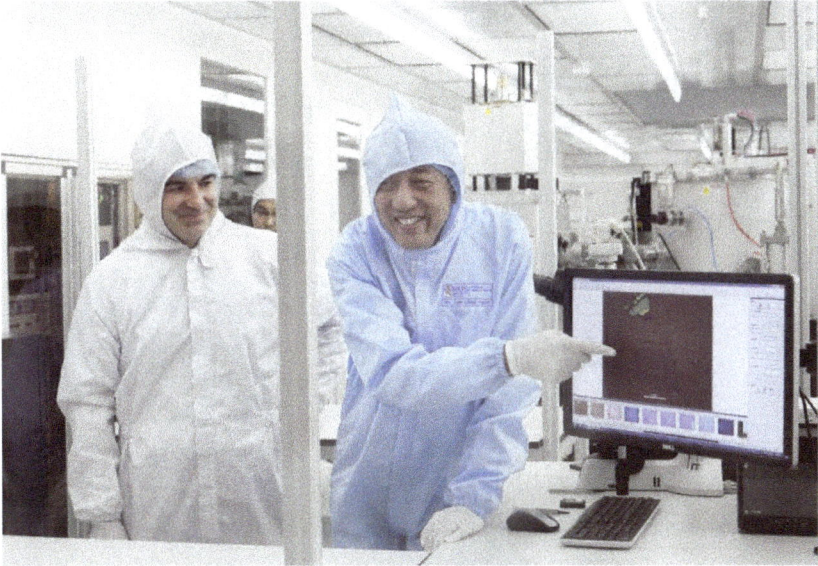

Fig. 9.   Prof Novoselov and DPM Teo at the GRC cleanroom after DPM Teo produced his first graphene flake under Prof Novoselov's guidance.

Fig. 10.   Inauguration of GRC Micro and Nano-Fabrication facility in June 2012. In the picture from left to right: FoS Dean Prof Andrew Wee, Prof Castro Neto, Prof Novoselov, Deputy President for Research and Technology Prof Barry Halliwell, NUS President Prof Tan Chorh Chuan and NRF CEO Dr Francis Yeoh.

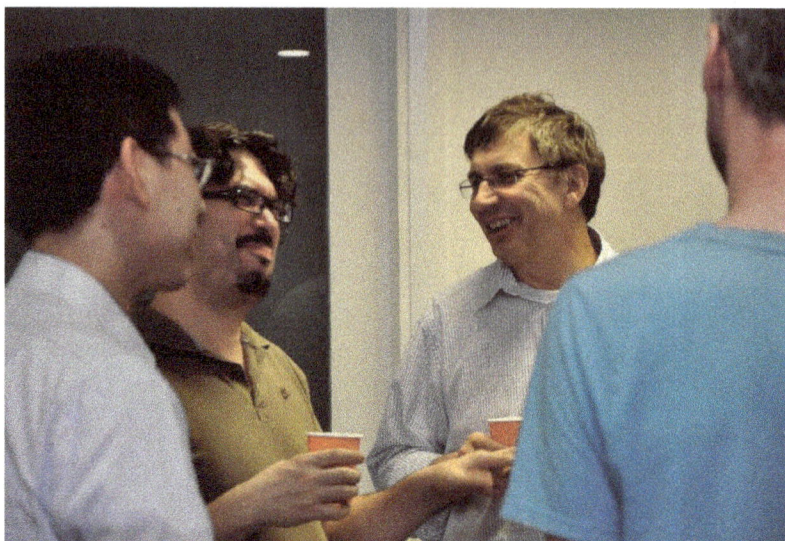

Fig. 11.   GRC Christmas party on December 2013: Prof Barbaros Oezyilmaz and Prof Geim.

materials was therefore a necessary condition for keeping a worldwide leadership role in materials science and technology. In fact, GRC was the first centre to publish on the electronic properties of phosphorene, a 2D allotrope of phosphorus, which has semiconducting properties, and can be transformed into a metal and even a superconductor under pressure.

Since 2010, researchers at GRC have been very successful in attracting funding and NRF has always been a major funder of GRC activities. In 4 years, GRC researchers were able to attract around S\$ 50 million in funding. As a result, hundreds of papers and invention disclosures were written in this area establishing NUS and Singapore as a hub for graphene research. Nevertheless, the costs of running a high tech facility with state of the art characterization, nano-lithography, and device development were very high. Grant funding could not be used to cover operating costs and hence, there was the important issue of the long term sustainability, and even existence, of GRC.

In 2014 NRF awarded NUS a Mid-Size Centre grant of S\$ 50 million over 10 years in order to create a natural progression for GRC to reposition itself as the Centre for Advanced 2D Materials, CA2DM. The new funding essentially allowed the continued operation of the high tech facilities that were created in the last 5 years and the retention of the skilled manpower that was needed for such endeavours. Furthermore, the Office for Industry and Innovation, under the leadership of Ms Tricia Chong (NUS Enterprise) was created to facilitate the interaction between academics and the industry.

Today, after 5 years of operation, CA2DM and GRC are centres of reference in the area of materials science and technology. Visitors from all over the world come to Singapore to work in CA2DM's facilities and interact with CA2DM researchers. In a recent *Nature Nanotechnology* editorial,[30] it was written:

> "Nanotechnology in Singapore is taken seriously. In 2010, the National University of Singapore provided S$40 million to set up the Graphene Research Centre, which boasts state-of-the-art fabrication facilities and is home to a faculty featuring several world-renowned experts in both experimental and theoretical techniques. In 2014, the centre expanded to become the Centre for Advanced 2D Materials, thanks to a S$50 million grant from the National Research Foundation (http://go.nature.com/ ykV7fZ). This move strengthened the technical capabilities and human resources of the centre and reflected the current trends in research now embracing all types of layered materials beyond graphene, such as transition metal dichalcogenides and black phosphorus."

At CA2DM we do not believe in predicting the future, we believe in inventing it.

References

[1] As quoted in *Growth, Dissolution, and Pattern Formation in Geosystems* (1999) by Bjørn Jamtveit and Paul Meakin, p. 291.

[2] http://www.physics.nus.edu.sg/~surface/

[3] National Science and Technology Board (NSTB), the predecessor of the Agency for Science, Technology and Research (A*STAR).

[4] http://ssls.nus.edu.sg/

[5] W. Chen, H. Xu, L. Liu, X. Y. Gao, D. C. Qi, G. W. Peng, S. C. Tan, Y. P. Feng, K. P. Loh, A. T. S. Wee, Atomic structure of the 6H-SiC(0001) nanomesh, *Surface Science* **596** (2005) 176–186.

[6] K. S. Novoselov, A. K. Geim, S. V. Morozov, D. Jiang, Y. Zhang, S. V. Dubonos, I. V. Grigorieva, A. A. Firsov, Electric Field Effect in Atomically Thin Carbon Films, *Science* **306** (2004) 666–669.

[7] Nature News 25 March 2009; *Nature* 458, 390–391 (2009).

[8] http://www.nature.com/news/2009/090325/full/458390a.html

9   T. Lin, GY Yu, ATS Wee, ZX Shen and KP Loh, Compositional mapping of the argon-methane-hydrogen system for polycrystalline to nanocrystalline diamond film growth in a hot-filament chemical vapor deposition system, *Applied Physics Letters*, 77, 17, 2692 (2000).

10 Chong, KF, Loh, KP, Vedula, SRK, Lim, CT, Sternschulte, H, Steinmuller, D, Sheu, FS, Zhong, YL, Cell adhesion properties on photochemically functionalized diamond, *Langmuir*, 23, 5615–5621(2007).

11 Zhong, YL; Chong, KF; May, PW; Chen, ZK Loh, KP, Optimizing biosensing properties on undecylenic acid-functionalized diamond, *Langmuir* 23, 5824–5830 (2007).

12 Ang, PK; Loh, KP; Wohland, T; Nesladek, M; Van Hove, E, Supported Lipid Bilayer on Nanocrystalline Diamond: Dual Optical and Field-Effect Sensor for Membrane Disruption, *Advanced Functional Materials* 19, 109–116 (2009).

13 Zhong, YL; Loh, KP; The Chemistry of C-H Bond Activation on Diamond, *Chemistry-An Asian Journal*, 5, 1532–1540 (2010).

14 Zhong, YL; Loh, KP; Midya, A; Chen, ZK, Suzuki coupling of aryl organics on diamond, *Chemistry of Materials*, 20, 3137–3144 (2008).

15 Xie, XN; Lim, R; Li, J; Li, SFY; Loh KP; Atomic hydrogen beam etching of carbon superstructures on 6H-SiC(0001) studied by reflection high-energy electron diffraction, *Diamond and Related Materials*, 10, 1218–1223 (2001).

16 Xie, XN; Wang, HQ; Wee, ATS; K. P. Loh, The evolution of 3 x 3, 6 x 6, root 3 x root 3R30 degrees and 6 root 3 x 6 root 3R30 degrees superstructuves on 6H-SiC (0001) surfaces studied by reflection high energy electron diffraction, *Surface Science*, 478, 57–71 (2001).

17 Soon, JM; Loh, KP, Electrochemical double-layer capacitance of MoS2 nanowall films, Electrochemical and Solid State Letters, 10, A250–A254 (2007).

18 Zhang, J; Soon, JM; Loh, KP; Yin, JH; Ding, J; Sullivian, MB; Wu, P; Magnetic molybdenum disulfide nanosheet films, *Nano Letters*, 7, 2370–2376 (2007).

19 Chen, W; Loh KP; Lin, M; Liu, R; Wee, ATS; Atomic force microscopy study of hexagonal boron nitride film growth on 6H-SiC (0001) — Editor's choice, *Physica Status Solidi A-Applications and Materials Science*, 202, 37–45 (2005).

20 Gao, Libo; Ni, Guang-Xin; Liu, Yanpeng, A. H. Castro-Neto, Kian Ping Loh, Face-to-face Transfer of Wafer-Scale Graphene Films, *Nature*, 505, 190–194 (2014).

21 Manish Chhowalla, Goki Eda, Kian Ping Loh et al., The chemistry of ultra-thin transition metal dichalcogenide nanosheets, *Nature Chemistry* 5, 263–275 (2013).

22 Jian Zheng and Kian Ping Loh, High Yield Exfoliation of Two Dimensional Chalcogenides using Sodium Naphtalenide, *Nature Communications*, 5, 2995 (2014).

23 Jiong Lu, Pei Shan Emmeline Yeo, Chee Kwan Gan, Ping Yu and Loh, KP, Transforming Fullerene Molecules into Graphene Quantum dots, *Nature Nanotechnology*, 6, 247–252, (2011).

24 Q. Bao, H. Zhang, B. Wang, Z. Ni, Candy H.Y.X. Lim, Y. Wang, D. Y. Tang, Loh KP, Graphene as broadband polarizer, *Nature Photonics*, 5, 411–415 (2011).

25 Loh KP, Bao QL, Eda G, Manish Chowalla, Graphene oxide as a chemically tunable platform for optical applications, *Nature Chemistry*, 2, 12, 1015–1024 (2010).

26 Jiong Lu, A. H. Castro Neto and Kian Ping Loh, Transforming Graphene Moire Blisters to Geometric Nanobubbles, *Nature Communications* 8, 823 (2012).

27 Jiong Lu, Kai Zhang, Tze Chien Su, A. H. Castro Neto, Kian Ping Loh, Order-Disorder Transition in a 2-D B-C-N alloy, *Nature Communications* 4, 2681 (2013).

28 Chen Liang Su and Kian Ping Loh *et al.*, Probing the Catalytic Activity of Graphene Oxide and its origin *Nature Communications*, 3, 1298 (2012).

29 Candy Su, Kian Ping Loh et al., A hydrothermal Anvil made of Graphene nanobubbles on diamond *Nature Communications* 4, 1556 (2013).

30 Editorial, *Nature Nanotechnology*, Vol. 10, October 2015, p. 825.

Chapter 23

Development and Progress of Marine Biology in the Last 50 Years

L M Chou

Introduction

The study of marine biology began soon after Sir Thomas Stamford Raffles founded modern Singapore in 1819. As an avid naturalist, the island's rich bio-diversity on land and sea occupied whatever free time he had from his primary duty of governing the colonial outpost as a trading port. Singapore was indeed a treasure trove of terrestrial and marine biodiversity. Being close to the equator and in a region now acknowledged as the global center of species richness, the island's natural history offered much in terms of species not seen previously or elsewhere. The potential of new species discovery was high and Singapore quickly became a focus of study by many naturalists from the developed world.

Raffles himself observed, collected and documented species and facilitated the search by other renowned naturalists to uncover the country's biodiversity. The term 'biodiversity' was not even coined then. Introduced only in the 1980s, 'biodiversity' has now become a buzzword. Research on species prior to the 1980s was classified mainly as natural history, more specifically taxonomy where new species are named, described, preserved and permanently cata-logued. The passion of the early researchers contributed greatly to the catalogue of Singapore's marine species richness. Most specimens of newly described species were shipped off to the British and other European museums in the early days. In 1849, the Raffles Museum was established to become a permanent repository of collected specimens that made up Singapore's natural history.

The biodiversity legacy continues right up to this day. The founding of the Raffles Museum maintained research interest at a steady momentum. This was boosted by the establishment of the Singapore campus of the University of Malaya and the formation of its Botany Department in 1949 and Zoology Department in 1950. Many of the earlier academic staff including Heads par-ticularly from the Zoology Department specialised in or had some interest in marine biology.[1] The formation of Nanyang University and its Biology

Department in the later half of 1950s contributed minimally with only one or two staff devoting some time to do research in marine organisms. While taxonomic studies continued, other aspects of marine biology such as ecology and physiology started to develop, all driven by individuals who were interested in understanding not only biodiversity but also ecological processes of the marine environment.

In 1962, the same year when the Singapore campus of the University of Malaya became an independent institution and operated as the University of Singapore, a boost to marine biology came from the formation of the Fisheries Biology Unit in the Zoology Department. It operated until 1973, offering the Diploma of Fisheries and the Certificate in Fisheries under the most able leadership of Professor Tham Ah Kow with his vast prior working experience in the Government Fisheries Department. Marine biology research established by the early pioneers provided a strong foundation for its expansion during the five decades from 1965 to 2015.

Marine Biology Since 1965

After Singapore became a sovereign nation in 1965, the study of marine biology developed at a slow pace for three decades and then more rapidly in the last twenty years. It was like a 1,000 m steeplechase with numerous and difficult obstacles packed into the first 600 m and fewer but by no means less challenging ones in the final 400 m. Personnel, funding and institutional capacity increased sluggishly and hesitantly until the mid-1990s before expanding more aggressively until the present. Research strength increased as more people were trained either locally or abroad and opportunities enlarged with the formation of many research institutes and centers focusing on marine science, including a few with a short existence before morphing or merging with others into more consolidated arrangements. Marine biology research has by today established an indelible contribution to our understanding of not just what is in the surrounding sea but how important marine biodiversity is in providing ecosystem services that benefits both the environment and society.

Personnel

Continued growth of marine biology research from 1965 was supported by a small core of academics and researchers from the Universities, Raffles Museum and the Primary Production Department (now Agri-Food & Veterinary Authority). The research expertise of many former Heads of the Department of Zoology and later the Department of Biological Sciences of the University of Singapore, which became the National University of Singapore (NUS) between 1971 and 1999 were in marine biology or fields related to it. Similarly, the research interest of many academic staff in those earlier years were wholly or partially focused on marine biology and they all contributed to supervising the research theses of many Honours, Masters and Doctoral students who had an interest in the subject.

From the Raffles Museum (later National Museum), ichthyologist Eric Alfred built up the fish collection, both freshwater and marine and expanded the repository. Marine biologists from the Marine Fisheries Research Department and also the Primary Production Department contributed to fisheries, marine aquaculture and marine environment research. Marine biology capacity also increased at the National Institute of Education from 2001. Among government agencies, National Parks Board (NParks) is today the biggest employer of marine biologists after it established the Biodiversity Centre in 2006 and extended its focus beyond the terrestrial realm to include marine biodiversity research and management. The Lee Kong Chian Natural History Museum (LKCNHM) established in 2015 and the Tropical Marine Science Institute (TMSI) opened officially in 2002 also have large core of marine biologists.

Other research centers and institutes affiliated to the universities (e.g. TMSI, DHI(Danish Hydraulic Institute)-NTU Water & Environment Research Centre and Education Hub (DHI-NTU Centre), NUS Environmental Research Institute, Centre for Environmental Sensing and Modelling, Singapore Centre on Environmental Life Sciences Engineering) all have some marine biologists among the research staff. Environment consultancy businesses also engage marine biologists but this started only in the 1990s when Environmental Impact Assessments were required of development projects. Moving into the 1990s, more young marine biologists trained locally or overseas joined the *workforce* as new appointment opportunities became available. Public attractions like the Underwater World Singapore, which opened in 1991, and S.E.A. Aquarium in 2012 have marine biologists on their staff. The pool of marine biologists has not diminished as more individuals acquired PhDs from either local or overseas universities.

Prospects for marine biologists have increased considerably in both the public and private sectors, and academia with the scope expanding in research, teaching and management. Local marine biologists do not confine their research to Singapore's marine environment but frequently extend the reach to the region and also international waters.

Institutional Development

Fisheries research and training was strengthened by the establishment of the Regional Marine Biological Centre under an agreement between UNESCO (United Nations Educational, Scientific and Cultural Organization) and the Singapore Government. It was officially opened on 19 March 1968 by Deputy Prime Minister, Dr Toh Chin Chye who was also Vice-Chancellor of the University of Singapore. The University contributed US$7,000 and UNESCO US$6,000 towards specialised equipment, which at that time was simply more superior microscopes, and space for a fully-furnished research laboratory provided within the Bukit Timah campus of the University. In addition, UNESCO committed an annual financial contribution of US$16,000 while the University made a matching contribution of the same amount to support administrative, research and technical staff. The Centre played a key role in analysing plankton samples collected by research vessels of participating nations engaged in the Cooperative Study of the Kuroshio and Adjacent Regions. This placed Singapore on the international map of marine biology for the next ten years until 1978 when the Centre was discontinued.

From the early 1980s, regional marine science started to gain prominence with the initiation of the Regional Seas Programme by the United Nations Environment Programme (UNEP). Marine biologists from NUS participated actively in the research activities of the intergovernmental Coordinating Body on the Seas of East Asia (COBSEA), which oversees UNEP's regional seas programme for East Asia. These included leadership roles in major projects such as the regional task force on climate change and the working group on training materials for coastal management. Regional involvement expanded towards the late 1980s with collaborative projects under the Association of Southeast Asian Nations (ASEAN) framework and COBSEA. The ASEAN marine science projects that were supported by Australia, Canada and United States of America, contributed immensely to capacity building of Singapore's marine biology.

The ASEAN-Australia Living Coastal Project (1987 to 1994) for example, established common monitoring methodology for assessing mangrove, seagrass, coral reef and benthic habitats, allowing for comparison and trend analysis of ecosystem health across the region. Participation in these projects helped to enhance the capability of NUS' Biological Sciences Department in habitat monitoring and assessment and was accompanied by the establishment of scuba diving facilities and relevant habitat sampling equipment. The improved marine biology capacity helped to increase regional and international collaboration. Individuals recognised for their marine biology expertise were also invited to join committees of international initiatives such as the Global Coral

Reef Monitoring Network of the International Coral Reef Initiative, Joint Group of Experts on the Scientific Aspects of Marine Environment Protection, and International Commission of Zoological Nomenclature.

Events during the 1960s signalled uncertainty to taxonomic studies. Most notable was the renaming of Raffles Museum to the National Museum of Singapore in 1965 and the shift in the museum's emphasis to history and culture in 1972. Its valuable natural history collection, of which an estimated 30% is marine was to be discarded as political leaders could not see any intrinsic value in a collection of preserved animals or how it can possibly contribute to economic growth. The collection was considered an economic burden as it needed prime space and a sizeable maintenance budget. Its iconic 14 m length Indian Fin Whale skeleton, for decades a majestic centerpiece suspended from the ceiling above the main stairwell was donated to the National Museum (Muzium Negara) in Kuala Lumpur, never to return and negotiations were started to sell or give the collection to interested institutions overseas.

Professor Chuang Shou Hwa, then Head of the Zoology Department made an effort to keep the collection within Singapore and particularly within the University but faced multiple obstacles and challenges.[2] The collection referred to as the Zoological Reference Collection was moved to temporary premises across Singapore three times sustaining loss and damage each time with unsuitable conditions of heat and humidity in these premises taking a further toll. It was only in 1987 that the collection was finally given a home in the National University of Singapore with suitable storage facilities and climate control that helped to keep the preserved specimens in top condition. It became the Raffles Museum of Biodiversity Research (RMBR). Marine biodiversity research took on a renewed vigor with more visiting scientists and students making full use of RMBR's facilities.

Today, RMBR has matured into a full-fledged museum of natural history, the Lee Kong Chian Natural History Museum where more of the reference collection including the marine component is proudly displayed for the public to admire and become connected with Singapore's natural heritage. In 2015, the intact carcass of an adult female sperm whale washed up near Jurong Island. The museum wasted no time in preparing the whale's skeleton and unveiled the 10.6 m long display in March 2016. This whale episode completes a full cycle of loss and recovery symptomatic of the fate of Singapore's marine natural heritage and research. The story with marine plant species is more tranquil. The Singapore Herbarium maintained by the Singapore Botanic Gardens remains the permanent repository of the plant collection, although marine species form a small component.

Field Facilities and the Tropical Marine Science Institute (TMSI)

Marine biology studies at NUS was supported by equipment and facilities built up over the years. This included a coastal research boat belonging to the Zoology Department in the 1960s and later in 1981, a recirculating seawater aquarium when the Department moved to the Kent Ridge campus. The boat was a small wooden one with a loud engine that belies its slow speed. Confiscated by the authorities for being used in illicit activities, it was given to the Department to support marine research. Named *Periopthalmus*, which is the scientific name for the mudskipper (also the logo of the Department), it remained a mouthful to the regulatory agencies. Field trips to the southern offshore islands were a whole-day affair as it took more well over an hour to reach for instance, Pulau Hantu and another extended hour to get back. There was no place in the boat to retreat to in order to avoid the loud engine and all users suffered from temporary deafness after each trip.

By the early 1990s, *Periopthalmus* reached the end of its serviceable lifespan. The Department replaced it with a somewhat larger fibreglass vessel, with a faster engine and named it *Mudskipper*. This cut travelling time by more than half each way, allowing researchers more time in the field. The vessel also had a winch to help with deploying and hoisting heavy benthic survey equipment, which was done by sheer human effort in the previous boat. Researchers were elated by the improved conditions and regulatory agencies were delighted with the name change. *Mudskipper* had to be decommissioned by 2010 and this time, replaced by *Galaxea*, a larger and faster aluminium-hulled vessel shared between TMSI and the Department of Biological Science. It has been in full use to meet the increased research demand. TMSI, which had its early beginning in 1996 as the Tropical Marine Science Initiative, started with 3 boats of its own but soon found them too much of a financial burden to maintain, eventually giving up two and combining with the Biological Science Department to share costs for *Galaxea*.

Marine biology field stations are considered essential to the advancement of research and teaching. Major overseas universities with strengths in marine biology have a field-based research station to enhance research. This has also been a need of the Zoology Department to advance its marine biology capability. As early as 1964, the Department was allowed to use Raffles Lighthouse as a modest marine station. A small make-shift laboratory with minimal equipment for water analysis was set up and some field experiments were initiated. However, the station did not last as the Indonesian Confrontation flared up and after it was over in 1966, security concerns made it necessary to close the station completely.

The search for an alternative site for a field station continued and thirty years later in 1996, the Department of Biological Sciences formed a Planning Committee to intensify the quest for a field station. Events proceeded quickly and with support from the then Singapore Tourist Promotion Board (now Singapore Tourism Board), St John's Island was identified as a suitable site for incorporating research and development facilities. Instead of a modest marine biology station, plans went ahead to establish the Tropical Marine Science Institute, with a multi-disciplinary research emphasis. The Institute's buildings on the island were completed and officially opened in 2002. Marine biology remains one of the Institute's major research strengths firmly embedded in two of its four research laboratories (Marine Biology and Ecology Laboratory, Marine Mammal Research Laboratory) and the Ecological Monitoring, Informatics and Dynamics Research Group.

The early development of TMSI is well documented in a commemorative volume published in 2012 to celebrate its 10th year.[3] But a steeplechase hurdle loomed not long after. In 2014, NUS made the painful decision to close the research facility at St John's Island because of mounting maintenance cost that made it financially unjustifiable. Support was sought from various government agencies that needed research data for policy formulation to help with the cost sharing. In 2016, The National Research Foundation conceptualised the *Marine Science Research & Development Programme* to consolidate Singapore's role in maritime development over the long term. This included transforming the research laboratories at St John's into a national research facility. This farsighted move will involve substantial upgrading to support high impact marine science research by all research groups in the country.

Research and Funding

Biology as a subject itself expanded and diversified into numerous fields of research as expertise, instrumentation, techniques and computing power improved. Similarly, marine biology advanced beyond taxonomy, ecology and physiology and from the 1990s incorporated the rapidly developing fields of ecosystem processes, genetics, evolution and the *omics*. Continued research in marine biology will increase Singapore's scientific capacity as we need a deeper understanding of marine ecological and ecosystem processes such as larval flow, spawning patterns, ecosystem response to local and global impacts and recovery so that the most efficient and effective management measures can be applied.

As the research scope widens, it adds onto the foundation laid down by early researchers. For example research on coral recovery was conducted as early as in the late 1960s. It consisted of gabions that consolidated rocks to provide a stable substrate for coral attachment. Another experiment made crude excavations on a sandy shore at Sentosa to create tidal pools that trapped water at high tide and did not drain completely before the next high tide. These pools were filled with live corals to see how well they can survive. Today, research on coral recovery continues but with improved techniques and stronger science. Coral restoration activities are now enhanced by nurseries and underwater epoxy. Tidal pools excavated on sandy beaches did not last because the sand redistributed with each tidal change. This was something that coastal engineers knew and it showed the importance of multidisciplinary inputs.

Scuba diving has become a necessary tool for underwater research. During the 1970s, scuba diving was a novelty and equipment was hardly available. Whatever was available was expensive. Wet suits were almost unheard of in Singapore and researchers could not stay warm underwater and resorted to wearing old long-sleeved shirts and pants as flimsy and psychological protection against cuts and stings. Some students resorted to diving in old pajamas. It was challenging for researchers then to look professional and they appeared like a motely crowd dressed in worn-faded apparel most of the time. The increasing popularity of scuba diving from the 1980s widened the market and equipment flowed in, giving researchers a wider choice of equipping themselves to optimise their research time underwater.

Improved equipment helped to advance field-based marine biology research by increasing efficiency. For instance, salinity could be measured instantly and directly with a probe. In the past, seawater samples had to be collected, taken back to the laboratory, chemically treated and titrated to obtain a reading of its salinity. Digital probes can now be used to measure temperature, dissolved oxygen, pH and other environmental parameters immediately. Even in the 1960s and 1970s, marine biologists were carrying mercury thermometers into the field to measure seawater temperature and litmus paper to measure pH. The downside was that students used these probes to simply record and accept whatever readings the instrument registered. I remember a group of undergraduates who did an ecology project at Sungei Buloh. They measured dissolved oxygen with a new digital meter, happily submitting a report showing that dissolved oxygen was over 20 mg/L and thought nothing of it. It did not occur to them that this was an abnormally high value and they did not question whether the equipment was functioning properly.

Research funding fifty years ago is miniscule compared to that available today. Project funding was in terms of a few hundred dollars and that curtailed the research scope and depth. The increase in research funding has increased project awards up to millions of dollars these days and with improved technology and multidisciplinary approaches, enabled more innovative research. The research landscape has improved dramatically for marine biology with funding available from government funding and development agencies, private sector and also external granting agencies. For example, a two-year reef restoration project in 2008 brought together researchers from the Biological Sciences Department and NParks with funding provided by Keppel Industries.

Education

It is interesting to note that despite the development of marine biology and its rapid expansion over the last two decades, none of the universities offer a dedicated marine biology degree course. A *Diploma in Marine Science and Aquaculture* programme was offered by Republic Polytechnic in 2014. It remains the only higher level marine science course and is geared towards developing skills for the aquaculture industry. For many of the basic or Honours biology degree holders, marine biology education is not comprehensive.

The situation became more challenging after 1996 when the Botany and Zoology Departments of NUS merged into the short-lived School of Biological Sciences before becoming the Department of Biological Sciences in 1998. Restructuring of the biology curriculum took place during this period influenced by the need to expand biotechnology. The biological sciences degree was replaced by a life sciences degree, reflecting the conversion of a significant slice of the previous curriculum to topics in biotechnology and cell and molecular biology. The biodiversity components of the previous biology curriculum were compressed such that graduates had little ability to identify plant and animal species. Another outcome is that students have essentially lost the skill of biological illustration. In the 1960s, biology students had to draw every preserved specimen that was displayed during the laboratory practical sessions. Regardless of the person's artistic ability, everyone had to capture an image of the specimen with pencil and paper. Line drawings were required and every biology student was capable of doing that. Today, students use a digital camera to shoot their way through. It won't be long before they start taking selfies with the specimens.

A reversal of this situation occurred over the last few years with a more balanced curriculum that included more biodiversity and aquatic biology modules. Students thus can opt for a specialisation in environmental biology, but there is still no marine biology degree course. Few modules related to marine biology were offered to Honours and graduate students. At Masters and

Doctorate levels, students can focus their research thesis completely on marine biology and quite a number have taken this path, but for the many students who desired a basic degree in marine biology, there was no choice but to study overseas.

Contribution of Marine Biology to Singapore

As an island nation located in a region recognised as the global hotspot of marine biodiversity, Singapore's marine biologists have played strong roles in the study of the rich living marine heritage and are making worthy contributions to science. Earlier publications were mainly taxonomic papers that described species new to science, but a former Science Dean critical of such publications described them as equivalent to stamp collecting. Publications in journals with an impact factor have increased geometrically and have gone beyond taxonomy and ecology. A Web-of-Science search using the keyword "Marine Biology OR Marine Biodiversity OR Marine Ecology AND Singapore" revealed no publications prior to 1996, six between 1996 and 2005 and thirty from 2006 to 2015. This trend is set to continue with the increasing capacity and diversifying fields of marine biology. Similarly, increasingly more books on marine biology have and are being written by local marine biologists dealing not only with Singapore's marine environment but also that of the region.

Many of the early and current marine biologists have or are serving in regional and international marine programmes and initiatives and a steady corps of marine biology graduates will help to meet the demand from the public and private sector as the region shifts its attention to developing the blue economy. Some of Singapore's marine biologists have been based overseas to serve in regional and international institutions such as the Global Coral Reef Monitoring Network and the Southeast Asia Regional Center for Global Change System for Analysis, Research and Training. One marine biologist served as co-leader of the ASEANAREAN Expedition Series, a private voluntary initiative that commenced in 1997 to document the marine parks of the ASEAN region. It was a major undertaking aimed at strengthening networking and developing research collaboration with marine science institutes, local communities and interested sectors throughout the region. The first expedition to Thailand resulted in the publication of the book *The Marine Parks of Thailand* in 1998. The second expedition to Indonesia in 1999 was the first Asian-led expedition to be given coverage by National Geographic Channel. Three half-hour documentaries were aired in September 2002 and a book *Marine Parks of Indonesia* was published in 2013.

Marine biologists from RMBR/LKCNHM and TMSI have not only participated in international marine biodiversity expeditions to various parts of the
world but also extended their capacity to deep-sea research. These include the
*Lumiwan 2008 Survey of the Deep Benthic Fauna of the South China Sea, 2009
Kumejima Marine Biodiversity Expedition* and the 2015 expedition to the 4,000 m
deep Clarion-Clipperton fracture zone of the Pacific Ocean.

The rise of marine biology from the mid-1990s can be correlated with the
government's apparent shift in management attitude towards marine conservation. Development without consideration of the marine habitats and biodiversity was the norm prior to the 1990s. Since then, development guidelines have
been incorporated to reflect the real essence of sustainable development.
Environmental Impact Assessments and real-time Environmental Monitoring
and Management Programmes are required of all development projects and
mitigation measures now involve species relocation and habitat rehabilitation
as the norm. Marine biodiversity conservation is fully entrenched in marine
development projects and requires significant contributions from marine
biologists.

Marine biology research has generated useful data that support many of
the major conservation initiatives. Earlier research was compiled to formulate
the Singapore Blue Plan in 2009, which called for more effective marine biodiversity management. It was submitted by civic groups to the government and
resulted in the *Comprehensive Marine Biodiversity Survey* programme, a systematic five-year study of Singapore's marine life. Led by NParks together with
researchers from tertiary institutions, NGOs and community volunteers, and
supported by funding from corporate groups, over 100 new species and 200
new records were scientifically documented. A significant outcome of marine
biology research is the establishment of the 40 ha Sisters' Islands Marine Park
in 2014. Apart from conserving marine biodiversity, opportunities for research
will be facilitated.

Civic groups and reputable blog sites (e.g. Wild Singapore, Toddycats,
Hantu Bloggers) have also played a prominent role in promoting awareness of
marine biodiversity to the general public. Various NGOs in partnership with
NParks and tertiary institutions are promoting citizen science and encouraging
the public to participate in research. The role of marine biology in sustainable dev-
elopment is further strengthened by the establishment of the Inter-Ministerial
Committee on Sustainable Development in 2008 under which the Technical
Committee on Coastal and Marine Environment, comprising relevant agencies
and academia draws on research data and experience. Through this committee,
Singapore adopted the Integrated Coastal Management (ICM) framework in

2009, labelling it *Integrated Urban Coastal Management* (IUCM) to reflect the relevance to its highly developed setting. The IUCM approach involves participation of different agencies in a combined effort to reduce coastal use conflicts and pave the way for a more holistic and mature management that is oriented towards sustainable development.

The study of marine biology remains crucial to marine environment management. Marine biologists were fully involved in the land reclamation dispute between Malaysia and Singapore in 2002.[4] More importantly, it demonstrated that marine biology when integrated with other disciplines becomes more effective in understanding marine environmental processes as a whole. It becomes more effective also when addressing emerging challenges such as climate change.

Marine Biology in the Next Fifty Years

Marine biology will remain relevant in the future and there is certainly a need for a marine biology degree course. Future research directions will be in applying biology to management, and how the ecological carrying capacity of the marine environment can be increased to allow further environmental and economic benefits. Singapore's marine territory is limited in space and constrained by the lack of an open sea. It is one of the world's busiest ports and competing uses for conservation, recreation, aquaculture, military training and industry are intense. To what extent can all these be accommodated, developed and managed without compromising the environmental health and securing it against climate change impacts? The IUCM framework is in place to help accommodate the needs of different sectors and agencies and optimise the use of the coastal area for greater environmental and socio-economic benefits

Two areas of research considered fundamental to management but are not well established are Ecological Carrying Capacity (ECC) and Strategic Environmental Impact Assessment (SEIA). The concepts have been promoted for some time now but assessment methodologies are not well developed. The importance of ECC estimation is that it helps to determine how impacts can be reduced and how the environment can be enhanced to accommodate multiple activities. Singapore's restricted sea space including its part of the Johor Strait will benefit from an overall SEIA compared with individual project-based EIAs as it facilitates identification of cumulative impacts from different activities, and determines the intensity and extent of each activity that can be permitted without overwhelming the ecological integrity of the area. Both ECC and SEIA require not just marine biology but a multidisciplinary involvement, and can draw on the strengths of the various research institutions.

Improving the Johor Strait environment as a whole requires the collaboration of Singapore and Malaysia but will clearly demonstrate transboundary management of a shared waterway. Both countries have adopted the ICM framework and can cooperate to improve the Strait's ecological carrying capacity, presently undermined by poor flushing, pollutant accumulation and depressed biodiversity. Replacing the Causeway with a full bridge will in time improve the Strait's ecological integrity and allow it to support more water-based recreation, fishing, aquaculture and biodiversity, which in turn will enhance environmental quality. Shipping should remain restricted as it is incompatible with the size of the Strait, constrains other sea activities, and increases erosion and accident risk. Both countries will benefit economically from the increased environmentally-compatible utilisation of the Strait where marine habitats can flourish under improved conditions and provide valuable ecosystem services. Currently, both countries lose a lot from the increasing frequency of environmental upheavals such as toxic plankton blooms that are detrimental to marine life. Left as it is, the Johor Strait can go in only one direction and that is towards ecological collapse.

The research need for marine biology keeps growing. Capacity in terms of research facilities and funding has improved and the corps of marine biologists has to be expanded to advance high-quality research. Singapore will do well with more of its own marine biologists and developing greater research strength that is relevant to sustainable development of the seas and oceans.

References

[1] Ng PKL, Wang LK, Corlett RT. 2011. History of biodiversity research. p 122–133. In Ng PKL, Corlett RT, Tan HTW (Eds) Singapore Biodiversity: An Encyclopedia of the Natural Environment and Sustainable Development. Editions Didier Millet, Raffles Museum of Biodiversity Research. Singapore.

[2] Tan KS, Chan ES, Teo S, Chou LM, Lam TJ, Taylor E. 2012. A short history of TMSI@ St John's Island. In Tan KS (Ed), Contributions to Marine Science, p vii–ix. Tropical Marine Science Institute, National University of Singapore. 168p.

[3] Tan, Kevin. 2015. Of whales and dinosaurs: the story of Singapore's Natural History Museum. NUS Press, Singapore. 266p.

[4] Cheong KH, Tommy Koh, Lionel Yee. 2013. Malaysia and Singapore: the land reclamation case: from dispute to settlement. Straits Times Press. Singapore. 128p.

Chapter 24

Tropical Marine Science Institute

E A Taylor

The Tropical Marine Science Institute (TMSI) is one of a kind — a groundbreaker. It arose from the need to have a multi-disciplinary research team to scientifically address the diverse yet interconnected problems and opportunities faced by Singapore's coastal waters. Singapore is an island nation on the equator, with one of the world's busiest shipping industries and has one of the most advanced coastal development strategies, and yet it's located in one of the richest regions in the world for marine biodiversity: how do you reconcile these factors to benefit the maximum number of people and accommodate their diverse interests?

The rationale leading to the formation of TMSI was based on the following: Singapore and Southeast Asia depend heavily on the sea for food, trade, economic development, and quality of life, but there were conflicting demands on sea space and coastal resources: land reclamation for new developments; seafood to supply the region's culinary culture; clean water for a myriad of purposes; outfall discharges; oil refineries, and port and residential developments. Marine science and engineering and the knowledge it would bring to the successful integrated management of the country's marine resources was therefore a critical discipline to ensure sustainable development. Singapore was well positioned on a number of levels to be one of the leaders in new knowledge and skills development, through teaching, research and practical application of the findings, and to be part of a global network in marine sciences and engineering which is particularly important since many marine problems and related land-based issues are trans-boundary.

Because the world's leading marine science laboratories were located in temperate or cold waters, we reasoned that our aspiration to be the best in tropical waters should not only be achievable but almost an obligation since there was an enormous gap to fill; and there was not only a national need for such R&D but strong international interest, subsequently born out by TMSI's many international collaborations. It was a big dream at the time but given all the fundamental positive aspects, we felt sure we could bring it off — all it needed was unflagging enthusiasm!

Until the mid-1990s, core competencies existed in the National University of Singapore (NUS) and external agencies but there was no multi-disciplinary critical mass in any one of these organisations. In 1993, the case for a TMSI began to be mapped out in response to a need for coastal research facilities by the Department of Biological Sciences headed by Professor Tom Lam, with support from Professor Chou Loke Ming, and a research scientist from the Department of Medicine, Dr Elizabeth Taylor who wanted to explore dolphin research, underwater vision in its widest sense and marine information technology, and who saw both the need and potential Singapore had for an integrated marine science and engineering programme. In order to do this and develop the necessary critical mass to support the proposal to establish a marine science institute, NUS needed to engage various government agencies such as NIE (National Institute of Education), AVA (Agri-Food and Veterinary Authority), ENV (Ministry of the Environment), STB (Singapore Tourism Board), EDB (Economic Development Board), MPA (Maritime and Port Authority) and SSC (Singapore Science Centre) to demonstrate there was a strong case and a national need. This was the first time there had been a 'bottom-up' proposal to establish a university level research institute and the proponents had to fight hard to realise their dream. Fortunately, they had very effective partners and mentors in Professor Lui Pao Chen and Professor Bernard Tan from the Department Physics, and NUS Deputy Vice-Chancellor, Professor Hang Chang Chieh. We started small.

Professor Tom Lam's research group had a long-standing interest in fish biology and aquaculture, and Professor Chou Loke Ming was very active in marine biology, particularly coral reef ecology. Joining them in 1994, Elizabeth Taylor established a Dolphin Study Group within the Department of Biological Sciences with the principle aim of studying dolphin cognition and communication. In January the following year a Physical Oceanography Research Laboratory was formed, headed by Professor Chan Eng Soon and hosted by the Department of Civil Engineering, and in 1996, the Acoustic Research Laboratory, headed by Dr John Potter was created and supported by research project funding from DSO (Defence Science Organisation) National Laboratories and hosted by the Department of Physics. (This was quite an unusual event since John had literally sailed across the Pacific Ocean from California in his own boat with his young family to join us!) The strategy was always to create win-win situations so that collaborating parties would gain from working together — and these three new laboratories were natural partners with each other and their host departments. By this stage, with groups from the Faculty of Engineering seen as essential complementary elements, Professor Hang Chang Chieh was pivotal in supporting Dr Taylor's concept that TMSI could not be established in any one faculty but had to embrace its multi-disciplinary elements and be viewed as a university-level organisation. We were grateful that NUS Vice-Chancellor, Professor Lim Pin agreed and understood the far-reaching potential of this approach.

The synergy between marine science and engineering, national defence and other national agencies such as the Agri-Food and Veterinary Authority (AVA — then Primary Production Department (PPD)), and the Singapore Tourist Promotion Board (STPB) enabled TMSI planning committee, headed by Professor Tom Lam to achieve unprecedented multi-agency research funding for the interdisciplinary research programmes being discussed between NSTB (National Science and Technology Board), EDB, DSO National Laboratories and NUS. This confluence resulted in the Tropical Marine Science Initiative being established in December 1996, directed by Prof Chou Loke Ming with Elizabeth Taylor as Deputy Director. A research home was found in one of the ex-British Army houses hidden away on pleasantly green and leafy Kent Ridge: it was was the first building of its kind to be used for university research. From this small base and nodes within the various NUS faculties, the Tropical Marine Science Institute was established in April 1998 after resolution of land issues regarding a groundbreaking development on St John's Island. It was the first university-level research institute. But not having the sparkling buildings and generous resources of other institutes, we had to be creative and for a few years operated largely out of laboratories and offices in stacked

By working in stacked containers on the NUS campus, TMSI got off to a rapid start.

containers — a decidedly unglamorous temporary solution but it worked and we had good views out to sea and a secluded location to unload field equipment after dive trips by researchers who's hair was still wet.

TMSI still needed a coastal research facility with easy access to the sea, in clean water with healthy coral reefs, beaches and inter-tidal flora and fauna, protected shallow water, and relatively deep but calm water nearby for core research such as underwater acoustics — not much to ask really! Coincidentally, the AVA required a new aquaculture facility, again with direct access to clean, calm and deep coastal water and there was potential for research collaborations. Thus, it seemed sensible for the two organisations to be sited next to one another. The STPB's Southern Islands Development created ideal conditions for both parties and in a happy coincidence, the STPB, and in particular Mrs Pamelia Lee, saw the opportunity for public education arising from co-locating both parties on St John's Island and she strongly supported the proposals. Marine science had been of particular interest to the late Chairman of the STPB, Mr Leong Chee Whye, and his vision continued to influence Mrs Lee's own interest in the development. TMSI created an education programme in 2002 run by Dr Taylor, and this foresight has been rewarded. In 2015, NParks established the Sisters' Islands Marine Park that includes St John's and Pulau Tekukor, and has set up a public visitor and education centre within TMSI's facilities. Not many NUS staff have had to start out buying and assembling their own furniture but this group has seen its hard work pay off and can look back and smile.

Back in 1998, it was the first time NUS had built an off-campus research facility — and not only off-campus, but offshore! It was quite an undertaking since services on the island were minimal, and was achieved *via* a loan from NUS of $6 million re-payable over 11 years. DSO National Laboratories provided funding of $4 million to support research infrastructure, and NSTB (now A*STAR — Agency for Science, Technology and Research) provided fundamental equipment for the research programmes they were funding. TMSI found an ideal location on St John's Island with strong and welcome support from STPB, which was responsible for developing the site. NUS's application was approved by URA and Land Office at market rate cost, so TMSI had to step up to the mark and find a way to re-pay this debt — not an arrangement for the faint-hearted. However, our funding level was deemed sustainable and we worked out a way to do it. We also managed to generate our own electricity and provide, mainly by desalination, the freshwater used in the laboratories, admin and dormitory block, and both raw and filtered seawater for the various aquaria that was drawn from about fifty metres out to sea and then pumped up a steep cliff over 25 metres high. We recognised the silver lining of the latter obstacle and the last undeveloped sea cliff in Singapore proved useful for the education programme and a research project on coastal biodiversity.

Transport to the island was a big issue but after some hair-raising blips, has been solved to the satisfaction of our staff and visitors. Travelling to work by boat and taking all your supplies with you is not as easy at it might sound, but, once they step ashore our staff are rewarded by the tranquil environment of the Southern Islands. It's always a pleasure to walk past a sandy beach with clear blue water on your way to and from your laboratory. Sometimes dolphins are seen just offshore: Indo-Pacific humpback dolphins or 'pink dolphins' and the more familiar bottle-nose dolphins which get everyone grabbing their cameras. Despite this natural setting, St John's Island is less than 30 minutes travel from the mainland — just a relative stone's throw away from the massive science and technology research infrastructure on mainland Singapore. Such a scenario is rare anywhere but even more so in the tropics, and the opportunities for novel research are immeasurable.

The daily commute to St John's Island is harder than it looks but arriving there is a reward in itself.

Why is marine science so important? Singapore began its current phase of rapid development as a trading node of the British East India Company, and the greater part of our history and fortunes has been carved as a trading port at this crossroads between the Pacific and Indian Oceans, and the South China Sea. The constraints of our land area on development appeared insurmountable, and yet, looking out to sea, the ocean spans as far and wide as the eye can see — a continuum limited only by our imagination and the means to venture forth. Shipping, trading and port operations are major pillars of our economy but bring related problems for environmental conservation, such as invasive species lurking in ship ballast water, oil and chemical spills, dredging and land reclamation that might smother coral reefs. Creating new opportunities while avoiding problems through marine science research appeared to be a promising way ahead. NUS Faculty of Engineering had staff undertaking research in oceanography, underwater acoustics (adopted from the Faculty of Science), and the new discipline of Marine Information Technology. The Faculty of Science maintained its strong tradition in marine biology and aquaculture, and the potential for greatly expanding marine biotechnology. Dr Taylor's concept of combining researchers from the Faculties of Engineering and Science into interdisciplinary teams would make breakthroughs not possible with conventional discipline-bounded teams. Fortunately, Professor Hang Chang Chieh and Professor Lui Pao Chuen, Chairman of TMSI's Management Board recognised and supported this new approach. In a way, it put us in no-man's land but that can be just the right place for discoveries and enterprise.

The two original NSTB funded research programmes — Marine Biology & Biotechnology, and Marine Aquaculture produced one of the main foundations for current research, and around two hundred and one hundred scientific publications, respectively, endorsed them as promising new research areas. Chan Eng Soon and Elizabeth Taylor proposed a follow-on but new, integrated research programme to study the marine environment that was fully funded by A*STAR. We were very pleased and the TMSI surged forward.

Marine Information Technology was moving in the same direction on a solid, different but always interacting track. This was a very complex area, which posed the question: how can you operate underwater in coastal zones in the most informed way? There were a lot of challenges and lots of opportunities.

In 2002, when the St John's Island's facility was officially opened by George Yeo, Minister for Trade and Industry, and NUS' campus laboratories were expanding rapidly, the main research thrusts were Marine Environment, Marine Food Safety & Security, and Naval Research. We needed to answer scientific questions about water quality and dynamics in relation to land reclamation, outfall discharge, oil and chemical spills, ballast water exchange, sustainability of benthic biodiversity with emphasis on coral reefs, inter-tidal zones, biofouling, toxic algal blooms, underwater acoustics and many other inter-related issues that were critical to environmental management. Application of the findings was just as important and we worked closely with partners such as MPA (Maritime and Port Authority of Singapore) and PUB (Public Utilities Board).

Opening of TMSI's Research Facility on St John's Island by B G George Yeo, then Minister for Trade and Industry, on 30 October 2002.

TMSI on St John's Island offers excellent research and teaching facilities.

In order to do this more effectively, TMSI created one of the first Marine IT developments in the world, something that's still right up there as an example of four-dimensional, real-time modeling. The Physical Oceanography Research Laboratory's (PORL) complex models and the Ecological Monitoring, Informatics and Dynamics (EMID) Group's data management systems are continually upgraded and refined with input from on-going field studies and NUS Centre for Remote Imaging, Sensing and Processing (CRISP). The resulting models of our coastal waters and influences on them from neighbouring regions are used by numerous national and international bodies. Complementing and extending the output and utility of this work together with that of the Marine Biology and Ecology Research Laboratory, the EMID group takes a cross-disciplinary, quantitative approach towards the overarching goal of sustainable resource management.

Half of Singapore is under the sea. Therefore, this is a very important research and development area for a land-starved country with huge investments in R&D that can readily be applied to the immediate geographical region and beyond. Safe aquaculture and desalination of seawater are critical survival projects, and Singapore has been quick to realise these objectives. TMSI develops and tests leading-edge marine technology for national defence and commercial system prototypes using integrated computer design tools. Elizabeth Taylor's original vision of multi-sensory integration for enhanced operational efficiency was at the forefront of TMSI and has led to technology that improves underwater vision and operations, and consequently safety, efficacy and performance of people working underwater. Underwater acoustic communication has been a strong growth area for the Acoustic Research Laboratory (ARL), which benefits from consistent funding support for this and their other projects such as dolphin sonar. The Marine Mammal Research Laboratory continues to inspire interest in dolphins and other marine mammals seen in Singapore waters, and works on underwater sensory perception — optical and acoustic (sonar) and marine mammal ecology.

In 2012, TMSI moved into its current upmarket home S2S — the gleaming green and blue glass, satellite dish-topped, 'Sea to Sky' building on the apex of Kent Ridge with all the vision the name and location affords. The purpose built research facility provides a very congenial base for both TMSI and its natural partner, CRISP. It still nestles in the leafy green top of Kent Ridge but provides a punchy shockwave to visitors.

S2S (Sea to Sky) Building housing TMSI and CRISP on Kent Ridge at the apex of the NUS Campus.

TMSI is not restricted to work in marine waters but now conducts scientific studies in freshwater bodies, and has contributed to national projects such as the Marina Bay Barrage and Ulu Pandan Reservoir. Neither is it restricted to the shallow water around Singapore but collaborates on deep-water exploration in the South China Sea.

In view of the value placed on close collaboration with NUS staff and students, our Directors and some members of staff held/hold dual appointments in the TMSI and a faculty. This arrangement facilitates interaction with students who are able to pursue a greater range of interests through internships, undergraduate research projects, and post-graduate degrees. International collaborators and students extended the range of projects that could be offered. TMSI's previous directors, Prof Chan Eng Soon (Dept. Civil Engineering) and Prof Peter Ng (Dept. Biological Sciences) inventively steered us through some tricky waters. The Institute's current director, Professor Wong Sek Man, a member of the Department of Biological Sciences is branching out to explore marine viruses. Dr Serena Teo is Deputy Director and has for many years worked on the economically important subject of bio-fouling and how to prevent this happening when it's not wanted and to encourage it when it is — a tricky balance; and Dr Tan Koh Siang is responsible for keeping our valuable facility on St John's Island running: maintaining electrical power generators was something he probably never bargained for when he said he was interested in studying marine molluscs, but the TMSI staff have proved they are a multi-talented and hardy bunch.

Our staff and students are so enthusiastic they've consistently fueled the heart of TMSI. Not only building their own desks, and maintaining power generators, but some willingly set out before dawn every day, and travel for an hour and a half on public transport, and end up jogging to catch a boat that they must not miss, heaving at its mooring in Pasir Panjang's unpredictable waters because they know that this level of enthusiasm is what really counts.

The announcement in 2012 by the National Research Foundation (NRF) and Maritime and Port Authority (MPA) of S$230 million of funding over five years for the development of facilities, infrastructure and research programmes for a new 'Maritime and Offshore' strategic research programme emphasised the commitment of the government to develop a scientific domain that is rapidly growing in importance. The St John's Island Research Laboratory, which has

TMSI's research vessel, Galaxea.

proved to be a valuable resource for both Singapore and international visiting scientists, is a beneficiary of the new NRF support and, as a new national research facility, the National Marine Laboratory hosted by the TMSI will provide a key research platform in a unique and important location. These new developments will offer avenues for bold new research initiatives built on firm foundations, and more opportunities for training the next generation of young and aspiring marine scientists. And let's not forget the intrinsic and consistently expressed interest of our young people. When they step outside their air-conditioned, artificially controlled environments, they are reminded they live on a tropical island on the equator with all the natural wonders that presents — and they want to preserve them. That's important.

Chapter 25

The Mechanobiology Institute — Defining a New Field of Science

Steven John Wolf, Low Boon Chuan and Hew Choy Leong

Introduction

In his 2013 address to the Comenius University in Slovakia, Dr Tony Tan, President of Singapore, described how Singapore now stands "at the cusp of a new phase of development that is based on knowledge and innovation". For a country just 50 years old, it is a remarkable feat to be able to claim a world-class research enterprise, and envision a future grounded in innovation and discovery.

As the country celebrates 50 years of independence in 2015, one of its greatest achievements is the establishment of a world-class research industry and higher education system that has pushed the boundaries of traditional science. By providing the infrastructure, funding, and freedom for researchers to explore new fields of science, Singapore has indeed become a leader in academic and innovative research and development.

The evolution of the higher education system, from comprising institutes grounded in fundamentals and traditional teaching methods, to institutes that encouraged the asking of questions that no textbook had the answer to, has been a major contributor to where Singapore is today.

This chapter explores a Research Centre of Excellence hosted at the National University of Singapore that is dedicated to understanding the emerging field of mechanobiology. Although still relatively new, the field of mechanobiology seeks to understand how cellular functions are driven by the integration of mechanical force, cellular geometry and biochemical activity.

The Mechanobiology Institute, Singapore (MBI) was built upon unique models that encouraged interdisciplinary research and provided an open-lab design where researchers from different professional backgrounds worked side-by-side, with shared facilities and infrastructure. Common to everyone at the institute was an overall drive to define and tackle a new field of science.

Although grounded in the principles of basic research, the discoveries made within the institute have proven to be highly translational. In less than a decade, new technologies have been patented, diagnostic tools brought to clinical trial, and several industry partnerships established.

In this chapter we look at how the institute was established, the model on which it was built and the research it conducts. Although younger than Singapore itself, the field of mechanobiology has already presented a new paradigm for those interested in biology, physics, chemistry and engineering. The influence of mechanical force on cell behaviour and growth states cannot be ignored, especially in diseases such as cancer, or in physiological processes such as wound healing. With continued support and through education and community engagement, MBI and Singapore are well placed to continue leading the way in this new and exciting field of research.

Establishing a New Institute in an Untested Field

On 1 January, 2006, the National Research Foundation (NRF) was established within the Prime Minister's Office. This government body was formed to oversee the direction Singapore would take in terms of research and development, and while its establishment formalised the importance of R&D to a growing economy, it also signalled a shift in ideas regarding the role of universities, both locally and internationally. To perform and become competitive on a global scale, the local universities had to reassess what it meant to be institutes of higher learning; they had to refocus their emphasis on traditional methods of teaching and incorporate research-based elements that could compete on an international scale.

To place the National University of Singapore and Nanyang Technological University ahead of other universities in the region, and cement the notion that research in novel and competitive programmes would boost their reputation as institutes of higher learning, a number of Research Centres of Excellence (RCEs) were established by NRF. These centres were co-funded by the Ministry of Education (MOE), and were aimed especially at creating new knowledge in highly specialised areas of study. Between 2007 and 2011, five RCEs were established. The Mechanobiology Institute, Singapore was established as the fourth RCE, with a commitment from NRF and MOE of S$150 million over 10 years. What was conceived was a research institute built from the bottom up where the overall themes and research directions would be defined by the work of specific research groups. This very much reflected the principles driving the field of Mechanobiology, where individual molecular or cellular processes directed the overall functions of the cell or tissue system.

Being hosted within the National University of Singapore proved synergistic. The provision of infrastructure and support eased the logistics of establishing a new facility, while MBI's association with NUS would ultimately enhance the research capabilities of the university, and contribute to its reputation as an internationally renowned centre of learning. Over time, this would contribute to Singapore's entrepreneurial economy.

A New Formula for Successful Research

The establishment of MBI was the culmination of a long and difficult task of acquiring funding and recruiting faculty members who had expertise in a range of fields, from pure cell biology to physics and computational sciences. Instrumental in this task was Emeritus Professor Hew Choy Leong, a prominent local scientist who was trained in the US and Canada. After spending more than 30 years as a professor at the Memorial University of Newfoundland and the University of Toronto in Canada, Professor Hew returned to Singapore in 1999 to head the Department of Biological Sciences at NUS, and served as the founding member and Director of the new Office of Life Sciences in NUS. It was a new mandate to establish strong, competitive and multidisciplinary research programmes that would propel NUS as a global and research powerhouse. The recruitment of world class scientists and the cutting edge research would be two of the minimal requirements.

To that end, Professors Michael P Sheetz from Columbia University, and Paul Matsudaira from the Massachusetts Institute of Technology (MIT) were appointed Director and Co-Director of MBI respectively. Being a pioneer of the field of mechanobiology, and with over 30 years' experience in biophysical sciences, Professor Sheetz brought unparalleled knowledge of the field, and was instrumental in setting the theme and direction in the initial proposal to establish the institute. His leadership was complemented with that of Professor Matsudaira, who brought expertise in tissue bioimaging and the cytoskeleton. A faculty was recruited that consisted of bioinformaticians, computational scientists, experts in signal transduction and cytoskeletal dynamics, the physics of cell migration, microfluidics and the engineering of microfabricated devices, cell adhesion and systems biology, nuclear dynamics, and even bacterial pathogenesis.

From 2006, this fledgling institute, originally known as the RCE in Mechanobiology, was given laboratory space in the Brenner Centre for Molecular Medicine at the National University of Singapore (NUS). With the support of Professor Judith Swain of the Singapore Institute for Clinical Sciences, the RCE was provided the infrastructure to support five laboratories, and approximately 50 personnel. The focus in these early days was on building

Fig. 1. Official Launch of MBI (left to right) Director of Projects of the National Research Foundation, Dr Lawrence Koe; Director of the Mechanobiology Institute, Prof Michael Sheetz; Permanent Secretary of the Ministry of Education, Mrs Tan Ching Yee; and NUS President, Prof Tan Chorh Chuan.

an imaging infrastructure that would later support the needs of biophysicists and cell biologists needing to explore the inner workings of cells with ever increasing detail. It was within this context that the first Mechanobiology Conference was held with an aim to identify the key research areas or themes.

In 2009, the Mechanobiology Institute, Singapore, was officially opened. The institute was housed within the newly built T-lab building at the National University of Singapore's Kent Ridge campus. Laboratories were spread out over levels 9 and 10, which were purposefully designed to be open spaces, without walls between researchers. Additional office space was provided for administration staff, facility managers and students on level 5 of the same building. By 2015, the number of principal investigators had expanded to 19, with an additional seven co-principal investigators. Dr G V Shivashankar was appointed as Deputy Director to provide scientific stability and growth, and Professor Hew Choy Leong served as Senior Advisor and Emeritus Professor. This brought the size of the institute to over 150 research staff and students, with around 50 facilities and support staff. A further 18 prominent researchers from around the world were recognised collaborators, and involved in regular visits to the institute.

A Model of Interdisciplinary Research

Seeking to understand how biological systems respond to physical stimuli is inherently interdisciplinary as it integrates methods that traditionally belong to the fields of biology and physics. For example, designing experimental systems that enable biologists to investigate the effect of force application, or control the microenvironment of individual cells, requires the expertise of engineers and computational biologists.

For this reason, MBI placed a strong emphasis on cultivating an interdisciplinary and collaborative research environment from the outset. Principal Investigators with a diverse range of interests and expertise were placed together under one roof, and encouraged to share ideas and hypotheses. Institute-wide events and social gatherings, a weekly seminar series attended by all research staff and students, and annual conferences provide valuable platforms to nurture collaborations between researchers of distinct disciplines.

As described by MBI Director Professor Michael Sheetz, "Small institutes provide good chances to bring together people with vastly different backgrounds and to encourage them to be adventurous."[1]

Within the first few years early collaborative efforts began to yield significant scientific discoveries and a culture was established where ideas could be freely discussed and innovative methods explored. With much of the research carried out at the institute pushing the boundaries of current methods, innovative techniques and new technologies used in the lab proved essential in answering many of the questions posed by MBI scientists. The success of such innovation was heavily reliant on open discussions between scientists coming from different perspectives.

To encourage continued collaborations, and to take full advantage of the wide range of skills and expertise shared amongst researchers, the institute adopted an 'open lab' concept. This was not only evident in the physical layout of the laboratories, where there were no physical barriers, partitions or walls in place between researchers, but also in the designation of lab space.

General lab benches were provided to all researchers using a random allocation system. This meant that benches would be shared by researchers from different labs. Colleagues working on distinct projects and under the supervision of different principal investigators found themselves to be neighbours at the lab bench despite coming from completely different scientific backgrounds. A physicist could be allocated bench space next to a biologist or a chemist and this ultimately aids in the casual sharing of knowledge and encourages less familiar experimental techniques to be trialled.

Centralised Facilities

Key to the success of the open lab concept is the centralised management of all day-to-day tasks. In a traditional research institute, individual labs are responsible for purchasing their own consumables, maintaining their own lab equipment, drafting their own protocol procedures, safety assessments and reporting on compliance on animal care and the responsible conduct of science, etc. Although necessary, these tasks can all detract from the time spent conducting research and can easily see a principal investigator bogged down by administration.

With the open lab concept, where consumables, chemicals and general lab equipment are shared, the administration and upkeep of the laboratories does not fall under the responsibility of any one lab or researcher. Instead a core team of dedicated staff provides the necessary support for all researchers, including the maintenance of lab equipment and consumable stocks, lab safety and training and much of the paperwork that ensures the institute meets the required safety standards.

This system of shared resources was further extended to include major research infrastructure, such as microscopy and microfabrication facilities, as well as a protein expression facility. Each facility is managed by PhD holders experienced in scientific research.

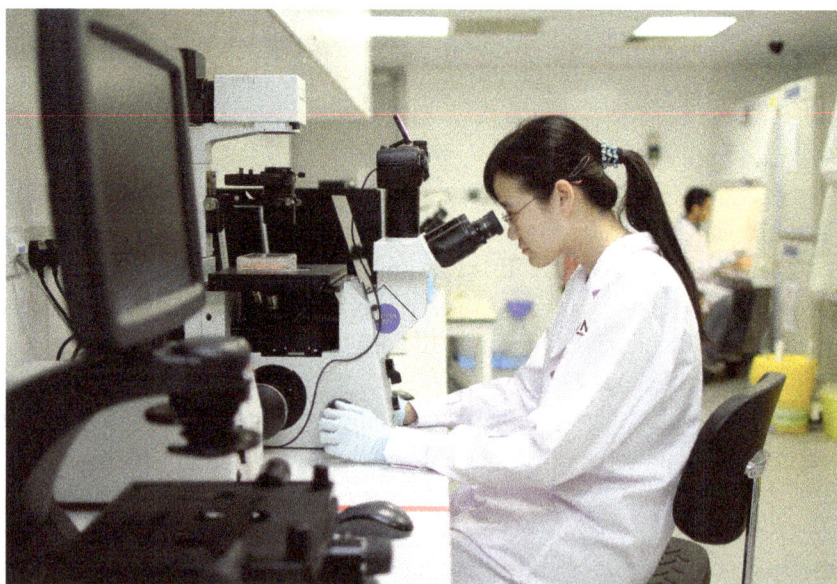

Fig. 2. The MBI open labs provide shared facilities, including several tissue culture facilities.

A microscopy core facility provides researchers with state-of-the-art imaging equipment that includes a collection of confocal fluorescence microscopes equipped with spinning disk units and chambers for live cell imaging. Two TIRF (Total Internal Reflection Fluorescence) microscopes are available allowing visualisation of membrane bound proteins and even single protein molecules. To further enhance MBI's imaging capabilities, a suite of super-resolution microscopes are also available. These newer systems, which include SIM, STORM and PALM microscopes, use the newest technology to provide superior resolving power and allow researchers to image cellular components such as actin filaments and individual adhesion complexes at high resolution.

A microfabrication facility is also available to MBI researchers, providing technologies that are used extensively in the lab. This facility, which is shared with the NUS Nanoscience and Nanotechnology Institute, comprises a class 10K cleanroom with a range of equipment including a scanning electron microscope, a thermal evaporator, a mask aligner and other equipment required to manufacture microfabricated devices. Furthermore, the microfabrication facility provides independent tools for in-house PDMS molding and mask design. Some of the devices commonly produced by the facility include

Fig. 3. The microfabrication facility at MBI.

micro or nanopillar arrays on which cells can be grown, defined geometrical patterns for protein stamping, microwells for growing cells and microfluidic devices. Finally, chips with a defined topography can also be produced.

Additional support units are also provided to free researchers from spending time on administration and other day-to-day matters. The IT core facility, for example, manages the storage and backups of all data generated at the institute. This unit also provides in-house image analysis expertise and works hand-in-hand with researchers. The Science Communications unit, on the other hand, provides editing and illustration services, produces media releases, and plans and runs outreach events. This unit also maintains a website that serves the mechanobiology research and student communities with descriptions of the cellular processes and molecular mechanisms relevant to mechanobiology. This website, known as MBInfo (www.mechanobio.info) serves as a hub for the field, and encourages contributions from the research community.

Establishing International Collaborations

To strengthen ties with the wider research community, and continue to develop international collaborations, the MBI has entered into a number of Memorandums of Understanding (MOU) that enable both shared funding and research opportunities.

In 2014, an official collaboration was established with the FIRC Institute of Oncology (IFOM), which is based in Milan, Italy. This led to the establishment of a Joint Research Laboratory (JRL) based at the Mechanobiology Institute and the IFOM-NUS Chair Professorship, which is held by Dr G V Shivashankar. A primary focus of the JRL is to establish how geometric constraints placed on individual cells, or on cells within tissues, affect the cellular response to cytokines.

A similar arrangement was established with the signing of a collaborative research agreement between MBI and the Centre National de la Recherche Scientifique (CNRS), Paris, France in 2011, and the establishment in 2014 of a joint international lab headed by Dr Virgile Viasnoff. This joint laboratory focuses on the mechanisms that regulate cell contacts and the formation of cell-cell junctions, mechanosensing, as well as the establishment of cell polarity. Other MOUs are also entered into with Waseda University, Japan; the National Center for Biological Sciences, Bangalore, India; the Weizmann Institute of Science, Israel and the University of Bremen, Germany. These collaborations led to regular student and faculty exchange, joint symposiums and joint research projects.

Further collaborations are ongoing, and are strengthened by the continued Visiting Professor programme at MBI. Through this programme, overseas collaborators of MBI may spend a month or more at the institute, initiating or continuing experiments, meeting with faculty and students and bringing expertise that is crucial to an interdisciplinary institute. The visiting professors have hailed from Japan, France, Germany, Israel, India, Spain, the United Kingdom and the United States of America.

To build further bonds with other researchers in the field, regular Mechanobiology themed conferences and workshops have been held. The first conference in October 2007 was to establish the themes of the institute. Since then, conferences were focused on specific themes such as cellular mechanics (2008), the dynamic architecture of cells and tissues (2013), the mechanobiology of chromatin and transcription (2012), and the mechanobiology of multicellular systems (2011). Additional collaborative conferences have been held together with the FIRC Institute of Molecular Oncology, Italy; Manchester University, UK; the Weizmann Institute of Science, Israel; Waseda University, Japan; the Institute for Molecular Biosciences, Australia; Japan Society for the Promotion of Science, Japan, and the Biophysical Society, USA, to name a few.

Fig. 4. The 5th Mechanobiology Conference, 2011.

Training Future Scientists

The MBI recognises the importance of training future scientists. To achieve this, MBI hosts a highly competitive postgraduate programme that places students in an established lab of their choosing for up to five years. During the first six months of their programme, students rotate around different labs to experience a range of disciplines and develop their interests on specific topics. As well as research and a doctoral thesis, students engage in coursework specifically designed for the field of mechanobiology. This curriculum includes four courses recognised by the National University of Singapore that introduce the principles of mechanobiology: MB5101- Cell as a Machine, which provides an in-depth analysis of biological functions, MB5102- Special Topics in Mechanobiology, which aims to train students in critical thinking and presentations, and MB5103- Trends in Mechanobiology. Most recently, a compulsory Bootcamp cum module MB5104- Integrative Approach to Understand Cell Functions, was also introduced. This module aims to equip new students with the core principles and techniques in biology and physics. Since the establishment of the institute, over 60 postgraduate students have enrolled in the programme, with the first batch graduating with PhDs in Mechanobiology in 2015.

Fig. 5. The first batch of MBI graduates with MBI Director Prof Michael Sheetz in 2015.

MBI also places a strong emphasis on the opportunities afforded to its postdoctoral candidates. By September 2015, just seven years after its establishment, almost 100 research fellows had been trained at MBI. A further four senior research fellows had been granted the opportunity to establish their own research programmes. In each case, an emphasis is placed to ensure the postdoctoral fellow receives strong interdisciplinary training that prepares them for an independent research career as principal investigator, in both the research and academic environment.

Engaging the Community

With the Singapore government providing solid support towards both infrastructure and research funding, MBI recognises the importance of continued community engagement and education. Whilst the recruitment of an established and expert faculty was beneficial in the initial stages of setting up the institute, encouraging students at high school, or undergraduate level to pursue science subjects and develop an interest in basic and translational research is essential in the long term. This is particularly important if Singapore is to continue as the research hub of Asia.

To this end, an outreach programme for school groups was initiated. During a visit the students are introduced to the field of mechanobiology through short seminars before touring the lab facilities. During each tour, the students take part in demonstrations on areas such as super-resolution microscopy, microfabrication, magnetic tweezer experiments and general biological research methods.

In addition, efforts are continuing within the institute to develop the online resource, known as MBInfo (www.mechanobio.info), which serves as a hub for all information, news and features on mechanobiology. This resource not only provides freely accessible information on the general concepts and specific molecular processes studied within the field, but it also allows members of the research community to redefine concepts and incorporate the most recent findings on a given topic. Although the resource is aimed at undergraduate or postgraduate students, as well as scientists in alternative disciplines, MBInfo provides anyone with an interest the chance to learn about the growing field of mechanobiology. This is possible due to the detailed explanations written by experts in the field, along with illustrations and animations designed to clarify the more complex descriptions.

Mechanobiology has Emerged at the Interface of Biology and Physics

At the heart of MBI, however, is its research. Driven by an aspiration to understand how biological systems integrate physical forces and mechanics, MBI continues to define the field of mechanobiology through basic research, and technological innovation.

All biological systems, whether individual bacteria cells, or large multicellular organisms like mammals, exist in 3D environments and are constantly under the influence of physical forces. Just as humans have sensory systems that determine the conditions of our environment (smell, sight, taste, hearing and touch) so too do cells. These systems allow the cells to detect and measure the physical or mechanical properties of their immediate environment. For example, cells may be compressed or stretched or even subjected to a flow of fluids over them. In many instances, the cells would need to test the stiffness of a surface or surrounding matrix. In some cases, the sensory systems within cells are as small as single proteins. However in most cases, they involve large multi-protein complexes that interact with networks of protein-based cables that span the entire cell.

Mechanobiology, by definition, explores how biological systems integrate mechanical or physical forces into their formation and function. To understand how a particular biological system forms and functions, researchers must understand the environment in which it forms, and the forces that define that particular environment. For this reason, the research at MBI integrates cell and molecular biology, as well as physics, and in particular, biophysics.

When cells are living as part of larger multicellular systems, such as in tissues like skin or muscle, then the cellular response to a given force must be coordinated with neighbouring cells. This requires systems of communication between cells that can transmit the mechanical signals. Such communication involves both biochemical and mechanical processes. The term used to describe the transmission of mechanical force within a cell, or through tissues, is **mechanotransduction**.

Importantly, the transmission of force is rarely one way. Cells will generate forces themselves and this facilitates cell migration, allows the cell to pull on the surface to test its stiffness, and to push or pull on its neighbouring cells in a tissue.

Basic Research Programmes

The research in MBI is structured around **four main research programmes**. Each programme approaches mechanobiology and mechanotransduction from a distinct perspective.

One programme, which is led by MBI Director Professor Michael Sheetz, looks at the **Molecular Mechanics of Mechanotransduction**. Researchers investigate how individual proteins are able to confer cells the ability to detect and measure the mechanical properties of their microenvironment. This requires the researchers to look at all regions of the cell, as mechanical cues can be detected from the cell periphery, where individual proteins can be stretched by force, to the cell nucleus, where DNA is tightly packed. These forces are then integrated through the specialised activities of individual proteins, or protein complexes. Of particular interest is the effect of stretching or compressing cells, on cell growth and proliferation. Cell stretching can occur, for example, during everyday activity and exercise. The forces generated from repetitive stretching events have been found to promote cell spreading, growth and proliferation.

Another programme, led by Professor Alexander Bershadsky, investigates the interactions between neighbouring cells, or between cells and specific proteins or molecules in their immediate vicinity. **The Cell-Matrix and Cell-Cell Mechanotransduction** programme reveals how properties of the extracellular matrix (which is a mixture of proteins and other factors that support cells in the body) can modulate cellular functions, such as migration, proliferation and differentiation. The internal architecture of the large multi-protein complexes that connect the cell to the extracellular matrix is dynamic; changing in size and composition in response to external as well as cellular forces. Work from this programme has also described how forces will induce the reorganisation of the cytoskeleton (which is a network of protein cables in the cell) and thereby direct changes to the cell shape, how cells move or how they divide.

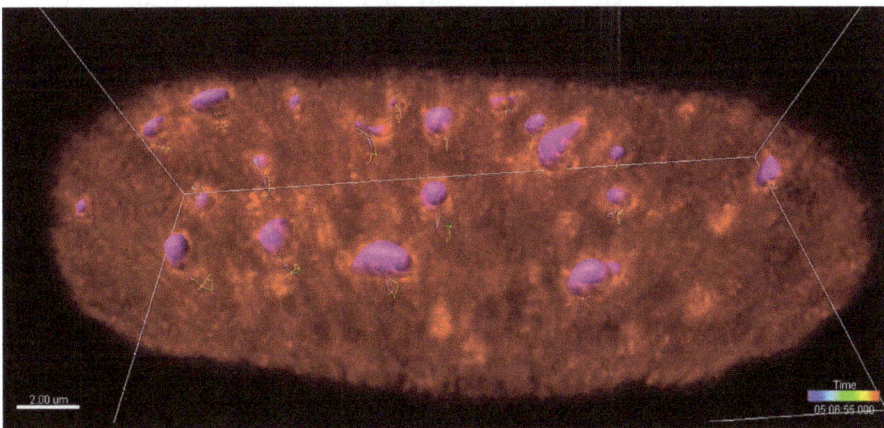

Fig. 6. Mechanobiology of Nuclear Dynamics and DNA packaging is investigated by the Shivashankar Lab.

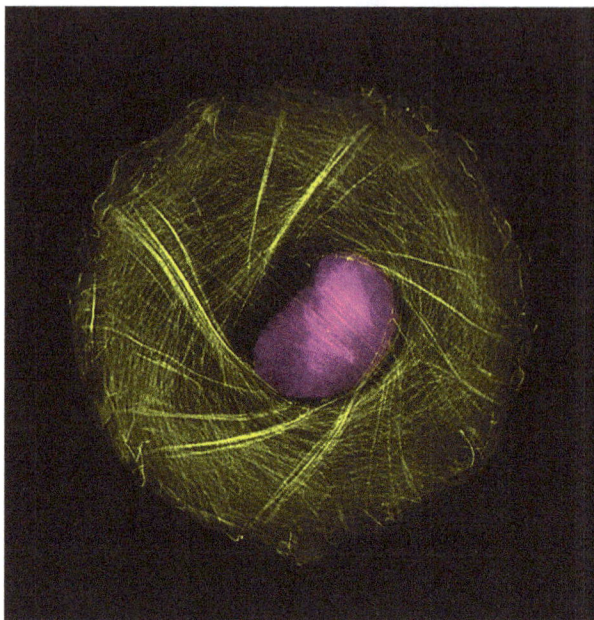

Fig. 7. The Bershadsky Lab researches how cellular behaviour is directed by the formation of the actin cytoskeleton (yellow). The cell nucleus is also shown in purple. Image Credit: Dr Tee Yee Han.

A third programme, led by Professor Paul Matsudaira, looks at **Mechanotransduction in Tissues**. How cell communities respond to the forces generated in tissues such as muscle or skin can be quite different to how they respond individually. Large scale forces are applied to tissues through every day activity, and cells must integrate these forces to both maintain tissue integrity, and repair tissues following injury or disease. Furthermore, mechanical cues can direct cell specialisation and organisation during organism development. The templates for tissue size and shape, as well as their organisation in the body, are set very early in development, and understanding these processes is integral in the development of new therapies for age related, or chronic diseases. Work from this programme has revealed several properties of the wound healing process. When wounds or gaps arise in sheets of epithelial cells, which line the stomach cavities, blood vessels, and skin, the cells surrounding the gap will work together to close the wound. This highly mechanical process can involve collective cell migration (where whole sheets of cells move together), cell contraction and the extension of cell projections in to the wound. The direction of forces into the surrounding environment, and the physical properties of the gap (size and shape) have all been found to influence how cells can repair the wound.

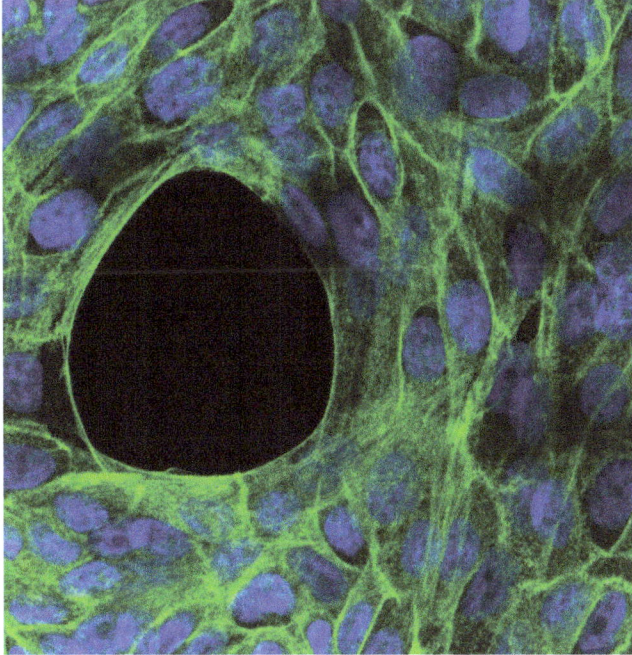

Fig. 8. Cells close a gap in an epithelial cell sheet. Image Credit: Dr Vedula Sri Ram Krishna and Prof Chwee Teck Lim.

Technological Innovations that Drive Today's Research

The fourth research programme at MBI is dedicated to the discovery, development and application of **Technology Innovation for Mechanobiology**. This research programme is led by Professor Chwee Teck Lim.

The methods by which researchers observe biological responses to physical stimuli is becoming increasingly quantitative. This means that new tools and techniques are required, both to control specific elements of the cell's environment, and to measure force exertion, accurately calculate changes in cell size and shape, and predict how a cell will respond under physiological conditions.

To achieve these innovative methods, MBI encourages biologists to collaborate with physicists, engineers and computational scientists.

Microfabrication technology has been used to produce devices that enable cell growth in environments with well-defined physical properties. For example, chips with defined topographies or patterns have been used to investigate how the properties of the substrate direct cell growth. In these cases the chip's surface chemistry, as well as its rigidity or stiffness may also be altered. Of course, being able to control the cellular environment is meaningless if the cellular responses cannot be observed. To this end, MBI researchers have focused on engineering a chip that enables single cells, or cell doublets (two cells joined together) to be grown in controlled conditions, and viewed on a microscope. Developed by Dr Virgile Viasnoff for use on conventional inverted microscopes, this technology brings optical capabilities similar to high end super-resolution microscopes. Termed so-SPIM, or 'single objective-Selective Plane Illumination Microscopy', the method has allowed MBI's researchers to view the formation and function of individual cell-cell adhesion complexes.

Other studies, particularly those aimed at defining the process of collective cell migration and wound healing, have benefitted from the design and production of substrates, such as microfabricated pillars, that allow cell movement to be monitored. These substrates also allow the forces generated during cell movement to be measured. Here, cells are grown on arrays of microfabricated pillars. As the cells move, the pillars are deflected. This deflection is observed and translated into a measure of force and direction.

Finally, an ongoing effort is underway to develop new diagnostic tools for cancer and other diseases. The work of Professor Chwee Teck Lim, for example, has focused on developing a microfluidics-based chip that isolates and captures floating cancer cells from blood samples. This work has continued through to the development of a chip that allows the captured cancer cells to be grown in the lab. These cell cultures retain many of the properties of those cells originally captured, which is important as researchers analyse the cells to determine why they became cancerous, and what properties could be targeted to prevent the formation of secondary tumours, or even stop the cancer cells from spreading.

Looking Forward

Mechanobiology presents a new paradigm to traditional fields of research, which largely treat biological phenomena as distinct from the physical sciences. It has become increasingly clear that all biological systems, whether single cells or whole organisms, possess ways to interact with their environment. Mechanical properties such as the stiffness of the surface under a cell, the flow of liquids over a cell, and even the push and pull from neighbouring cells, direct the basic cell functions regardless of whether the cells are living independently, or as part of a tissue or organ. These same mechanical cues will even influence stem cell differentiation — which is the process whereby stem cells become a specialised cell type. To appreciate the importance of mechanics on cellular functions, the perspective from which researchers approach biological or physiological phenomena will need to change. Cellular functions can no longer be viewed independently of the physical environment, and it cannot be assumed that cell fates or functions are determined purely through chemical or genetic means.

The success of the Mechanobiology Institute is therefore dependent on the continued sharing of ideas and expertise that is inherent in interdisciplinary research. Since its establishment in 2009, MBI has come a long way in its contribution to Singapore's research and development efforts. A new paradigm has been presented and is being accepted throughout the international research community. With improved technology and new methodologies, the molecular processes that facilitate mechanosensing, mechanotransduction, and the cellular response will continue to be defined by researchers at MBI. This will undoubtedly lead to further advances in medical diagnostics, and translate to new technologies and industry partnerships. With Singapore's aspiration to become a world renowned research hub and continue to pursue further innovations, MBI is well placed to continue leading the world in the study of mechanobiology.

For more information on MBI, please visit mbi.nus.edu.sg. For more information on the science behind the field of mechanobiology, please visit www.mechanobio.info.

Reference

[1] Sheetz MP (2014) Can small institutes address some problems facing biomedical researchers? Mol. Biol. Cell, 25:21, 3267–3269; doi:10.1091/mbc.E14-05-1017.

Chapter 26

Mathematics Education in Singapore (1965–2015)

Lee Peng Yee

Introduction

Singapore is a city state with a population of five million. We often say that the size of a problem changes the nature of the problem. For example, being a city state, we could make changes more quickly if necessary. Over the last 50 years, we have moved from having no industries before independence in 1965 to having industries and more after independence. In Singapore, education is regarded as an investment rather than a social service. In schools, mathematics is a compulsory subject up to Grade 10. In this chapter, we shall look back and record events concerning teaching and learning of mathematics in the past 50 years. We shall look at the K to 12 mathematics curriculum and its implementation vertically in time and horizontally in scope.

In perspective, there are two major events. One event is *Math Reforms*, a worldwide movement, in the '70s. It generated a series of changes in school mathematics till today. It is not only the change in content but also the localisation of syllabus and textbooks among others. Another event is the publication of primary school mathematics textbooks by the Curriculum Development Institute of Singapore (CDIS), Ministry of Education, Singapore, in the '80s. It made a great impact on mathematics teaching in Singapore. For a comprehensive reference to the years 1945–2005, see the two conference papers by Soh Cheow Kian,[1] Lee Peng Yee,[2] and references therein. The missing period, namely the years 2005–2015, will be covered in this chapter. As far as school mathematics is concerned, the highlight during the period is the mathematics syllabus implemented in 2013. It contains *learning experiences*, which makes explicit the learning processes. It includes *mathematical modelling*, which encourages students to learn mathematics in context. In the first six sections, we give a chronological story of the 50 years. In the last five sections, we elaborate on some important aspects of the story.

After Independence

Singapore was a British colony for about 150 years. It became independent in 1965. After World War II or 1945, there were not enough school premises. A school would normally run two sessions, namely, a morning session and an afternoon session. More precisely, a school was used by two different sets of students — one in the morning and another in the afternoon. So students had to wake up before sunrise for the morning session. For the afternoon session, students reached home after sunset. This went on for over 50 years. Eventually, the schools adopted a single session, but students still have to go to school as early as before. This is because the social structure such as school buses and family life was conditioned by the system for two generations. It was hard to change. Some schools did try to start schooling later in the morning, and it did not work out. Consequently, for all these years till now, teachers and students do not have a decent lunch at a decent time. Life goes on.

After independence, we were short of teachers. Also, not all teachers were qualified. It was much later in 2013 that we started training teachers mainly for replacement only. Primary school and secondary school teachers were trained differently. Grade 12 graduates were trained to teach in primary schools. For secondary schools, teachers required university degrees. This requirement was met only in the mid '80s. Classroom teaching was traditional. Teachers followed textbooks closely. Lessons consisted of mainly examples and exercises.

Singapore was and remains a multi-racial society. There are Malay, Chinese, Indian and other races. At the time before the late 70s, we had English and Chinese schools, and also Malay and Tamil primary schools for different communities. In English schools, the medium of instruction was entirely in English. In Chinese schools, the medium of instruction was in Chinese though students learned English as a second language. The same applied to Malay and Tamil primary schools. There were more Chinese schools than English schools. They ran independent of each other. In Chinese schools, students spent more hours in mathematics than those in English schools. In English schools, the pass mark was 50 (out of 100), whereas in Chinese schools it was 60. It was a general belief among parents that Chinese schools' students did better in mathematics than those in English schools.

The first local mathematics syllabus was drafted in 1957 and announced in 1959 before independence. It contained the mathematics syllabus from year 1 (Grade 1) to year 13. It was printed in English and Chinese. This was the first attempt to go local. Before that, everything was imported — syllabus, textbooks, and also teachers. It took about 20 years after independence for all schools of different language streams to merge into one, using English as the medium of instruction and a common syllabus. In the mid '60s, the Advisory Committee on Curriculum Development (ACCD) was formed within the Ministry of Education, Singapore. As a consequence, the secondary school mathematics syllabus, *Syllabus C*, was introduced in 1968. It started the so-called *Math Reforms* in the following decade. We shall describe in the next section.

Math Reforms in the '70s

It was a world trend. It started in Europe and North America in the '60s. It came to Singapore in the '70s. It was generally agreed that the three landmarks in mathematics education up to the 20th century are Euclidean geometry, calculus, and pure mathematics. In pure mathematics, rigour was an important element. A group of French mathematicians called themselves Nicolas Bourbaki, published a series of books called *Elements of Mathematics*. The books had a great impact on mathematics, and later mathematics education. Consequently, a new syllabus was proposed for the schools. In Singapore, it was called *Syllabus C* for secondary schools (Grade 7 to Grade 10). It was introduced in 1968, and revised in 1973. The primary school mathematics syllabus (Grade 1 to Grade 6) was also revised in 1971 and again in 1976, resulting in a reduction of content. We called the new syllabus, *Modern Math*.

The major change was in the content, more so in secondary school than in primary. The concept of "set" was introduced in Grade 7. In geometry, the term "a line segment" was used in addition to "a line". Also, "measure of an angle" was used in addition to "an angle". Different notations for minus 3 (–3) and negative 3 ($^-3$) were adopted. Commutative law, associative law, and distributive law were highlighted. Most of these terms disappeared afterwards. For good reasons, the term "a line segment" stayed. So did "Venn diagram". In short, *Modern Math* was structural and formal.

A serious casualty was geometry. Teaching geometry was for hundreds of years nothing but teaching proofs. This approach was abandoned to a great extent. A certain amount of transformation geometry was introduced. Consequently, school geometry was neither transformation geometry nor classical geometry. This went on till the end of the reforms. In fact, the matter has not been fully resolved even now. An attempt was made to introduce the concept of transformations into primary schools. Hence, tessellation was inserted in the primary school syllabus. It stayed in the syllabus for a long while. It was never an important part of the syllabus.

Mechanics was gradually replaced by statistics. Consequently, we lost two rich topics, namely geometry and mechanics, which were rich in content and in examination questions. One gave us proofs, and another showed us applications. Statistics was also in the primary schools in form of pictograms. Statistics was probably the only topic in modern math that was not of the 19th century. To show applications, linear programming was included. To make it available to secondary school students, a problem in linear programming was restricted to two variables so that it could be solved using the coordinate plane. More precisely, to find the maximum point in a linear programming problem is to construct a convex polygon and then find the corner point where the maximum occurs. It was a fashion at the time to teach advanced mathematics at an elementary level, even though the method used to solve a problem is not the one used in the real situation as in the case of linear programming. Tessellation is another such example.

The teaching approach had changed somewhat. For example, worksheets were used. Some teachers used Cuisenaire rods (colour blocks) to teach four operations. Discovery method was introduced in the west. By discovery method, we mean teachers did not give students a formula, for example, distributive law. Students were supposed to discover the formula by themselves. Normally, it would take one full lesson to teach the distributive law. The discovery method did not catch on in Singapore. However, it did influence what happened later on. Problem solving was in the centre of a syllabus in the 1980s and thereafter. In a way, problem solving was nothing but guided discovery.

Foreign textbooks were adapted for local use. The first locally-produced textbooks appeared in 1969.[3] Due to new content in the new syllabus, massive teacher training took place. This was the beginning of going local. More precisely, we had our own syllabus. We produced our own textbooks. We paid greater attention to teacher training and professional development of teachers. Meanwhile, two events took place in the educational scene:

- Science and mathematics were taught in English starting from 1976 but it was not fully implemented until 1984.
- Mathematics at the O-level (Grade 7–10) was made compulsory as an examination subject as of 1974.

From 1974 to 1977, more secondary schools switched to *Syllabus C* and the failure rates increased. It was not true that schools in the west all turned to teaching modern math. For example, it was known that only one third of the schools in the UK adopted modern math. In Singapore, almost all secondary schools, if not all, adopted *Syllabus C*.

Again, it was a world trend. Students performed poorly in algebraic manipulation after adopting the so-called modern math. In the west, there was a call of "back to basics" at the *International Congress on Mathematical Education* (ICME) 1980 in Berkeley, USA. As before, we followed. Since we did not go as far as some other countries, we also had less to change. *Math Reforms* may be negative for many countries. In Singapore, it hastened the localisation of mathematics syllabus, textbooks, and teacher training. In other words, there was a positive aspect of the *Math Reforms*. We shall describe in the next section what happened afterwards.

Back to Basics

As mentioned before, for those countries (including Singapore) adopting modern math, the standard fell. More precisely, examination results dropped. So we took a U-turn, known as "back-to-basics". As it happened, the time coincided with the implementation of the New Education System (NES) in Singapore in 1979. Under NES, students were streamed into two and later three streams. The objective of streaming was to reduce the attrition rates in the school system. The cause of attrition rates was partly due to the change of the medium of instruction in schools. Modern mathematics was another part of it. However, some years later, streaming helped to retain students in schools for 10 years and more.

At the time, some students left schools at Secondary Two (Grade 8). The "set" concept was not relevant to their jobs, for example, in the marine industry. Trigonometry was more practical. Hence, a new syllabus, called *Syllabus D*, was adopted in 1981. Modern mathematics was moved to Secondary Four (Grade 10). There was simple trigonometry in Secondary Two (Grade 8). *Syllabus C* was phased out progressively. Do not forget that when a new syllabus was adopted, there were still students using the old syllabus and taking examinations based on the old syllabus. So it would take two to three years for the new syllabus to replace the old syllabus entirely. Similarly, there was a new mathematics syllabus for primary schools. The changes in primary school were not as much as in secondary school.

Concerning streaming, it is easier to explain in terms of school leaving examinations. At the end of Primary Six (Grade 6), there is the Primary School Leaving Examination (PSLE). There are two versions of the examination: one based on the Standard Mathematics syllabus, and another on the Foundation

Mathematics syllabus. The latter is a subset of the former. Pupils who offer Foundation Mathematics spend time revisiting some early topics in their final two years. In other words, we adopt *differentiated syllabuses*. At the end of Secondary Four (Grade 10), there is a national examination. Again, there are two versions: O-level and N(A)-level. Similarly, we adopt differentiated syllabuses for the O-level mathematics. Here "O" stands for "Ordinary" and "N(A)" for "Normal (Academic)". As far as content is concerned, the N(A)-level is part of the O-level.

In the O-level examination, students normally take eight examination subjects. The minimum number of examination subjects is six. Following the British system, students could offer two subjects in mathematics, namely, Mathematics and Additional Mathematics. In other words, some students may study more mathematics than others. Though not required, most science students in the universities would have done two mathematics subjects in their secondary. As time went on, another N(T)-level was added. Here "N(T)" stands for "Normal (Technical)". The N(A)-level was formerly called N-level. The original idea was that those N-level students could join the O-level students if they had done well. Such transfer was encouraged though not common. Also, there was a group of students finding even the N-level too hard for them. Finally, after many years the N-level was divided into N(A)-level and N(T)-level. The N(T)-level syllabus is in a sense not linked to the O-level syllabus. Teaching approach of the N(T)-level also differs.

A new syllabus required new textbooks. Ministry of Education (MOE) set up a special unit called CDIS. It produced a series of mathematics textbooks for primary schools. It was used in schools for over 20 years. It was not the textbooks alone that make a success. It was the support given by MOE and teachers, and the time taken for the textbooks to mature. We shall give more details later. The series was known as *CDIS Textbooks*. This is probably the second major event in mathematics education in Singapore. The impact was tremendous in terms of syllabus design and teaching approach. It helped make a name for Singapore internationally.

In the educational scene, scientific calculators were introduced into secondary schools (Grade 7–10) in 1982. It was used in Secondary Two (Grade 8) first, then Secondary One (Grade 7). We were conservative in the use of calculators. Graphic calculators (GC) were introduced in junior colleges (Grade 11–12) much later in 2006. The GC is not popular at the university level. The use of calculators was introduced at Primary Five (Grade 5) only in 2008 and at Primary Six (Grade 6) in 2009. To be precise, when we say "use", we mean we use calculators in national examinations.

The publication of *CDIS textbooks*, together with streaming, was an important part of the back-to-basics programme. We had achieved finally the full localisation of textbooks, et cetera. For the next two decades till now (2015), the significant changes were mainly in the teaching approach and the design of syllabus, among others.

New Initiatives

It started in 1997 from a speech in parliament by the Minister for Education, Singapore, on his vision for the future educational system. The speech was titled *Thinking Schools, Learning Nation*. The three initiatives were: national education, information technology, and thinking. By national education, we refer to good citizenship and nation-building. Here a learning nation refers to adult education outside the school environment. In what follows, we shall confine to school mathematics.

In simple words, thinking means more learning and less teaching. We wanted our students to take greater initiatives in their own learning process. We wanted them to put more emphasis on the process, and to learn mathematics, if possible, in context. In fact, they had done it all along. The difference was that now we had to make it explicit. In order to do that, the first move was to reduce the content by 10%. The reduction referred mainly to secondary schools. Also, teachers changed their teaching approach. For example, they referred to materials outside the textbooks more when planning a lesson.

When we changed the teaching approach, we also needed to change the assessment tools. The test questions were often structured and hence became longer. Out-of-syllabus takes a different meaning. For example, as long as a question is within the ability of students though not in the textbooks, it is not regarded as out of syllabus. We cannot abolish the examination, at least not now. But we can downplay it. MOE introduced the Integrated Programme (IP) in 11 schools in 2004. Informally, we call it "through-train". The IP allows students to go through secondary schools (Grade 7 to 10) and the A-level (Grade 11–12) in six years without taking the O-level examination. Hence, students have more time to study. In fact, this was the case before the merging of different language schools. The programme proved to be successful. After 2010, seven more schools, hence a total of 18 schools, offered the IP. Similarly for primary schools, some academically-strong pupils can be admitted to secondary schools before the PSLE examination. Obviously, the IP changed the teaching schedule for those students involved. In other words, there was no need to spend time preparing for the O-level examination.

Concerning technology, the first move by MOE was a gift of 100 personal computers for every primary school. The instruction was to switch on the computers. Then three five-year master-plans (1997–2002, 2003–2008, and 2009–2014) followed. Initially, 22 schools were designated as demonstration schools. Next, around 100 schools were involved. Eventually, all schools were involved. The introduction of technology in the classroom was gradual. Step one: switch on. Step two: use in the classroom. Step three: involve the curriculum. The objective was to enrich and transform learning environments. It is now a common sight in schools that teachers use PowerPoint and other softwares in classes.

In short, the content did not change much. The change was in the teaching approach. In a sense, we are still working within the pentagon framework, which was introduced in 1990. We will cover more fully the components of the pentagon later.

Mathematics Syllabus 2013

The pace of life is faster now. Something that seems to be important now may not stay. However, we know *Mathematics Syllabus 2013* will stay. The full title is *Mathematics Teaching and Learning Syllabus 2013*.[4] The year 2013 refers to the year of implementation. It is a revision of an earlier syllabus 2007, which is supposed to last for 6 years. This syllabus is different from the previous syllabuses in many ways. It is designed as one document describing an overview of the curriculum from the primary level all the way to pre-university level (A-level). So did the first local syllabus in 1959, though not the others. The new syllabus is implemented progressively year by year. That means, the first PSLE based on the new syllabus will take place in 2018, and the O-level in 2016. This was not the case for the 2007 syllabus. For example, the O-level examination based on the 2007 syllabus took place in two years after its implementation. The new schedule allows teachers and students to be better prepared for the implementation of the new syllabus.

The 1959 syllabus was descriptive. It listed the items to be covered in the syllabus. The 2013 syllabus was more detailed, and in some ways, more prescriptive. In fact, the syllabuses in the past 35 years were almost always prescriptive. A prescriptive syllabus would state what was included and sometimes also what was not included. It spelt out in detail at the operational level what students should go through. We emphasised the process and put mathematics in context. In the 2013 syllabus, we introduced the term *learning experiences*. The content and learning experiences were listed in two columns side by side and by level. In one column listed was the content. On the other parallel column, actions that we wanted our students to experience were listed under the phrase: "students should have opportunities to". This was to make explicit the processes we want students to experience.

To help teach mathematics in context, *mathematical modelling* was introduced. Basically, we posed a question in context and solved it. Teaching materials were produced by teachers. They were designed to be used within curriculum time.[5] We should be able to say something when the revision is due in 2019. In fact, the review will begin two to three years before 2019. The changes, if any, will take place only in 2019. Also, by then we may know to what extent attitudes and metacognition (two components in the pentagon) will be involved in the syllabus and teaching. For example, the processes component is more involved now, compared with the 2007 syllabus. The implementation of the 2013 syllabus was monitored within MOE.

In the educational scene, a new policy from MOE was *every school a good school*. We implemented differentiated syllabuses. The purpose was to provide each child with an education best suited his ability. We do have elite schools and neighbourhood schools. Now the idea is to provide each child with an equal opportunity if we can. In addition, special schools were established and catered for different needs. Here are two examples. NorthLight School was set up in 2007 for pupils who did not pass PSLE. SOTA (School of the Arts) was set up in 2008 for students who take one of the four art subjects in addition to other secondary school subjects. The four art subjects are music, dance, media (including film making) and visual art. There are other such schools. In terms of mathematics teaching, we now have different kinds of students learning mathematics under different environments, using different approaches and for different objectives. There is another new initiative. That is, the government is participating formally for the first time in pre-school education in Singapore. Furthermore, there is a learning support for primary one (P1) and primary two (P2) pupils. Not every kid has attended kindergarten before entering schools. Making P1 and P2 as a single unit for teaching helps to overcome the learning difficulty at that level.

In summary, we make explicit what students need to learn in the syllabus. The assessment landscape is changing. For example, in national examinations, candidates are required to answer all questions. This was not the case in the early days. In those days, there were options in an examination paper. Alternative assessments were suggested. So far in this chapter, a history of 50 years was given vertically in time. In what follows, we shall go through the history horizontally in scope. The scope covers curriculum design, textbooks, teaching approach and teacher training.

The Pentagon Framework (1990)

The pentagon framework was announced in 1990.[6; p. 15] The sides of the pentagon represent the five components: *concepts, skills, processes, attitudes,* and *metacognition.* In the centre is *mathematical problem solving.* It describes the principles of teaching and learning mathematics in schools. Initially, we concentrated on concepts and skills. Later, when teaching approach became prominent, we spent more time on processes. Only recently, we talked about attitudes and metacognition. The pentagon has provided the continuity of curriculum design and classroom practices. The continuity lasted 25 years.

There are different versions of the pentagon. Let us compare pentagon 2001 with that of the latest, in 2013. The pentagon 2007 is basically the same as that of 2013. For concepts, in addition to "numerical, algebraic, geometrical, and statistical", we added "probabilistic and analytical". This shows the importance of randomness and formal reasoning. We deleted "mental calculation" and added "spatial visualisation" under skills. It shows what is needed and what is not. We also added "reasoning, communication and connections" under processes. Finally, we added "beliefs" under attitudes due to the latest research in the area. To understand the role of beliefs in teaching, look at the following catch phrase: *if you think you can catch the bus, you'll run for it.* The changes to the pentagon mentioned above reflected the refinements we made to the pentagon. Most likely, there will be no changes till 2019.

In syllabus 2001, there were 13 heuristics for problem solving under processes for the secondary level, and 11 for the primary level. Teachers could hardly remember all of them, not to say use them. If we check through some teaching materials, we may find that some commonly used heuristics are: draw a diagram, make a list, guess and check, and act it out (including "work backwards"). So in syllabus 2007, the heuristics were clustered into four categories. The four are: to give a representation, to make a calculated guess, to go through the process, and to change a problem. Further, in syllabus 2013 mathematical processes is a strand in addition to the three existing content strands (algebra, geometry, and statistics). We call it *process strand.*

The pentagon framework remains relevant after so many years. One component, namely processes, is now upgraded to a strand. Due to recent researches on affective domain and metacognition, we can see in the near future more involvement of the two components, namely attitudes and metacognition, in mathematics teaching. Also in research, for example, measuring tools for metacognition.

CDIS Textbooks in the '80s

Other than *Math Reforms*, the series of CDIS textbooks is another highlight in the past 50 years. CDIS stands for Curriculum Development Institute of Singapore. It was set up by the Ministry of Education, Singapore, in the early '80s. One of their aims was to write textbooks. It took six persons, including one chief editor, working full-time to produce a series of primary school textbooks in five years. Textbooks were tested in schools, and then revised. Teachers were trained to teach the materials, in particular, the model method. It took a few years before the textbooks were used effectively in schools. As far as textbook writing is concerned, it set a standard format for other Singapore textbooks to follow. Also, a set of heuristics cards was produced for classroom use. Later, it was taken over by other available teaching tools.

That was the first time the model method was introduced formally. It was made public in 1978 at the *Fourth Southeast Asian Conference on Mathematics Education* held in Singapore.[7] Due to *Math Reforms*, algebra was introduced in primary schools. Later it was out of primary schools. As a compromise, we took algebra out, but kept algebraic thinking in the form of model method. Briefly, the model method is nothing but a pictorial way of solving simple simultaneous linear equations in two unknowns. In fact, it was taught in Chinese schools at Secondary One (Grade 7) in the '50s as arithmetic. Model method was designed to make an easy entry to algebra. It is not meant to replace algebra. Eventually, algebra should be the core.

The comments made by others outside Singapore about the textbooks are as follows.[8; p. 21] The books cover fewer topics though each with greater depth. The books contain less noises. There is more direct instruction, as compared with some other countries. Concrete, Pictorial and Abstract (CPA) is an underlining figure in the textbooks. In fact, CDIS published another set of textbooks for the N-level. It did not take off like the primary school textbooks. Do not forget that the N-level has a smaller population of students. At that time, the popular secondary school textbooks were published commercially. A-level (pre-university) and junior colleges (JC) do not really use textbooks. Teaching approach at JC is closer to the university than to schools. For schools offering IB programme (International Baccalaureate Programme), they use imported textbooks. The CDIS textbooks ceased publication afterwards to give way for commercial publications.

It is true that no other textbooks in Singapore were produced and implemented in the way the CDIS textbooks were. It remains a challenge to produce textbooks like the CDIS textbooks. We must keep in mind that publishing a textbook should be a slow process. A good textbook takes time to mature. We must also be aware that textbooks play a different role now.

Gifted Education

The gifted education programme (GE programme) is an extension of NES, which was mentioned in *Back to Basics*. The GE programme was first implemented in 1984 with four classes in two primary schools and five classes in two secondary schools. Each class had 25 students. The first-batch of students constituted about 0.25% of the cohort. It is generally agreed that, for administrative purpose, 1–2% of the student population can be regarded as gifted. The key factor of the programme is enrichment, and not an accelerated programme. The emphasis is on thinking processes and social responsibility, and not just the knowledge alone. This is also the reason why the programme was housed in regular schools. It is known internally that every GE teacher has in their possession two huge files of teaching materials. Teachers were trained, well-supported, and monitored by the GE unit (later called GE branch). The programme moved slowly and cautiously, as stated in one GE report.

The enrichment programme did not continue in junior colleges (Grade 11–12). In its place, is the Science Research Programme (SRP), where a small group of selected first-year junior college students carry out some minor research projects with university lecturers. They are asked to write the reports at the end of the projects and present them at a seminar. This programme started in 1988 and is ongoing. The seminar is now called a congress. The IP as mentioned in *New Initiatives* is part of the GE programme. There are other such mentorship programmes. The GE programme was reviewed after 10 years in 1994. Consequently, it opened up to more students. Furthermore, some projects become school-based.

The GE Branch has served the schools beyond the GE programme. In fact, the GE branch provided professional development for teachers. The experiences learned also benefited other units.

Differentiated Syllabuses

Streaming led to differentiated syllabuses. At first, it served to reduce attrition rates. Later it helped retain students in schools for at least 10 years. Normally, we have a standard syllabus. Then we have an alternative syllabus for those students who do not need so much or cannot manage so much. We say we adopt differentiated syllabuses. In fact, the syllabuses for the GE programme or for students of the arts are also differentiated syllabuses. Differentiated teaching in a class was practised in the west during the *Math Reforms*. It means the class uses one book but students work on different exercise books. Our approach to apply differentiated syllabuses is a compromise.

Teaching approaches may be different due to differentiated syllabuses. For example, use of calculators could be to speed up computation in one syllabus, but in an alternative syllabus it is to avoid heavy computation. The former is to push students to do more whereas the latter is to allow students to do less.

After O-level, students can join either junior colleges (JC) or polytechnics (poly). This is also a form of differentiated syllabuses. Currently, the ratio of the two groups of students is about 7:10. More students go to poly. In the past, those who joined JC would go to universities. Those who joined poly would probably not. Now more poly students will go to universities. In the past, poly graduates could attend some remedial classes to qualify for universities. Now they may take an intervention programme in poly before entering universities. The intervention programme prepares students for universities whereas reme-dial courses repeat what students did not do well enough at poly. The educa-tional system is evolving to keep up with the times. Mathematics education cannot be otherwise but be part of it.

There is a practice in schools labelled SoW or Scheme of Work. It gives a day-to-day schedule of lessons to be delivered in class throughout the year. It includes examples to present and exercises to assign. Syllabus is a rough plan. SoW is syllabus in operation. Almost every school has its own SoW. Normally every teacher in the school follows it. Publishers use it as an essential reference for writing textbooks. The official documents describe the differentiated sylla-bus. SoW describes the corresponding differentiated teaching approaches and differentiated testing. It is an important component of school teaching. We do it but we do not often talk much about it.

The implementation of differentiated syllabuses is also an extension of NES, mentioned in *Back to Basics*. It is here to stay. In fact, syllabus used in NorthLight School is also a differentiated syllabus, though not officially docu-mented. As expected, the school has SoW. In other words, the idea of differen-tiated syllabuses has already taken an extended meaning.

Teacher Education

Teachers' Training College (TTC) was established in 1950. It was and still is the only teachers' college in Singapore. It is now called National Institute of Education (NIE), attached to a university. It means research in mathematics education is now an integrated component of NIE. Before 1982, teachers would teach in a school in the morning and attend courses at TTC in the afternoon. Now teacher trainees attend full-time pre-service training at NIE. For the story of TTC turning into NIE and engaging in research, refer to a book by Chen Ai Yen and Koay Siew Luan.[9] We shall confine to certain recent issues in teacher education.

Currently, teacher trainees spend one year at NIE for training after they graduate from universities. In 2017, they will spend one year and four months at NIE so that they have more time for teaching practice and attend more courses. Eventually, all teachers including primary school teachers will be university graduates. It is not the case yet.

For the primary level, teachers are trained to teach three or at most four subjects. A recent policy to be implemented is to specialise in only two subjects, like in the secondary schools. For the secondary level, every teacher should have two teaching subjects. It is common though not necessarily preferable that the two teaching subjects could be mathematics and English. This happened due to the shortage of English teachers when mathematics and science were taught in English during the '70s and '80s. Furthermore, after independence there were predominately more science students than arts. This will change in due course.

For primary teachers, they are required to take a course on subject knowledge (SK) or pedagogical content knowledge (PCK). There are notes prepared by NIE lecturers but not a published book on the subject. The issue is whether SK should be a course for teaching or a course for teachers. Different countries do it differently. There are two books available on pedagogical knowledge,[10,11] one for secondary and another for primary. Informally, they are known as the *green book* and the *yellow book* respectively because of the colour of their covers. From the guidelines given to the authors, we can see the content in the two books. They are the practical issues: syllabus (standards) and concepts, teaching approaches and ideas, common errors and teaching difficulties, and sample activities and test items.

As stated above, there are unfinished jobs in the area of teacher education.

Conclusion

In summary, the 50 years can be divided into five periods as shown below. For easy reference, figures are rounded.

- 1965–1970: After Independence
  Building up the hardware such as school premises, and beginning of local syllabus and its implementation.

- 1970–1980: Math Reforms in the '70s
  Modern syllabus leading to the localisation of syllabus, textbooks, and teacher training.
- 1980–1995: Back to Basics
  Streaming of students to reduce the attrition rates, and publication of CDIS textbooks.
- 1995–2005: New Initiatives
  Making explicit the teaching approach with emphasis on the process and teaching in context.
- 2005–2015: Mathematics Syllabus 2013
  Bringing in learning experiences and mathematical modelling.

Again in summary, the major events in the educational scene include:

- Improvement of school facilities.
- Merging of different language schools.
- Streaming of primary and secondary school students.
- Use of technology in the classroom.

In mathematics education, we experienced the changes as stated below.

- Mathematics is made compulsory up to Secondary Four (Grade 10).
- Mathematics is taught and examined in English.
- Adopting differentiated syllabuses.
- Introduction of the pentagon framework.
- Publication of CDIS textbooks.
- Making explicit learning experiences in Mathematics Syllabus 2013.

It is expected that mathematics education will keep changing. However, there are things that are fundamental and do not change so fast. Multiple approaches to teaching will stay. There is no best way in teaching. A crucial factor is teachers. For any reform, it can move only as fast as teachers can move. Also, a teaching approach that makes high demand on teachers is not scaleable. We need to find our own way to solve our problems. We need institutional memory so that we do not repeat the same mistakes. It is too early to discuss e-learning now, whatever e-learning means.

Mathematics is also changing. A recent discussion is to redefine mathematics. Perhaps this is more relevant to universities than to schools. It may take some years before it reaches schools. Another discussion is that future jobs may occur in the area of STEM (science, technology, engineering, and mathematics). If so, what kind of mathematics? Do we know? Are we willing to find out?

When we look forward, what are the challenges? Looking at three papers of three different periods in the past, we pick one challenge which was dominant at the time. In 1959, it was to have a common secondary school mathematics textbooks following one syllabus and for all schools.[12] In 1972, it was to train teachers to teach modern math with new content.[13] In 2005, it was to adopt multiple teaching approaches including the use of technology.[2] Look at things around us. Mathematics is taught not only in the mathematics departments at the universities. In fact, there are more and more subjects which require mathematics as a prerequisite. Look at what happened in the past years. Now in 2016, the challenge could be to teach different mathematics in many different ways and to different groups of people.

References

[1] Soh, Cheow Kian (2008). An overview of mathematics education in Singapore. In Z. Usiskin and E. Willmore (Eds.), *Mathematics Curriculum in Pacific Rim Countries — China, Japan, Korea and Singapore* (pp. 23–37). Charlotte, North Carolina: Information Age Publishing.

[2] Lee, Peng Yee (2008). Sixty years of mathematics syllabus and textbooks in Singapore (1945–2005). In Z. Usiskin and E. Willmore (Eds.), *Mathematics Curriculum in Pacific Rim Countries — China, Japan, Korea and Singapore* (pp. 85–94). Charlotte, North Carolina: Information Age Publishing.

[3] Teh, Hoon Heng (1969), Editor. Chen Chuan Chong, Koh Khee Meng, Ng Peng Nung, Shee Sze Chin, Tan Sie Keng, Yap Kim Yew. *Modern Mathematics for Secondary Schools*. Singapore: Pan Asian.

[4] Ministry of Education, Singapore (2015). *Primary Mathematics Teaching and Learning Syllabus 2013* and O-Level *Mathematics Teaching and Learning Syllabus 2013*. http://www.moe.gov.sg/education/syllabuses/sciences/

[5] Lee, Ngan Hoe, and Ng, Dawn Kit Ee (2015), Editors. *Mathematical Modelling: From Theory to Practice*. Singapore: World Scientific.

[6] Wong, Khoon Yoong (2015). *Effective Mathematics Lessons Through an Eclectic Singapore Approach*. Singapore: World Scientific.

[7] Kho, Tek Hong (1987). Mathematical models for solving arithmetic problems. In *Proceedings of Fourth Southeast Asian Conference on Mathematical Education* (ICMI-SEAMS), pp. 345–351. Singapore: Institute of Education.

[8] National Mathematics Advisory Panel (2008). *The Final Report*. U.S. Department of Education. http://www.ed.gov/MathPanel/FinalReport/

9 Chen, Ai Yen, and Koay, Siew Luan (2010), Editors. *Transforming Teaching and Inspiring Learning: 60 Years of Teacher Education in Singapore* (1950–2010). Singapore: Institute of Education.

10 Lee, Peng Yee, and Lee, Ngan Hoe (2009), Editors. *Teaching Secondary School Mathematics: A Resource Book*. Singapore: McGraw Hill.

11 Lee, Peng Yee, and Lee, Ngan Hoe (2009), Editors. *Teaching Primary School Mathematics: A Resource Book*. Singapore: McGraw Hill.

12 Lee, Peng Yee (1959). On secondary school mathematics syllabus. *Bulletin of Mathematical Society, Nanyang University*, No. 1, p. 24, p. 37.

13 Lee, Peng Yee (1972). Mathematics curriculum in 1990. *Seminar for Senior National Administrators*, RECSAM, Penang, Malaysia.

Chapter 27

The Formation and Transformation of Science Centre Singapore

Lim Tit Meng

The Beginning (1968–69)

In accordance with the functions of the Science Council of Singapore as defined in Sections 8–10 of the Science Council of Singapore Act, 1967, the Council made a recommendation to the Ministry of Science & Technology in 1968, to establish a Science Centre in Singapore. Accordingly, the Council appointed a Special Committee in late 1968 to carry out preparatory work. The Committee was chaired by Mr Ronald Sng Ewe Min (Science Council) with three members comprising Mr Sng Yew Chong (Ministry of Education), Mr R A Shelley (Hume Industries (F E) Ltd) and Dr Bernard Tan (University of Singapore).

On 20 November 1969, Dr Lee Kum Tatt, Chairman of the Science Council submitted to Dr Toh Chin Chye, the then Minister for Science and Technology, the views and recommendations of the Special Committee to set up a Science Centre in Singapore. The historical document carried the report by a UNESCO Advisor Miss M K Weston (of Science Museum London) who came to Singapore during late September 1969 for three months to work with the Special Committee to draw up proposals to set up the Centre.

In Miss Weston's Report, the main benefits for the Centre were summarised as follows:

- It would greatly complement the science and technical education programme (which were receiving urgent attention then).
- It would provide the adult population a quick and greater understanding of technology and its part in Singapore's Industrialisation Programme.
- It would stimulate in the younger generations interest for a career geared to industry and to overcome the past preference by parents for white collar rather than blue collar occupations.
- It would find favour with industrialists from the recruitment aspect and they may feel more disposed to contribute to a semi-permanent exhibition.

- It might attract a proportion of foreign visitors who tended to visit such institutions when they were away from home.

It's amazing how these benefits spelt out more than four decades ago are still relevant today in our nation that thrives on knowledge and an innovation-based economy. While the focus then was on developing human resource for industrialisation, today we continually need to develop and deploy talents in Science, Technology, Engineering and Mathematics (STEM) professions, which are in great demand for our post-industrialisation growth.

The comprehensive proposals by Miss Weston suggested two permanent exhibition halls with one devoted to Industrial Technology and the other to Communications and Transport. A third hall would be dedicated to Special Exhibitions which could also be used to stage school science fairs and other events. Other suggested services included a planetarium, a library, an observatory, an amateur radio transmitting station, workshops, stores, classrooms, lecture theatres or an auditorium, and a restaurant.

The recommended sites for the Centre by Miss Weston in consultation with the then Department of Urban Renewal were assessed, and three were put forth in order of preference as follows:

- Tanglin Golf Link (opposite the Botanical Gardens).
- Fort Canning (next to the King George V Park).
- Kent Ridge (now the National University of Singapore).

Interestingly, the Government did not take up any of the recommended sites but decided to build it at Jurong instead. I remember when I first visited the Centre as a student and I had to walk on a mud track because it was built in an area where roads were not yet fully developed. The mud track later became a proper road and named, Science Centre Road. It was literally 'in the middle of nowhere' and many people have considered the Centre as being 'very far away', even till now!

Science Centre in the 1970s

The Science Centre Act was established in 1970 by the Singapore Government. The Centre was aptly given a mission as follows:

> *"To promote interest, learning and creativity in science and technology through imaginative and enjoyable experience and contribute to the nation's development of its human resource."*

The Centre was to function as a Statutory Board which was originally under the then Ministry of Science and Technology. When the Government subsequently restructured the Ministry, the Science Centre Board then came under the purview of the Ministry of Education.

After the Government decided to build the Centre, a design competition was called and the entry by architect Mr Raymond Woo was selected. An exhibition of architectural designs for the Science Centre was held at the Victoria Memorial Hall on 22 December 1971. The award-winning design depicting a spacecraft which had landed on a surface later inspired several 'copycat' buildings along the Jurong Town Hall Road.

In 1972 when the building's plan and layout for the Centre were finalised, Dr Toh Chin Chye wrote a farsighted message which contained the following paragraphs:

> "Establishing a science centre may appear to be a paradox when in some indus-trialised countries there is a prevailing mood that science has not been able to solve the problems of their societies and that technology has been responsible for the pollution of cities and the environment.
>
> "In Singapore, however, it is not less but more knowledge of science and technology that we need, particularly so when the livelihood of the people of the city-state will increasingly depend upon our ability to develop new talent and skills to operate industries based on technical expertise. However, it will be a mistake to focus our sole attention on man-made machines and forget about the living world. For this reason there will also be a division to show biological processes and ecosystems. These two divisions, the Life Sciences Division and the Division of Science and Technology, will together form the basis for the working of the Science Centre.
>
> "By striking such a balance we hope that planners, administrators and city dwellers will learn how to create an environment in which the population of Singapore can live in harmony in what will become an industrialised society."

These words of wisdom have withstood decades of national develop-ment. In fact, the fast pace of change in Singapore has much to do with how we stay true to the core value of science and technology, and apply the power of STEM to create solutions for our survival. For example, Singapore now has a growing biomedical industry derived from life science research and applications; and it is reputed as a forerunner in sustainable and liveable city development.

The first Chairman of the Science Centre Board, Mr Wee Cho Yaw, stressed that an important task of the Centre was to prepare, train, and motivate our youth to meet the challenges of modern science. An equally important task was to inform our citizenry of the critical problems facing the world and point out possible solutions. His message still resonates well today.

While the construction took place, a team was assembled headed by the first Director, Mr Kenneth V Jackman who came from the Lawrence Hall of Science, USA. Under him were nine staff comprising an Executive Secretary, an Executive Officer, two Assistant Education Officers, one Chief Technician, three curators and a graphic artist. They operated out of an office in Friendly Hill at Depot Road. The Board's office was once temporarily accommodated in the main office of the National Museum at Stamford Road.

The Government allocated a sum of $15 million towards the cost of the building project, exhibits and ancillary equipment. On 18 August 1973 when the construction actually started, a time capsule containing examples of Singapore's burgeoning industries and flourishing social infrastructure was buried to mark the occasion. The building was completed in October 1976. By then, Dr R S Bhathal had become the Director of the Science Centre and he was to lead the team from Friendly Hill to the Jurong campus.

On 10 December 1977, the Singapore Science Centre was officially opened by Dr Toh, after taking nine years of studies, consultations, visits to other science centres and construction from the point of conception to opening. It was then the very first contemporary museum of science and technology in Southeast Asia.

Mr Tan Keong Choon took on the Chairmanship of the Science Centre Board and in his address at the official opening of the Centre, he began his message with the following statement:

> "Singapore has no natural resources and for us to survive as a progressive nation, we need to rely on advanced technology for our continued growth. Hence there is no other alternative open to us but to educate our youth in the principles and applications of science and technology, so that we can up-grade our science based industries."

This sentiment is still valid in today's Singapore after 50 years of nation-building. It is indeed the core reason why Science Centre is still important for our nation.

Science Centre 10th Anniversary: Expansion

The Centre with its interactive exhibits and diverse enrichment programmes grew to become an institution beloved by both the locals and foreign visitors. It has been described as one of the best science museums in the world by Nobel Laureate, Sir George Porter, and many others who have visited. The visitor

Fig. 1. The original Science Centre (top) was built in the middle of nowhere in Jurong (bottom).

numbers grew significantly from 233,697 in 1978 to 578,550 in the Financial Year 1986/1987.

A recollection from Mr Yong Jian Yi typified the young visitors of those days:

"We were on a field trip to meet with a British palaeontologist on the subject of dinosaurs. It was the first experience that I had of meeting a scientist and the impression had stayed with me since. Although the topic was broadly about the

mating habits of dinosaurs, we came to the topic on the origins of humankind. There is a certain protrusion on the back of our skull, which as informed by the professor, was also found on the skulls of Neanderthal. This similarity might point towards our common ancestry. Diagrams were presented and heads were compared; but unfortunately since my head was the only exception that day, we had no final conclusion. That day was the day when I truly became interested in Science and its processes."

When the Centre celebrated its 10th Anniversary on 10 December 1987, the Omni-Theatre was officially opened by then President of the Republic of Singapore, Mr Wee Kim Wee. It was part of the expansion plan to develop the Centre into a Science Complex, an idea mooted as early as 1980. The Omni-Theatre was built to show OMNIMAX movies and Planetarium shows. Costing a total of $18 million, the project was jointly financed by the Ministry of Education, the Singapore Tourism Board and funds donated by the private sector. It took four years of planning before the project got off the ground.

The building of the 22.15 m high Omni-Theatre designed again by Raymond Woo and Associates, was launched on 2 August 1985 with a ground-breaking ceremony officiated by Dr Tay Eng Soon, then Minister of State for Education. Construction began on 1 April 1986 and completed on 20 July 1987.

The Omni-Theatre had a 284-seat amphitheatre with a massive hemi-spheric screen tilted at a 30° angle to the horizon. The screen wraps over the audience to cover 80% of the hemisphere giving the audience the sensation of being completely engulfed in the picture. Today it remains the only immersive dome theatre in Singapore and is one of the largest in Asia.

The Science Centre Observatory sitting next to the Omni-Theatre was opened in 1989 by Dr Tony Tan, then the Minister for Education. It is situated at 1.3342° N 103.7357° E, and is 15.27 m above mean sea level. The main tele-scope, a generous gift from the Japanese Government to Singapore, is a 40 cm Cassegrain reflector of a combined focal length of 520 cm. Its equatorial mount was designed for Singapore's unique location which allows constellations of both the northern and southern hemispheres to be observed.

The Observatory became an integral part of the Centre with many visitors young and old coming to learn about astronomy and space science. One of the astronomy highlights made possible through the Observatory programme is described in the following account by Ms Chan Sue Li:

"I remember clearly the 2004 Transit of Venus (partially because the last passing was almost 200 years before) as typically astronomy events were conducted at night, but on 8 June 2004, it was the first time we did an afternoon session because

Fig. 2. Original Science Centre logo (left) and the revised logo with Omni-Theatre incorporated (right).

of the passing timing. Altogether 13 viewing stations were set up throughout the Science Centre grounds from Kinetic Garden to the greens outside the two observatories. Thousands of people came within that two to four hours and they queued under the terribly hot sun. It was very telling of their passion and dedication..."

As Chairman Mr Tan Keong Choon wrote in one of his reviews, the Centre entered a new and dynamic phase in the 1980s with a more active role in the popularisation of science and the creation of a 'science culture', with an increasingly important role in advancing 'science literacy'.

Because of the addition of the Omni-Theatre complex, the Science Centre changed its original logo which was a red stylised SSC arranged as three-hexagons stacked like the aerial view of the main building, into one with a dual representation of the Singapore Science Centre and the Omni-Theatre depicted at the top half of the graphic. The overall graphic was a stylish version of the letter 'S' denoting science. At its centre was a globe symbolising the Earth. The graphic element was in blue, the main colour associated with sky, water and gas.

Science Centre 20th Anniversary: Revamp

When the Centre turned 20 years old in 1997, it saw another major revamp under the leadership of the then Science Centre Board Chairman BG(NS) Lee Hsien Yang. In the 1990s, the Ministry of Education started the concept of a 'thinking' school which further heightened the function of the Centre to nurture critical and creative thinking in our young.

With a development grant of $38.68 million from the Government for its Revamp project, the first in the history of the Centre, it was then able to upgrade the exhibitions in its galleries and set up an Annex Building to host world-class exhibitions, and also build the outdoor Kinetic Garden.

The Kinetic Gardens were specially opened on 31 December 1999 to usher in the Millennium with a countdown party. The Annex Hall was officially opened by Minister Teo Chee Hean on 4 November 2000 to commemorate the re-launch of the Science Centre. The first world-class exhibition brought into the Annex was on Dinosaurs.

The revamp brought the Centre to a new growth phase and the Centre's educational programmes became very popular. However, due to space and manpower constraints, it was able to accept only 50% of the bookings from schools. More than 95% of our schools joined as institutional members with 130,000 students participating in the science enrichment programmes. The number was a far cry from the original 40,000 students target set by the UNESCO Advisor in 1969 when the proposal to set up the Singapore Science Centre was put up.

The Centre in the 1990s established itself as the key player in promoting mass-based science education with many partners working closely. The events in collaboration with organisations such as the National Science and Technology Board (NSTB), National Computer Board (NCB) and Economic Development Board (EDB) included the Innovation Festival and Singapore Robotic Games. These were in addition to the long-term partnership with the Shell Companies in Singapore which has sponsored the Singapore Youth Science Festival since the inception of the Centre.

The National Junior Robotics Competition (NJRC) was inaugurated in 1999. It spurred students into learning and loving everything robotics, and spawned an industry of robotics suppliers and training programmes in the nation. Year on year, the NJRC never failed to intrigue and engage the masses of students, with many working adults and professionals volunteering as judges. The success of NJRC made the Centre a founder of the World Robotics Organization (WRO) which now has more than 50 member countries in the global network. The champions in NJRC each year get to take part in the inter-national competitions organised by WRO hosted in different parts of the world. This success story is one of the indicators of how the Centre has become very reputable and respected in the international arena.

It is a fact that the Centre has successfully popularised robotics in Singapore and even the region. Dr Pang Kian Tiong who works for the Centre thinks robotics programmes should continue and even grow further at the Science Centre with the following rationale:

> "When I joined the Centre (at the end of 2003), one of my first assignments was to "help out" in robotics programmes. I soon learned about the NJRC which was

very much still rising in popularity with local schools. I was also to help in the inaugural FIRST Robotics Singapore (FRS); including many other less major robotics competitions and events. It wasn't difficult to see that these activities are vital in early STEM exposure. The players are almost always incredibly energised, enthusiastic, and learning experientially! It is therefore important to continue such multi-disciplinary activities at the Centre."

Science Centre's Third Decade: Fortification

When the Centre entered its third decade post 1997, the Millennium was just around the corner. NSTB was restructured as the Agency for Science, Technology and Research (A*STAR), and Singapore developed One-North to strengthen our capacity in knowledge and an innovation-based economy. With the setting up of the Biopolis and the worldwide life science revolution, the Centre went into a collaboration with the Cold Spring Harbor Laboratory's Dolan DNA Learning Centre in New York to set up the DNA Learning Lab in Singapore. It became an instant success and has enriched more than 250,000 students since its opening in 2003.

A*STAR also commissioned the Centre to run outreach activities by sponsoring a whole new department named 'Science Upstream'. The concerted efforts to promote science to the public then took on a new playing field, and the Centre literally made its presence felt in public areas such as shopping malls and, libraries, as well as on national television through the highly competitive National Science Challenge for secondary schools.

The Centre opened the first-of-its-kind Waterworkz outdoor play area in 2005 which inspired many other attractions and shopping malls to build similar facilities for their guests. It was not a surprise that the Waterworkz helped to bring in even more visitors to the Centre and it is still one of the most popular spots especially on a hot and sunny day. It even won an award from the Singapore Tourism Board (STB) for contributing to tourist experience.

Snow City adjacent to the Centre originally built by ST Engineering and operated by NTUC Income was transferred to the Centre because NTUC Income failed to sustain its business. The Centre assumed full ownership in 2007 and has made it viable ever since. The Snow City later become a supplementary science enrichment experience to the Centre especially with the importance of educating visitors on global warming and climate change issues.

The Centre marked its 30th Anniversary with a special event welcoming the 16th million visitor. Under the leadership of Chairman Mr Wee Heng Tin, a new brand positioning exercise was conducted and the vision statement of

Fig. 3. The current Science Centre logo.

'To be a world class science centre' was changed to a new aspiration proclaiming the Centre as a place 'Where science befriends and transforms the minds of millions'. The Science Centre logo was also reinvented into one with a contemporary look with the words 'Science Centre' forming a forward rolling circle portrayed in trendy bright orange. The official name of the Centre became Science Centre Singapore.

Science Centre's Fourth decade: Transformation

A new partnership with the Defence Science and Technology Agency (DSTA) and the DSO National Laboratories was forged in 2008. The collaborative efforts saw the introduction of the 'Amazing Series' competitions in subsequent years. These competitions which were organised for students of all ages to spur their interests in STEM and unleash their creativity through the making of 'amazing' science-based machines and interactive exhibits quickly gained popularity.

The Science Buskers Festival was launched in 2008. It is a competition for those aged 7 to 70 to demonstrate science through performance and showmanship. It was initially modelled after the Singapore Idol competition and was later modified into a science busking contest in a shopping mall. During the competition, contestants do a "show-and-tell" on any science topic, and judging is based on audiences' votes and judges' scores. It serves as an excellent platform to spread the key message that science is intriguing and fun!

In 2010, the Centre set up yet another innovative laboratory named CRADLE: Centre for Research and Applied Learning in Science. To facilitate working with students and youth passionate about science and technology, the Singapore Academy for Young Engineers and Scientists was registered with a clubhouse in the Centre. The acronym of SAYES was meant to prompt members to say 'Yes' to a career in engineering or science. It is run by a Council of young leaders elected by its members.

When the Centre marked its 35th anniversary, DPM Teo Chee Hean was officially introduced as the Patron. At the celebration dinner in 2012, he presented a special token of appreciation to thank all past corporate sponsors who had contributed to the success of the Centre. The token was a fire tornado lamp designed, built and patented by the Centre. It was a miniature version of the 6 m tall fire tornado exhibit engineered entirely in-house which also had the honour of gracing the Youth Olympics Games hosted in Singapore in 2010 as the pinnacle of the Olympics Fire Cauldron.

Over the years, the Centre also witnessed a rise in the number of young parents and their young children visiting Science Centre. As the Centre matures in school outreach and enrichment, it was becoming obvious that the pre-schoolers would need a specially-dedicated space for them to explore and experience science. Another significant transformation thus became possible when the Ministry of Education granted funding to build a new wing for children aged 8 and below. The KidsSTOP™ then took shape in the Omni-Theatre building and was officially opened on 10 June 2014 by then Minister for Education, Mr Heng Swee Keat.

During the official opening, Chairman Ms Tan Yen Yen shared the philosophy behind KidsSTOP™ in her message:

> "It's a milestone in the history of Singapore, as our nation's first science centre for children **where every child gets to Imagine, Experience, Discover and Dream.** Children learn by doing. They feel, think, touch, see and move as they employ creative problem solving skills. KidsSTOP™ aims to arouse and foster children's natural curiosity by providing such experiences. We are glad that Science Centre Singapore is able to provide this unique facility to ignite that journey."

The KidsSTOP™ designed for purposeful play was an instant success with kids and young parents. The membership signed up far exceeded the original target set. More importantly, KidsSTOP™ also attracted collaborative projects and partnership with many pre-school service providers and kindergartens, especially with the Early Childhood Development Agency (ECDA).

Another bold initiative by the Centre in 2014 came about when the Minister for Education wanted to make every school a good school. MOE then commissioned the Centre to offer STEM Applied Learning Programme (ALP) in at least half of all the non-elite secondary schools. A new unit called 'STEM INC' was therefore set up to work directly with schools.

The name 'STEM INC' was deliberated to convey the meaning of: (1) students incorporating STEM knowledge into real-world problem solving and (2) students beginning to see STEM as an enterprise through which they can create wealth or a future for themselves.

STEM INC provides interactive and hands-on experiences within domains such as engineering and robotics, information and communications technology and programming, food science and technology, environmental science and sustainable living, materials science, health science and technology, transportation and communication, and simulation and modeling. The Curriculum Planning and Development Division of the MOE works closely with STEM INC which deploys curriculum specialists and STEM educators to train teachers and work with them to co-develop and co-teach STEM lessons. STEM INC also coordinates an Industry Partnership Programme where relevant companies or corporations are matched with specific schools as industrial collaborators.

The Curriculum Specialists employed by STEM INC were selected based on their working experience in either a research institution or an engineering industry. Lessons and hands-on activities designed by them are to bridge conceptual understanding of the curriculum to applications in real-world scenarios. For example, students might apply what they have learnt about biology, electronics, computer programming, and microcontroller technology to build an electronic heartbeat sensor. It is very reassuring that within three years of operation, more than 60 schools have joined the STEM ALP.

Reinvention Strategy

Like how science advances, the Centre is constantly evolving — creating and presenting innovative ways to spark interest and ignite curiosity in science. The Centre has also become a preferred partner for collaborations, and has grown and deepened its alliances with local and international agencies who are in the business of promoting science and technology. In order to strengthen the leadership role in science outreach and promotion, a transformation and reinvention has to take place amidst ongoing discussions on moving the Centre to a new site.

The Centre has declared itself to be an Edutainment Champion and positioned itself as a thought leader in the Science Centre agenda with the following four slogans:

- **The Science Centre is an Alternate Classroom**

The Centre offers a wide range of science enrichment workshops in specialised teaching laboratories, and lecture demonstrations or problem-based explorations in the exhibition galleries. They aim to complement the school science curriculum with hands-on and inspiring learning experiences. Learning opportunities are also infused within the many competitions and events organised by the Centre, often involving professionals from the academia and industries to enhance authenticity and credibility. As an informal learning institution, the Science Centre is not bound by curriculum and convention, and this gives opportunities to experiment with innovative ideas to make the teaching and learning of science more fun and engaging than what could be done within a traditional classroom. As an alternate classroom, the Science Centre creates an 'education ecosystem' to bridge the gap between formal and informal science education which is engaging, effective and empowering. The number of students engaged in the Centre has exceeded 230,000 annually in recent years.

- **The Science Centre is a Preferred Partner**

The Centre is widely acknowledged as a centre of excellence for edutainment, and a desired partner for science and technology promotion. Agencies such as the Agency for Science, Technology and Research (A*STAR), Defence Science Organisation (DSO), Economic Development Board (EDB), Infocomm Development Authority (iDA), National Environment Agency (NEA), National Library Board (NLB), National Parks Board (NPB), National Research Foundation (NRF), Public Utilities Board (PUB), Workforce Development Authority (WDA), Early Childhood Development Agency (ECDA), Singapore Tourism Board (STB); the Singapore National Academy of Science (SNAS), the Science Teachers Association of Singapore (STAS), the Singapore Association for the Advancement of Science (SAAS); the Institution of Engineers Singapore (IES), and all institutes of higher learning, collaborate with the Centre on a regular basis to promote STEM. Many corporations and companies such as the Shell Companies in Singapore, SONY, Abbott, Panasonic, Cerebos, National Instruments, ArtScience Museum, Resorts World Sentosa, Microsoft, Autodesk, etc, also sponsor or partner the Centre for their Corporate Social Responsibility efforts or for joint-promotional events.

The Centre is also in collaborations with many overseas science centres or museums. Notable ones include the Exploratorium in the USA, the Natural History Museum in England, the Ars Electronica in Austria, Questacon in Australia, Universcience Le Cite in France, and the Mind Museum in the Philippines. The Centre is a founding member of the Asia-Pacific Science Centre Network (ASPAC) and also an active member of the Association of Science Technology Centres (ASTC). The key leaders of the Centre are also represented in their Executive Council or Board of Directors.

- **The Science Centre is Without Walls**

The Centre has championed the Singapore Science Festival (SSF) as a powerful and effective platform to bring science to all corners and to be experienced by all walks of life in the country. It highlights how science has a positive impact on our lives, transformed our economy, and continues to be central to our national plans on research, innovation and enterprise.

In 2015, the SSF involving multiple agencies and partners from the public and private sectors (total 47 partners with 55 events) saw close to 180,000 participants overall. The survey feedback for the exhibitions held at VivoCity revealed that 97.97% agreed or strongly agreed that the event showed that science can be fun; 95.96% agreed or strongly agreed that the event helped them understand science better. More significantly 95.67% agreed or strongly agreed that the event encouraged them to learn more about science.

The SSF has gained an international reputation recognised by overseas counterparts. The Centre has received many invitations to take part in similar festivals in countries such as Abu Dhabi, China, Indonesia and Thailand. The counterparts in Beijing even made their China City Science Festival a sister-festival with SSF.

- **The Science Centre is Community-Owned**

While the Centre grows in stature and competence, the level of science literacy and expertise in our community is also growing rapidly. People are now more educated, and very well connected in the new Millennium, and the Centre has recognised the power of co-creation with the diverse talents present in the community. There are now co-creation spaces in the Centre in the form of a Tinkering Studio, the Einstein Room and the Eco-garden. But more significantly, by tapping on the worldwide Makers Movement which started in the USA in 2005, the Makers Spaces are not just confined within the Centre.

In 2012 the Centre hosted the first Mini Maker Faire in Singapore. Within a short span of three years, the Singapore Maker Faire was elevated to a full Maker Fair status in 2015. The Founder of Maker Media Inc Dale Doughety even came to give a keynote address on how the Maker culture has redefined learning. The event saw a dramatic growth from 5,000 participants in 2014 to 12,000 in 2015, with an increase from 19 workshops in 2014 to 60 in 2015. This growth was of great significance because when the Faire started in 2012 only 20 Maker groups and 1,000 participants took part. A book telling the stories of Maker families in Singapore entitled 'Busy Hands Happy Hearts' was published with a Foreword written by Minister Vivian Balakrishnan who is a Maker himself.

The Centre is also an incubator for experienced teachers and educators for the development of STaR Kits® (Science Teaching and Resource Kits). These kits are designed to support the teaching and learning of science through inquiry-based activities. Targeted at primary and secondary school levels, the kits are used to develop inquiry process skills in students and to stimulate discussion as they construct understanding about science concepts or ideas.

With the advent of digital technology, the Centre has also installed a new digital planetarium projector system in the Omni-Theatre, and opened a new gallery called E3: E-mmersive Experiential Environments in 2015. These digital tools and virtual reality systems open up new co-creation channels for the Centre to work on visualisation projects with native and foreign experts. In due course, the Omni-Theatre with the digital planetarium, as well as the E3 gallery, will become experimental playgrounds for students and research scientists who deal with big data analytics and visualisation to showcase their output. The Centre strives to become a creative powerhouse to innovate communication platforms to narrate the wonders and beauty of science. This new direction is especially important to engage the growing generation of 'EPIC' learners who are 'Experiential, Participatory, Image-driven and Connected to the world'.

Fulfilling the Science Centre Mission Through Endearing Means

As Singapore is proud of the achievements over 50 years of nation-building, the Centre has also many proud initiatives that are endearing and impactful. Among the most cherished are the Young Scientist Badge Scheme and the Singapore Science and Engineering Fair.

Young Scientist Badge Scheme

The objectives of this scheme are to stimulate interest in science activities among primary school students aged 7–12, enabling them to carry out self-directed activities in areas of interest in science, and to provide opportunities for students to develop initiative and creativity. The badge scheme was launched in 1983 and more than a million badges have since been awarded.

In its 30th year, the cards were revamped and incorporated more inquiry-based activities, through a partnership with the Singapore Association for the Advancement of Science (SAAS), the Science Teachers Association of Singapore (STAS), Singapore National Academy of Science (SNAS), and the Academy of Singapore Teachers (AST). Some cards also involved the zoos, nature parks, facilities of the Public Utilities Board (PUB) and the Housing and Development Board (HDB) as these are places where students may also carry out investigative activities. There are now more than 17 different Young Scientist badges, which can be earned by primary school students.

There is high awareness of this scheme and many adults and students could recall working on their activities in primary school, especially with their family. The scheme has certainly inspired many budding scientists. One good example is Dr Jonathan Loh, an A*STAR scholar and later a multi-award winner who had started his passion in science when working on 12 different badges. One of Singapore's atmospheric scientists, Dr Koh Tieh Yong, also professed how his interest started when working on badges for astronomy, meteorology and geology.

Singapore Science & Engineering Fair (SSEF)

The SSEF is a national competition affiliated to the prestigious Intel International Science and Engineering Fair (Intel ISEF), well regarded as the Olympics of science fair competitions. SSEF is open to all students between 15 and 21 years of age. They submit research projects covering all areas of science and engineering which are judged by professional scientists and engineers.

Feedback from SSEF participants found that they valued the opportunity to work closely with mentors and friends or classmates on a project of interest. Important learning points for them included qualities of a good researcher, exposure to the work of a researcher, use of research equipment that was different from school labs, forming a hypothesis and designing the experiment to test the hypothesis. Close to 80% of respondents in a 2012 survey indicated they planned to pursue a career in science research, medicine or engineering as opposed to business or non-science careers.

One example among many is Dr Loh Xian Jun who presented his research project as a 17-year-old in SSEF and went on to read Bioengineering when he went to the university. He became a research scientist studying general polymer characterisation techniques with the potential for such novel materials to be used for the delivery of drugs or as scaffolds for living cells. Another candidate Mr Gao Guangyan who took part in SSEF when he was 15 years old, later went on to study Engineering and graduated from MIT in 2015. While he was in MIT he set up a company to make Tesla Coils and other electronic gadgets. He kept a close affiliation with the Centre, and when he returned to Singapore upon graduation, he even built a 'Singing' Tesla Coil for the Centre!

Concluding Remarks

The Centre with a history spanning close to 40 years, is an icon of Singapore and is one of the top science centres in the world. It has been voted as a cherished institution by Singaporeans in the SG50 Heart Map.

Over the decades, the Centre has benefited from the leadership of many Chairmen, namely Mr Wee Cho Yaw (1970–76), Mr Tan Keong Choon (1976–88), Mr Michael Yeo Chee Wee (1988–97), BG(NS) Lee Hsien Yang (1997–2004), Mr Willie Cheng (2004–07), Mr Wee Heng Tin (2007–12) and Ms Tan Yen Yen (2012 to present), supported by their respective Board Members. The Centre has attained its fame and status through the dedicated teamwork of a passionate crew lead by a succession of Directors, namely Mr Kenneth V Jackman (1970–76), Dr R S Bhathal (1976–82), Professor Leo Tan (1982–91), Dr Chia Woon Kim (1991–95), Dr Chew Tuan Cheong (1995–2010), and Associate Professor Lim Tit Meng (2010 to present).

The Centre has been winning many local and international awards because what it offers to the visitors is unique, fun, engaging, endearing and meaningful. Affirming the Centre as an excellent Edutainment Champion, the Singapore Tourism Board has bestowed three times upon the Centre the Singapore Experience Award in 2010, 2011, and 2012, consecutively, for Best Enrichment Experience. The Asia-Pacific Science Centre Network (ASPAC) has given the Centre at least three awards thus far, including an award for Creative Science Exhibition in 2013 for 'Fire Tornado', and two for Creative Science Communication for 'ArtChemist Show' in 2013 and for 'Sex Cells: A Science Musical' in 2015. The international Giant Screen Cinema Association has given the Omni-Theatre an IMAX Best Booth Award in 2012, and to the Science Centre two other awards for Best Movie Programme in 2011 and 2012, respectively. Several family and lifestyle magazines have their readers voting for the Centre as one of the best attractions in Singapore in

Fig. 4. The current Science Centre complex in Jurong East.

recent years, and the Centre was recognised with the TripAdvisor's Traveller's Choice Award in 2015. The Singapore Science Festival 2014 received the Bronze Award at the Marketing Awards 2014 for 'Best Government Sector Event' and another Bronze for 'Best Government Sector PR Campaign' after competing against big-budgeted events submitted from other Asia-Pacific countries. In 2015, the Science Centre-NEI joint-venture exhibition 'HBX: Into the Human Body Experience' brought in two trophies while the Family World honoured the Centre with the Best of the Best products and services award.

Whether the Science Centre gets to finally move to the 'promised land' or not, it shall remain faithful to its original purpose articulated in the 1969 proposal, and continue to be driven by the noble mission as stated in the Science Centre Act, 1970. As Singapore celebrates 50 years of national development, the Science Centre is not forgotten as a pivotal institution that has inspired generations of nation builders both past and present, and will press on to inspire many more in the future!

References

Proposals for the Setting Up of a Science Centre in Singapore (1969).

Singapore Science Centre Official Opening Souvenir Programme (1977).

Science Centre Board Annual Reports (1970/71, 76/77, 77/78, 86/87, 87/88, 97/98, 2007/08).

Chapter 28

The Lee Kong Chian Natural History Museum*

Kevin Y L Tan

Introduction

While the Lee Kong Chian Museum of Natural History, which opened in March 2015, is one of Singapore's newest museums, its collection is one of the oldest. Much of its heritage collection was once part of the natural history collection of the Raffles Library and Museum (later the National Museum). This chapter tells the remarkable story of how this collection — threatened with destruction on so many occasions — survived and how this museum was established. This is a story of how a small group of individuals, sensing the importance of Singapore's natural history legacy, battled against all odds not only to preserve and protect the collection, but to also establish Singapore's first dedicated natural history museum.

A Rafflesian Legacy

The establishment of Singapore's first museum was accidental, rather than planned. When Sir Stamford Raffles laid down his plans for the Singapore Institution (precursor to Raffles Institution), he included in it a plan for a comprehensive library as well as a museum. However, it was not till 1845 that a private library, known as the Singapore Library was established on the premises of the Singapore Institution. In January 1849, Temenggong Daing Ibrahim presented the colonial government a gift of two ancient coins which were placed in the Reading Room of the Singapore Library. These coins — now lost — formed the kernel of what was to become the museum wing of the Singapore Library. As the Library was a private enterprise, its success and viability depended greatly on the commitment and contributions of its members. Alas, by 1874, the Library

---

*I have relied heavily on my *of Whales and Dinosaurs: The Story of Singapore's Natural History Museum* (Singapore: NUS Press, 2015) for this account.

was broke and the Straits Government was invited to take over the Library, pay off its debts and transform it into a public library.

The late 19th century was an age of learning and of collections and museums, and it was no different in the Straits Settlements. On 28 March 1874, the Straits Legislative Council resolved to set aside $10,000 towards the formation of a Library and Museum, and thus was born the Raffles Library and Museum. James Collins was engaged as Economic Botanist and Librarian. Due to space constraints, Collins was unable to get the museum off the ground and spent most of his time managing the Library. It was only in 1877, after the Raffles Library and Museum moved to a newly-completed block at the Raffles Institution, and when Dr Nicholas Belfield Dennys was appointed that the museum got underway. One of his first efforts was to obtain the carcass of a female rhinoceros that had died at the zoo at the Botanic Gardens in August 1877. The hide was found to be in bad condition but the skeleton was 'very neatly mounted'[1] and put on display at the end of 1877. Dennys acted as Curator of the Museum till 1882. During that time, the Museum grew tremendously and large crowds thronged its galleries to see the natural history collection. Between 1882 and 1895, the Museum was managed by Arthur Knight (1882–1887); William Ruxton Davison (1887–1893); and George Darby Haviland (1893–1894); and Thomas Quin (1894–1895).

Davison, a well-known ornithologist oversaw the move of the Library and Museum from its temporary premises at Raffles Institution to its own building on Stamford Road (today's National Museum of Singapore building) in 1887. It was during his tenure that the natural history collection grew, with numerous gifts of specimens from visitors and well-wishers, and it was Davison who first effected a collection policy for the Museum, determining that the Museum should contain specimens from areas demarcated by the 50-fathom line set down by the great naturalist Alfred Wallace in 1859. Davidson himself was involved in several collecting expeditions, the longest of which was a two-month trip to Pahang in 1891. Unfortunately, the death of his wife drove Davison to despair and drink and eventual suicide in 1893.

George Darby Haviland, then Secretary of the Committee stood in temporarily as Curator. Prior to his arrival in Singapore in April 1893, he had been the second curator of the Sarawak Museum and put his experience to good use, and improving the quality of display and curation in the Museum. The management were keen for Haviland to take up the position permanently but he demurred on account of the paltry salary the Museum offered. He resigned in January 1894 and his duties were taken over by Thomas Quin while the Committee searched for a permanent curator. The man who was eventually hired was Dr Karl Richard Hanitsch (1860–1940), a German naturalist with a doctorate from the University of Jena and who had taught in Britain. He took up the post on 1 July 1895.

Becoming the Raffles Museum

Hanitsch was to become the Museum's longest serving head, occupying the post for some 24 years. He was young and energetic and prioritised 'overhauling, cleaning and arranging the large collection of stored zoological specimens'.[2] Hanitsch's long tenure — first as Curator, and then Director of the Raffles Museum (the title having been changed from Curator to Director in 1908) — as well as his scientific training enabled him to organise and grow its collections at an unprecedented pace. In 1906, when a much-needed extension was added to the Stamford Road buildings, Hanitsch overhauled its entire exhibition space, moving its entire zoological collection to the new building. Hanitsch gave the highest priority to the zoological exhibits partly because of his professional interest, but mainly because he knew that it was this collection that attracted the greatest number of visitors. Most of the local population were still illiterate and were less attracted by the Museum's ethnographical material, but there was something primeval and palpable about gazing at these magnificent specimens of nature at such close range that had people returning to the Museum again and again.

During his long tenure as Director, Hanitsch also maintained a healthy relationship with many of the region's museums as well as those in India and London. Through this network, he traded surplus specimens and bought new ones from other museums to add to the zoological collection. He was particularly concerned to 'complete' as much of each collection as possible by having a comprehensive set of specimens. It was quite clear that by this time, the importance and reputation of the Museum's zoological collection was becoming quite significant. This is evident from the fact that each year, numerous scientists, naturalists and scholars visited the Museum to study its collections. In 1919, Hanitsch retired to Oxford at the age of 59 where he continued doing research till his last days.

While Hanitsch had certainly made the Museum popular and attractive, he did not give much attention to the reference collection. It was his replacement, Major John Coney Moulton (1886–1926), who in a whirlwind of activity from 1919 to 1923, turned the Museum into a truly scientific institution. Moulton was both a scientist and military man. At the age of 22, Moulton cut his teeth as Curator of the Sarawak Museum, but left to join the army in 1914 when war broke out in Europe. When he assumed the Directorship of the Museum, Moulton decreed that the reference collection be reorganised and that all specimens be relabelled to the latest classification standards. He divided the Museum's material into three collections. The first was the Exhibition Collection, consisting mainly of mounted specimens displayed for popular instruction; the second was the Reference Collection, which should be on as large a scale as possible, but carefully classified, named and stored away for consultation by scientists or advanced students of natural history. The third was the Duplicate Collection, which should be stored away for use as replacements for damaged specimens or for exchange with other museums. Beyond being a mere repository for unique specimens, Moulton felt that the museum should also be a centre for research, and this required making the Reference Collection known and available to those who would study it.

Halcyon Days

In the four short years that Moulton was Director, he transformed the Raffles Museum into a scientific institution of great regional importance. He was ably assisted by Valentine Knight as Curator and Frederick Nutter Chasen as Assistant Curator. By the time he resigned his post in 1923 to become Chief Secretary of Sarawak, Moulton had thoroughly overhauled the museum's collection, re-catalogued all its holdings, and weeded out duplicate specimens for disposal or exchange. Moulton also established a great working relationship with two scientifically-driven curators in Kuala Lumpur — Herbert Christopher Robinson and Cecil Boden Kloss, respectively the Director and Assistant Director of the Federated Malay States (FMS) Museums. Indeed, it was Kloss who took over as Director of the Raffles Museum upon Moulton's resignation. When Robinson retired as Director of the FMS Museums, Kloss became concurrently Director of both the FMS Museums and the Raffles Museum, a post he held till his own retirement in 1932.

It was during this inter-War period that the zoological collection of the Raffles Museum really grew, thanks in large part to Robinson, Kloss and Chasen — three of the most important naturalists of the region during that era. It was during Kloss' tenure as Director that a major exchange of specimens was effected between the Raffles Museum and the FMS Museums. In 1927, the entire mammals and birds collection of the FMS Museum was transferred to Singapore in exchange for Singapore's insects collection. At first blush, this wholesale transfer of collections seems highly irregular and defies logic. However, further probing reveals that Robinson, Kloss and Chasen — all in

Singapore — were keen to undertake a thorough study of the mammals and birds collection and would have preferred to have the collection in Singapore rather than in Kuala Lumpur. At the same time, H.M. Pendlebury, an entomologist, had been appointed to replace Robinson as Curator of the Selangor Museum and he too, would have preferred to study his insect specimens in Kuala Lumpur. With this transfer, the Raffles Museum's collection of birds and mammals grew dramatically. When Kloss retired in 1932, he was replaced by Frederick Nutter Chasen (1896–1942). Chasen had risen through the ranks and was the Museum's Director at the time Singapore fell to the Japanese in 1942.

The Interregnum of War

Throughout the inter-War period, the Directors of the Raffles Museum — especially F.N. Chasen — regularly transferred type specimens to the Museum of Natural History in London. The biggest of these transfers — over a thousand skins — took place sometime around 1935, shortly after Chasen had finished working on his magnum opus, *A Handlist of Malaysian Birds* (1935). It was not clear if Chasen was planning another massive transfer of mammal specimens after he completed his *Handlist of Malaysian Mammals* (1940) when the Pacific War broke out. In any case, Chasen, fearing for the safety of the type specimens, arranged for them to be moved to the Herbarium at the Botanic Gardens where he believed they would be safer. When the Japanese invaded Singapore, the specimens were still in the Botanic Gardens, but Chasen, who escaped Singapore by boat, and taking with him some of the more valuable type specimens, perished when the *HMS Giang Bee* was sunk by a Japanese destroyer in the Banka Strait.

When the British surrendered to the Japanese on 15 February 1942, the Museum was bereft of its top leadership. Chasen had perished at sea while his deputy, Michael Wilmer Forbes Tweedie (1907–1993) had been transferred to the Royal Air Force in 1941. The Museum was thus nominally left in the charge of its Archivist, Tan Soo Chye. In the days before the British surrender, Dr Carl Alexander Gibson-Hill, the Acting Medical Officer, was appointed to take charge of the museum and library. While everyone was fearing for their lives, Eldred John Henry Corner (1906–1996), Assistant Director of the Botanic Gardens, took it upon himself to try to protect the Museum from the rampaging Japanese troops.

Through a series of near-miraculous encounters, Corner managed to convince the Japanese that the museum and library be protected against pillage.

As it happened, the Imperial Army had despatched a Dr Hidezo Tanakadate from Saigon to Singapore 'to investigate useful minerals'. However, Tanakadate — who only had a nominal rank of sub-lieutenant — had other ideas. He saw himself as the protector of cultural heritage and proceeded to take over the Museum and Botanic Gardens. Because of Tanakadate's connections — he had been classmates with General Yamashita, and his father-in-law was the well-known Baron Tanakadate — no one questioned his authority to do what he did. Tanakadate remained in Singapore for only thirteen-and-a-half months but because of him and Yoshichika Tokogawa (1886–1976) — more famously known as 'The Marquis' — the Raffles Museum and Library and the Botanic Gardens emerged from the Japanese Occupation relatively unscathed.

Becoming the National Museum

When civil government was restored in 1946, the old Straits Settlements (of which Singapore was a constituent part) was abolished and Singapore became a separate Crown Colony. Michael Tweedie returned to the Museum as Director, and Carl Gibson-Hill as Assistant Director. The next decade was a period of calm and transition for the museum. With Tweedie at the helm and Gibson-Hill as his deputy, there was a sense of continuity as both men had been involved with the Museum before the Japanese Occupation. It was a return to the pre-War routine of relabeling, rearranging and collecting. It was almost as if nothing had changed. However, things were very different outside the museum walls. Singapore was inching towards self-government and the British were under increasing pressure to 'localise' or 'Malayanise' the civil service. In 1957, Eric Ronald Alfred (b 1931), a local zoology graduate from the University of Malaya in Singapore, joined the museum as Curator of Zoology. Alfred became the first local person to hold the post of Curator, signalling a shift in the political climate. Tweedie retired in 1956 and was succeeded by Gibson-Hill, who was the last expatriate Director of the Museum. When Gibson-Hill himself retired in 1963, it was Eric Alfred who took over as Acting Director.

In 1957, the administration of the Raffles Library and Museum separated through the enactment of the Raffles Museum Ordinance and the Raffles Library Ordinance In 1960, the museum ceased to be known as the Raffles Museum and was renamed the National Museum and placed under the Ministry of Culture. At the time of its creation, the National Museum inherited the entire collection of the old Raffles Museum, which included the ethnological collection and the natural history collection, as well as its general collection of art works and historical artefacts of Singapore and the region. However, the government had new plans for the museum. The museum, for so long a centre of scientific learning, research and education, was now going to be mobilized for the nation-building exercise that would follow, with a corresponding shift away from the natural sciences to history, culture and anthropology.

Very shortly after the National Museum's creation, the government decided to employ the museum for scientific education as well and a Science and Technology Museum Committee was established. Then, in 1961, the government announced plans to establish a science and industry museum in a section of the National Museum to 'popularize science by holding public lectures using visual aids, films and television, thereby making more effective the acceptance of the museum as an instrument of education'.[3] Carrying the endorsement of both the Minister for Education and the Minister for Culture, it was felt such 'a museum would be of immense value to the cause of scientific and technical education for the people of Singapore' and would 'stimulate the economic life of the nation through the display of the nation's products and processes in the economic field'.[4] Eventually, the idea of situating a science and technology museum within the National Museum would grow into something bigger than the museum itself.

By the time Singapore achieved independence in 1965, the National Museum was beginning to look dated and old, leading to public calls for the government to 'modernize' it. People had started making unfavourable comparisons between the National Museum and the Muzium Negara in Kuala Lumpur which had been built in 1963. Interestingly, Woon Wah Siang, Permanent Secretary to the Ministry of Culture, declared that Singapore's National Museum was 'second to none in Malaysia' on account of the quality of its 'study collections' and the research generated from scholars working in them.[5] Woon further informed the public that a committee had been set up in 1964 to consider plans to modernize the museum's public galleries 'so that more exhibits can be effectively displayed in modern show-cases'.[6] Several development plans were prepared but none were realised.

Science and Technology or Natural History?

The increasing emphasis on science and technology as key drivers of industrialization led to the National Museum being transferred from the Ministry of Culture to the Ministry of Science and Technology in 1968. Suddenly, a whole new vision appeared. The National Museum would now be used to push the science and technology education programme. In December 1968, the government announced that the National Museum was to be transformed into a Natural History Museum at a cost of $245,000.[7] In 1969, after a brief survey, and following the renewed interest in Singapore's history on the occasion of the celebrations for the 150th anniversary of Singapore's founding, the ministry decided to transform the National Museum into a pure Natural History Museum. In response to a question from J.F. Conceicao, MP for Katong, the then Minister for Science and Technology Dr Toh Chin Chye replied:

> … the Museum, as it stands at the moment, is almost a junk-shop containing a hodge-podge of exhibits. You enter one room in which you see snakes and monkeys and you enter the next room and you see old Chinese furniture and sarongs.
>
> Obviously the Museum needs a new policy and, accordingly, a Gallup poll was conducted among Museum visitors to ascertain their preferences for the various categories of exhibits in June this year. The results showed a general preference for exhibits of flora and fauna. In view of this preference and of the limitation of funds which makes it impossible to build up the Anthropological Section of the Museum, it has now been decided to develop the National Museum into a Natural History Museum.[8]

According to Dr Toh, the museum had sought the expert opinion of Dr A.N.J. Vandehoop, the Museum Advisor from the United Nations Educational, Scientific and Cultural Organization (UNESCO) back in 1957, as well as the Science and Technology Museum Committee that had been established in 1961.[9] The space allotted for public galleries would be increased from 35,000 sq ft to 52,000 sq ft, and when completed, would include special exhibits on archaeology, porcelain, pottery, woodwork, furniture, boats, brassware, cloths and fabrics, coins and medallions, weapons, metalwork, musical instruments, primitive peoples, mammals, birds, reptiles, amphibians, fishes, and invertebrates. The plan also included improvements to the museum's research facilities, which would be open to university students and other research workers.[10] In 1969, Eric Alfred was again asked to prepare a plan to transform the National Museum into a Natural History Museum. This lengthy proposal, which consisted, in the main, of plans for a physical renovation of the museum's premises, once again made the case for better research facilities and care for the Reference Collection which then numbered some 70,000 specimens.[11]

In the meantime, a parallel development was causing great confusion among the museum staff and the Ministry of Science and Technology officials. In March 1970, it was announced that a 'Popular Science Centre' would be built at an estimated cost of S$5.2 million.[12] A site had yet to be selected but the government had set aside $1 million in the development estimates for this centre.[13] In July 1970, an official from the Ministry of Science and Technology announced that the Science Centre would have two divisions: the first was a 'technology division to show the workings of machinery, rockets, telephones, production of paper, oil refining and the like'; and the second was to be 'a life sciences division portraying man in his natural habitat at various stages of evolution, different types of animals, birds, vegetation, and even detailed models of man's body structure.'[14] The official added that, 'the National Museum with its large and diverse collections can contribute to this project' and that it was probable that 'the centre would be housed under the same roof as the museum, since this would foster greater administrative efficiency.'[15]

In September 1970, the government created the Science Centre Board by enacting the Science Centre Act.[16] The main objective behind the Science Centre was to promote science and technology through a museum/gallery site. In projecting the importance and popularity of the Science Centre, Dr Toh estimated that the Centre would attract an average of 340,000 visitors annually, more than three times the number at the National Museum.[17] Part of the plan was to move the National Museum's natural history collection to the new Life Sciences Division of the Science Centre while the Museum's anthropological and ethnological collection would remain at the Museum which would be 'returned' to the Ministry of Culture for the establishment of an art gallery as well.[18] By the end of 1972, it became clear that the Science Centre would not be housed within the premises of the National Museum. Instead a new $9.5 million building would be constructed in Jurong to house the Centre. All talk about transforming the National Museum into a Natural History Museum suddenly evaporated and the Science Centre was beginning to have difficulty deciding on what to do with the National Museum's natural history collection even though it was stated that its display would be 'modernized'.[19]

Carving Out the Natural History Collection

On 31 March 1972, the National Museum was transferred from the Ministry of Science and Technology back to the Ministry of Culture, who would 'convert the back portion of the public galleries into an art gallery and will also run the Museum as an Anthropological and Art Museum.'[20] The museum's collection was split up with effect from April 1972 when the Ministry of Culture resumed control of the National Museum and its building in Stamford Road: it would keep its ethnological, anthropological and art collections and divest itself of its natural history collection.

The natural history exhibits in the museum, including the reference collection that was now officially referred to as the Zoological Reference Collection (ZRC), would be taken over by the Science Centre Board. Pending the completion of the new Science Centre in Jurong, the Board would occupy a portion of the museum and continue exhibiting the natural history exhibits at Stamford Road.[21] For a time, it was known as the Science Centre at the National Museum. This collection of exhibits would 'eventually be enlarged and depicted in a modern systematic way in the new Science Centre'.[22] Shortly after its creation, the Science Centre also established a Schools Service to provide scientific programmes for teachers and students, and this too, was located at the museum's premises on Stamford Road. It later moved to Friendly Hill, off Depot Road in October 1972.[23]

Exactly what caused the radical shift in plans is not clear but may be surmised. The following facts are indisputable. First, the government, visitors and curators felt that the museum, venerable though it was, needed to be modernised and updated. Second, the government had determined that science and technology was of paramount importance, both in terms of education and in terms of it being a key driver of industrialisation. This led, among other things, to the creation of the Ministry of Science and Technology. Third, the government was prepared to invest generously in technical education and to this end, it was felt that a brand-new, ultra-modern facility — the Science Centre — would be built. The combination of these three factors led government planners to turn the National Museum into 'an Anthropological and Art Museum'[24] that promoted 'the art, culture, history and anthropology of the Southeast Asian region.'[25]

The split in the collection meant that the ZRC would eventually have to move out of the Museum's premises at Stamford Road. Because a new and spacious Science Centre would be built, most people simply assumed that the entire ZRC would move there and form the core of its natural sciences exhibits. However, what most of the planners did not realise was the vastness and value of the ZRC that lay hidden in locked drawers, shelves and rooms. What everyone knew were the mounted zoological specimens on display, so dramatically epitomised by the skeleton of the Indian Fin Whale, the gigantic seladang, G.P. Owen's crocodile and the giant leathery turtle. But these were just the proverbial tip of the iceberg.

The ZRC, scientifically important as it was, simply could not be exhibited in its entirety and the Science Centre Board was not about to waste precious exhibition space to store the scientific collection. Pondering this problem, former Assistant Director of the National Museum, Eric Alfred felt that a 'split' in the collection was a practical necessity:

> This so-called splitting up of the collection was an idea that I thought up. I said this Science Centre was going to come up, and they've got nothing in store. So what actually was split up was not the Reference Collection. It was the mounted exhibits that people knew. I said, 'Let all of these go to the Science Centre' … So it all went! … The Science Centre was going to be only exhibition, no reference. And here was a ready, available zoological collection, which made them very happy.[26]

The Science Centre was happy to have these iconic mounted specimens for display since it was an exhibition, rather than a research centre. These specimens were eventually moved to the Science Centre in July 1976 in preparation for its opening in 1977.

The Zoological Reference Collection 1970–1977

With the display collection going to the new Science Centre, what was to become of the ZRC? According to Professor Bernard Tan — one of the original four members of the Science Centre Committee that had been formed in 1969 — the Centre never once considered taking over the Reference Collection:

> The Science Centre was initially going to be [about] merely physical science and engineering. That was the initial idea. It was Dr Toh Chin Chye's idea. He gave us parameters. That's why we called in the London Science Museum to help. They were basically a physical science and engineering and mathematics museum; a museum of engineering and industry. Somewhere along the way, the fate of the Raffles Museum got caught up in this. It was decided that there was to be another wing. I was slated to be the director of the physical science wing. Eric Alfred was going to be director of the life sciences wing, which would include the collection. The whole thing got delayed because of the incorporation of the life sciences … So, on Eric Alfred's side, the $64,000 question was: 'Do we make a modern centre or a 'museum' kind of centre?' So we decided, 'No', we didn't want all that collection. We'll make it a centre that educates people on the life sciences; a 'hands-on' type of centre. That's when the whole collection was jettisoned and in fact there was some uncertainty as to its fate and as to who would take it.[27]

The ZRC's situation was dire, and there was 'even talk about selling it or donating it'.[28] News of this moved quickly within the international museum fraternity. On 9 July 1971, E.C. Dixon of the British Foreign and Commonwealth Office's (FCO) Overseas Development Administration wrote to Mrs Joyce Pope of the British Museum, confirming what he heard from Dr D.N.F. Hall — the FCO's Fisheries Adviser — concerning the future of Singapore's 'Natural Museum':

> One incidental scientific item which came my way was the future of the National Museum. As the old Raffles Museum, the Museum in Singapore became world famous. Deposited in it there are very many TYPE specimens from the region, particularly of birds.... I understand there is a move to convert the National Museum into a cultural museum, and to remove completely the natural history section with no clear idea what is to become of it, and in particular the type specimens. A new Raffles Museum would perpetuate a distinguished name.[29]

Pope discussed the matter with her colleagues at the Natural History Museum and contacted Michael Tweedie — who had retired as Director of the Raffles Museum back in 1957 — about the likelihood of *type specimens*[30] being in the ZRC. Tweedie confirmed that no holotype specimens were 'held there'. In any case, Arthur Percy Coleman, Secretary of the Museum thought it prudent to discuss the matter with Dr Ron Hedley, Deputy Director of the British Museum who 'felt that Keepers might wish to know about the possibility that the collection at the Raffles Museum might be disposed of' and if they

> ... felt it would be useful, we might approach Lord Medway, who is in Kuala Lumpur, to make appropriate enquiries and perhaps to advise on what was available. This we would do through Lord Cranbrook.[31]
>   I would be grateful if you would let the Director know by 20 August whether you would be interested.[32]

This must have excited the various Keepers somewhat because Lord Cranbrook was indeed approached. His son, Lord Medway, had in the meantime, resigned from his position of Senior Lecturer at the University of Malaya in Kuala Lumpur and had gone off to New Hebrides, where he was not contactable.[33] In 1971, Dr Anthony J. Berry, of the School of Biological Sciences at the University of Malaya, spoke to Eric Alfred and 'to people in Singapore University' and determined that the Museum's collections would officially be handed over to the University of Singapore and that 'the University is being

given the necessary funds and other provision for handling them properly.'
Berry concluded:

> My general impression is that fairly good care is being taken to house the col-
> lections suitably (perhaps even better than hitherto) and that the collections
> are certainly not 'going begging' or in dire need of rescue.[34]

Exactly what happened 1970 and 1971? As noted above, the initial alert
about the collection's fate came from Dr D.N.F. Hall, the FCO's Fisheries
Adviser in July 1971. Hall was then involved with the key 'fish' people in the
region, including Associate Professor Tham Ah Kow and was quite aware of its
tendentious fate. Unknown to most people, it was Tham who brought the
University of Singapore into the equation.

Tham Ah Kow (1913–1987) was a fisheries expert who, in 1962 established
the Fisheries Biology Unit at the Department of Zoology at the University of
Singapore and became its first Director. From 1968 to 1978, under an agree-
ment with UNESCO and the Singapore Government, a Regional Marine
Biological Centre (RMBC) was established at the Department with Tham as its
Director.[35] Tham would serve as the Centre's Director till his retirement in
1973. He would be succeeded by Chuang Shou Hwa. It was in the context of
Tham's work at the RMBC that he became engaged in the subject of reference
collections. In 1969, there began a discussion on the feasibility of establishing a
Fish Taxonomy Unit in Singapore and whether a reference collection should be
created by RMBC. Dr Raoul Serène, the UNESCO Marine Science Regional
Expert for Southeast Asia,[36] stated that while it was 'not the main aim' of
RMBC 'to establish a Reference Collection, its 'activities will encounter a tremendous amount of obstacles' in its absence.'[37] In the course of these discussions, Tham informed Serène that Singapore's natural history collection would
be split into two units:

> … one unit in charge of the exhibition of Natural History with perhaps empha-
> sis given to the educational aspect; training of teachers, popular education,
> mass media for natural history, etc; one unit in charge of scientific reference
> collections of natural history.[38]

Serène questioned the wisdom of splitting the collection up but maintained that as it was 'an internal matter'; he was not qualified to officially comment as UNESCO expert.[39] Tham wrote to Hoe Hwee Choo of the Ministry of
Science and Technology advising against taking the Fish Reference Collection

out of the ZRC.[40] In April 1970, at a meeting to discuss a proposed Liaison Bureau and the wisdom of separating the Fish Reference Collection from the ZRC, Tham met with several notable marine biologists and Dr Toh Chin Chye, who was concurrently Minister for Science and Technology and Vice-Chancellor of the University of Singapore. At this meeting, Toh told Tham to prepare a report on the ZRC, which they referred to as the 'National Museum Reference Collection'. It was submitted in May.

Nothing happened for another year until Tham received a note from Hoe in March 1971. She informed him that although a 'decision on the fate of the Museum's Reference Collection and Reference Library' had not yet been made, 'there is a possibility that the Collection may be accommodated at the new campus' of the university in Kent Ridge. That being the case, Tham was advised to bring this to the attention of the University of Singapore Development Unit (USDU) so that floor space requirements could be factored in.[41]

A few days later, Tham penned a note to Reginald Quahe, Deputy Vice-Chancellor of the University of Singapore, attaching Hoe's letter and giving Quahe a background to his involvement in the discussions and how he came to draft the report for the minister. Tham prefaced his note by stating that as he was not a Head of Department and that he had no standing to communicate directly with USDU. He asked Quahe:

> If you think that the University of Singapore can entertain the proposal that the National Museum Reference Collection should be transferred to the University, could you be so good as to transmit the floor space requirements to the USDU. The estimated expenditure to be incurred in keeping the National Museum Reference Collection at the University of Singapore is of the order of $50,000 per year. However it may be possible for the Government to give a grant to the University to cover this extra expenditure.[42]

Just as Quahe was digesting the implications of Tham's note, he received a note from Hoe, dated 30 March 1971 in which Dr Toh Chin Chye, as Minister for Science and Technology, 'directed that the Reference Collections and Reference Library of the National Museum ... be transferred to the custody of the University of Singapore.'[43] Accordingly, the USDU was informed of this decision so that sufficient space could be set aside to accommodate the collection.

Dr Toh's directive came as a great shock to the university authorities who scrambled to find a way to accommodate the collection. At this time, the university occupied a beautiful but small campus on Bukit Timah Road. It was a campus that had been built for no more than 2,000 students but was now bursting at its seams with five times that number. Indeed, there was barely enough room for classes on campus. There was simply no more space to accommodate the ZRC. Dr Lim Chuan Fong of the Department of Zoology, who had been assigned to represent the University in negotiations on the ZRC tried to delay the transfer of the Collection till after the University's new Kent Ridge Campus was ready but the National Museum could not wait. The ZRC had to be moved out by 31 March 1972, a day before the Museum was transferred from the Ministry of Science and Technology back to the Ministry of Culture.

In the meantime, the fate of the ZRC was attracting the attention of internationally-renowned scholars who had worked on the Collection before. The world's eyes were anxiously fixed on how Singapore was going to deal with the collection. There were rumours that the Collection would be split up between the University of Singapore and Nanyang University, with specimens to be used for teaching rather than research. Eventually, Professor Chuang Shou-Hwa, Acting Head of the University of Singapore's Zoology Department, and Professor Nan Elliott, Professor of Physiology at Nanyang University 'came to an agreement to house the Zoological Reference Collection (or Raffles Collection), as it quickly came to be known, at the University of Singapore.'[44] It is unclear exactly how this decision was to be effected since the University of Singapore campus on Bukit Timah campus was already groaning at the seams. What we do know is that sometime between September 1971 and March 1972, a decision was made by — either by the university or by the Department of Zoology — to temporarily house the collection in five Romney huts that had previously served as workshops of the British Ministry of Public Building and Works' Department of the Environment. These properties had recently been handed over to the Singapore government as part of the British military withdrawal from Singapore.[45] Chuang was obviously not pleased with being given this added responsibility and did little to prepare the Zoology Department for the arrival of the collection and it was left to other stalwarts, like Professor Desmond S. Johnson to advertise for a Curator to look after the Collection.

Among the applicants for the position of Curator was Mrs Yang Chang Man, a former student of Tham Ah Kow. Yang, a biology graduate from Nanyang University, had been a teacher before signing up with the University of Singapore's Zoology Department in 1966 to complete a Diploma in Fisheries. Thereafter, Yang remained at the University of Singapore to pursue an MSc degree, which she obtained in 1972 for her study of the copepods of Singapore's waters. In 1968, when Tham Ah Kow became the Director of the Regional Marine Biological Centre (RMBC), he persuaded Yang to join his unit and arranged for her to transfer her Public Service Commission bond to the University. For the next four years, Yang's main job was to study, analyse, curate, and maintain about 60,000 plankton samples collected from the South China Sea.[46] She knew from Tham that the university would soon take over the Zoological Reference Collection and that they would be advertising for a Curator in the local newspapers in March 1972. Tham — who obviously thought very highly of Yang — strongly encouraged her to apply for the position; she joined the Department of Zoology as Curator while working at the RMBC concurrently.[47]

Up to this time, Yang had never seen the museum's Reference Collection. On her first day of work in July 1972, she reported to Chuang and was told to see Eric Alfred at the National Museum. Yang recalled what an 'eye-opening lesson'[48] it had been for her, opening the countless drawers of specimens, huge boxes of skins and skulls, and bottles of specimens preserved in formalin. The experience was overwhelming and she was stunned and both by the value of the collection and the enormity of the tasks ahead.

Yang had no experience in curatorship or in managing a century-old historical collection, but her scientific training as a marine biologist came in useful. The task was huge but not impossible when broken up into steps. The first step was to create an inventory. There were catalogues of crabs, molluscs, and fish, and some lists of holdings in the museum but these were of no use to Yang since her task was not so much an audit, but a full stock-take at the National Museum premises before the collection was transferred to the university. Everything had to be painstakingly taken out, examined and recorded. This she did by hand with the help of two technicians, Lim Keng Hua and Augustine Raphael, and the lists were religiously typed up by her clerk, Woo Lee Wu. It took Yang and her team over six months to do a complete inventory of over 126,000 reference specimens,[49] and by the time they were done, it was already early 1973, almost a year after the initial deadline set for removing the collection from the museum. At the end of the process, Yang and her team found that they were custodians of 126,000 specimens, of which 4,700 were type specimens (mostly paratypes).

The next task was to move and set up the ZRC at Ayer Rajah. As Chuang had taken no steps to renovate the five Romney huts — which were like airplane hangers with some abandoned machinery and huge doors — Yang was shocked by their bare state. There was neither shelves nor fans, the zinc roofs were leaking, there was no ceiling or furniture and the doors were damaged. She got the Estate Office to fix them and install ceiling fans, and personally went out to order angle-irons for the Department Technicians to build the shelves necessary to house the collection.

Once the shelves were ready, Yang and her team had to pack the collection. There was no proper packing material available and the team scavenged whatever they could find. Her budget for the whole move was a paltry $750, which she used sparingly. In all, the moving of about 60 lorry loads of specimens took almost a year, and did not include the books from the museum's reference collection.

Unfortunately, the Romney huts were terrible for the collection. The zinc roof made the huts like ovens and temperatures indoors reached 43°C in mid-afternoon and then plummeted to about 23°C at night. These huge fluctuations were deleterious to the specimens, but this was the best Yang and her small team could do. They spent most of their time maintaining and repairing the specimens, and trying to keep out the dastardly insects, mites and mould. The conditions were far from ideal but at least the collection was safe. The ZRC spent almost five years at Ayer Rajah and was visited by a number of scientists and researchers, some of whom became good friends with Yang. The first visitor to the ZRC at Ayer Rajah was Lord Medway, who was extremely concerned about the fate of the collection. Many of these early visitors were to prove instrumental in the next fight to save the ZRC.

Moving To Bukit Timah

In 1976, Yang was told that the ZRC had to be relocated; the land on which the Romney huts stood was required for the new university hospital building. At this time, Yang came in direct contact with the University's Deputy Vice-Chancellor, Reginald Quahe, who broke all protocol and called her directly, enquiring about the ZRC. Quahe, a commerce graduate and life-long university administrator, truly appreciated the value of the Collection. CM Yang recalled: 'He gave us so much hope, and at one point, offered a location on the new campus for the ZRC.'[50] Ironically, at a meeting to relocate the ZRC, Quahe offered the University's Estate Office Workshops at Kent Ridge Campus for housing the collection. This is the same site that would eventually be occupied by the Lee Kong Chian Natural History Museum. Nothing came of this offer because Quahe died of a heart attack in his office on 14 September 1977. The offer was eventually rejected by the Estate Officer whose decision was supported by the new Deputy Vice-Chancellor, Professor Choo Seok Cheow. It was never mentioned again.[51]

Not only were no premises made available at the university's new campus to temporarily house the collection, the original plans for a new building to house the ZRC at Kent Ridge were also squeezed into oblivion in 1977. The urgent need for more and more space to accommodate an anticipated spike in student numbers at the new campus simply meant that the planned space allocation for the ZRC continuously 'tumbled down the priority list.'[52] In late 1977, Yang was informed that the original plans for the building to house the ZRC at the Kent Ridge Campus had been scrapped. Apparently, Lim Chuan Fong, the Head of the Zoology Department, had informed the university's Development Unit that the ZRC need not be housed at Kent Ridge. With Lim's assurance, the Development Unit quickly pushed the ZRC off its planning radar and the collection was once again in jeopardy.

In the meantime, Yang was given a list of vacant government buildings to consider as possibilities for ZRC's new home. Together with her laboratory technicians H.P. Wang and Lua Hui Keng, Yang visited and inspected military buildings at Sembawang and in Portsdown Road, the PUB pump house at Ayer Rajah; university messes at Dalvey Road and Dalvey Estate, and several other places. None of the buildings could accommodate the entire collection. Moreover, all the buildings were in need of repair and renovation. After discussions with her Head, Lim Chuan Fong and the Estate Office, it was decided that the ZRC would be split up and housed at the Dalvey Road Mess, a space at Manesseh Meyer Block, and the old Arts Block. The plankton samples were still at RMBC at that time. They were then moved to the basement in anticipation of their transfer to Japan in 1978.

Back from the Brink: 1977–1987

Even as C.M. Yang was in discussions with Reginald Quahe over where the Collection might next be moved to, she decided to appeal to the many friends she had made while caring for the Collection to help save the ZRC. If well-known scientists could weigh in on the importance of the collection, perhaps its chances of survival would be heightened. One of the first to write in to the university about the collection was someone Yang had not met at that time: Professor Francis John Govier Ebling (1918–1992) from the Department of Zoology of the University of Sheffield. He was, at the time, External Examiner of the Department of Zoology at the University of Singapore, and had heard about the fate of the ZRC from other faculty members. Ebling dashed off a memorandum to the Department, stating how the specimens in the ZRC were of 'irreplaceable scientific value' and expressed the hope that the University would 'do its utmost to preserve this unique historical collection, and to display it worthily.'[53]

Yang's staunchest ally in the fight to save the collection was Dr David Roderick Wells, from the Department of Zoology at the University of Malaya. Wells had gone to Kuala Lumpur on a Malayan Commonwealth Graduate Scholarship in 1961 and was a former PhD student[54] of Lord Medway.

He became aware of and interested in the collection 'when Lord Medway started working on a series of handbooks on Malayan birds,'[55] and met Yang when she first took charge of the collection in 1972. Wells believed this Collection to be of tremendous importance from a scientific point of view but was also realistic enough to know that the only way to make oneself heard in an economy-obsessed state was to put a monetary value to the collection. Thinking back to the desperate days of 1970s, Wells recalled:

> I was down here and there was a panic. Oh my god, the government is going to hand out several of these reference skins to every school and that was going to be the end of it. Yang Chang Man had taken over, as a very young girl, with this enormous task. It's her and her alone we have to thank, for having this collection at all. So she said, 'What can we do? What can we do?' So I said, 'We know they think it's worthless, so I will write to the heads of some of the world's major museums and get them to write to the Singapore government, saying that they heard that the museum was being disbanded, and would the government be interested in a bid for purchase.' So I got Dr Les Short in the American Museum of Natural History and David Snow … from the British Museum …
>
> I knew all these people. So I said, 'For God's sake, cook up a letter that makes it look like you want to make a bid for it. You are interested in taking it because it is of world importance.' We wanted to make the Singapore government realise that: (a) this is recognised globally as a collection of importance; and (b) nominally, it might be worth some money. That was just my idea. Other people said, 'No, you wouldn't make an impact whatsoever, they just weren't interested.' *But*, it certainly coincided with the dropping of the scheme to disperse the whole thing.[56]

Wells' strategy was brilliant. He fired off letters to three major museums in Europe and America — the American Museum of Natural History; the Field Museum of Natural History; and of course the British Museum (Natural History) in London. To his friend Dr Lester Leroy Short (b. 1933), Lamont Curator of Birds at the American Museum of Natural History, Wells wrote:

> Trouble in Singapore. The government plans to build a road through the present accommodation of the old Raffles Museum collections. Unless an adequate monetary (as distinct from scientific) value can be put on them, I understand from the Curator no adequate alternative set-up will be provided. I wonder therefore if you could give us a figure for the insured value of the AMNH birds, say, per 1,000 skins or other convenient unit? Everything in Singapore has its price and the fact that the collection there is virtually untapped biologically will, alone, impress no-one important.[57]

A similar letter went off to Dr Jack Fooden (b 1927) from the Division of Mammals at the Field Museum of Natural History in Chicago, and to Ian Courtney Julian Galbraith (b. 1925), then Head of the Sub-Department of Ornithology at the British Museum (Natural History). Yang, who knew Galbraith from earlier correspondences, also wrote to him, asking him if he could give an estimate of the value of the collection for the purposes of insurance. Galbraith sent around a memorandum to his colleagues in the other departments, request-ing price estimates for a collection. This dumbfounded them since they had never been faced with such a request; and in any case, the British Museum did not insure its collections since this was considered government property.[58] Reginald William Sims (1926–2012) Head of the Annelid section of the British Museum (Natural History) offered perhaps the most sensible answer:

> It is of course impossible to estimate the financial value for these *Oligochaete* collections since the only yardstick would be the price for any other collection, and I have no knowledge of such collections being offered for sale.[59]

The American curators were less nonplussed by Wells's and Yang's requests. Dr Short offered the following guidelines on pricing:

> It is exceedingly difficult to place a value on items many of which are utterly irreplaceable. The collection there contains many such specimens. One might figure some scheme such as the following: a) all type specimens that are sent out (and this is not a usual practice) are insured for $500.00 to $1,000.00, and this would be an appropriate value for each of them; b) to all unique specimens of historical-archival importance, I would assign a value of perhaps $100.00 apiece; c) for data-bearing specimens from areas in which one cannot obtain replacements (i.e., the forests are gone, or the area is politically impossible) I would guess a value of perhaps $50.00 each; d) to all other specimens bearing data, a value of $10.00; and e) to specimens with little or no data $5.00 apiece. None of these estimated values are to be taken as truly indicative of worth, for one could argue that a unique specimen that could not be replaced should be valued higher, at thousands of dollars!
>
> I have given you one set of figures treating 'values'. In practical terms, collections that are *sold* today, and they rarely are, do not come up to the value just given. It is in fact standard practice to purchase specimens comprising a collection at $3.00 to $5.00 a piece.[60]

Yang continued to collect estimates from other quarters — on mammals from Lord Medway; and insects and other arthropods from the Field Museum of Natural History — and compiled a one-page table of estimated value of the collection. It came to a staggering $8,747,853.00.[61] Based on an insurance premium of 50 cents per $1,000 value, the insurance premium for this collection alone would exceed $4,000 per annum. This was the 'bill' Yang would present her bosses. Unfortunately, the economic and financial argument, compelling as it was, did not help secure a proper space for the ZRC on the new campus — the authorities thought that the collection had little to contribute to Singapore's development.

Holding the Fort ... Amidst Shifting Sands

The years 1977 to 1979 were a nightmare for the ZRC. The new university campus at Kent Ridge — a part of which included the Ayer Rajah compound that the collection was forced to vacate — would not be ready till 1980. In the meantime, the Bukit Timah campus was so overcrowded that classes were being run in the most appalling conditions. Where would the ZRC go? As a temporary measure, the ZRC was split up into parts and stored in various places, some in a dreadful state of disrepair. The splitting up of the collection meant that access for curation, maintenance, research and study was severely limited, if not well-nigh impossible in some cases. Worse still, the conditions, especially at Dalvey Mess, were very humid, and the buildings were frequently infested with termites. Fortunately, the stout teak cases housing the specimens prevented them from being destroyed. In fact, only Yang and her assistant, Lua Hui Keng were able to work in those dreadful conditions to safeguard the collection. Oftentimes, they had to deploy very strong and poisonous insecticides and fungicides to keep specimens in good condition. Despite these harrowing conditions, Yang and her staff doggedly and stoically provided the special care required for the delicate specimens. Most importantly, they did everything in their power to continue to make them available for study by visiting researchers and scientists. What kept them going was the importance of this scientific collection, which had to be cared for and preserved for the future generations. And of course, there was the hope of the collection gaining a permanent home.

In 1979, this hope shattered when the university had a space crisis and rumours spread that the collection would have to be thrown out by the new university term in July 1979.[62] The government had decided to merge the Chinese-medium Nanyang University with the University of Singapore to form the National University of Singapore. As a prelude to the merger, Nanyang University students joined University of Singapore students in attending courses taught in English under a Joint Campus scheme. This led to an influx of even more students onto Bukit Timah Campus. Worse still, the new campus in Kent Ridge was nowhere near completion and every department was asked to justify its use of space. Classes were given top priority and every available space at the Bukit Timah campus would be converted to pedagogical use. Roland E Sharma, then Acting Head of the Zoology Department, recalled:

> ... it was at this time that the Collection had to contend with ever-increasing demands on its allotted space, due to an increase in student numbers resulting from a Joint Campus scheme with Nanyang University. With pressing need for additional lecture and tutorial rooms it became evident that once again, the Reference Collection would have to go.[63]

It was Sharma who obviously bore the brunt of these pressures. A pleasant, avuncular man not prone to confrontation, Sharma felt deeply about the collection but was also powerless to fight for its preservation in Singapore.[64] In desperation, he began thinking of 'arranging for its acquisition either *in toto* by a major Museum or University overseas, or its fragmentation and dispersal amongst various institutions.'[65] News spread that the insect, fish, bird, and mammal collections would be dispersed to several Malaysian universities and institutions, including the University of Malaya, the Universiti Kebangsaan Malaysia, and Universiti Sains in Penang. He even once asked Yang if she would be prepared to 'follow the collection and work in Kuala Lumpur.'[66]

Sharma felt extremely conflicted by this situation since he was well aware that the disposal of the collection 'would incur the loss of an academic heritage closely associated with Singapore and held in trust for the future' and that would in fact be 'contrary to a UNESCO Recommendation for the Protection of Movable Cultural Property, which had been adopted the previous year (1978) and one to which Singapore was a signatory.'[67] Things did not look good. In March 1979, a distraught Yang wrote to Dr Short again informing him of the possible transfer of the collection to Malaysia. It would, she thought, 'be more safe to keep this collection in well-known museums that would preserve it for science.'[68] Yang's international contacts continued to press the Singapore authorities for news of the ZRC and even offered to absorb it into their own collections.

Enter the Malayan Nature Society ... and Nancy Byramji

In many ways, Sharma's hands were tied and he was too low in the university hierarchy to do more.[69] However, Sharma was also the Chairman of the Malayan

Nature Society (Singapore Branch) and it was through the Nature Society that a 'flank attack' could be undertaken. Sharma, who had long been involved with the Society, became the Chairman of the Singapore Branch in July 1977. His committee comprised some of the most important names in Singapore's scientific community, including Dr Popuri Nageswara (PN) Avadhani and Professor Anne Johnson, Chairman of the Nanyang University's Department of Biology, who was the society's Honorary Secretary. Anne Johnson was the widow of Desmond S. Johnson and was as passionate about the ZRC as her late husband had been.[70]

The committee had long been concerned with the fate of the ZRC but could do little more than to quietly engage the government officials privately to persuade them of the value of the collection.[71] Johnson was particularly upset over Sharma's decision to dispose of the collection to the Malaysian institutions and resolved to do something about it.[72] She decided to alert a young journalist who was a member of the society — Nancy Byramji — and get her to write an article about it.[73] Johnson provided Byramji with the background to the crisis the collection was facing[74] and on 29 April 1979, there appeared on the front page of *The Straits Times*, Byramji's article, entitled

Save our heritage: Priceless Raffles Collection may end in the dustbin unless $70,000 a year is found.[75]

Byramji had sounded the alarm bells about the university's space crunch and the 1 July 1979 deadline in the most dramatic fashion. Her long article detailed the history of the collection and how the tensions of space and money led to its plight. In her article she quoted 'a zoologist' as well as 'a biologist' as saying that the collection 'is one of the best and most complete collections of tropical fauna in the world' and another senior biologist querying 'Why has no space been allocated for this obviously valuable collection at the Kent Ridge Biological Sciences complex?'[76] In 2014, Byramji confirmed that the source in this article was in fact Sharma and that the quotes had all come from him.[77]

At this point, the situation may be summed up as follows. The University of Singapore's Zoology Department was instructed to get rid of the collection by 1 July 1979 so that room could be made for students in the coming academic year. Byramji tried getting the university to justify its actions but the 'helpful' spokesman said 'she could not comment on the situation.'[78] The biologists at the Nanyang University were keen to take over the entire collection but it did not have the money; its Director-General was 'believed to have rejected the offer because Nantah's biologists cannot justify to their administrators the

$70,000 extra, and the additional staffers, that will be needed to keep the collection.[79]

Off to Nanyang University

Byramji's article had its desired effect. Readers and other journalists chimed in.[80] Less than a week later, there was a breakthrough. At a meeting on 22 May 1979, between Mrs Lu Sinclair (Registrar of the University of Singapore), Sharma, Yang, and Johnson, Sinclair offered the Nanyang Campus as the site for the ZRC's new home.[81] It is not clear what transpired between the authorities and the universities during this time but in July 1979, it was reported that talks were underway between the University of Singapore 'and interested parties to keep the century-old Raffles Collection intact.'[82] The *Sunday Times* editorial of 19 August 1979, revelling at having broken the news through Byramji's article, heaved a sigh of relief but offered an unusually critical and introspective perspective of the whole episode:

> But, amidst this moment of relief, there creeps in a feeling of shame, even fear of a sort. First, let us take the shame ... shame that a valuable part of our scientific lore was bandied about the way it was. The Raffles Collection was treated shabbily of late, and that is shameful....
>
> Why, one must ask, did this priceless collection pass from hand to hand as if it were a leperous object that would taint the hand that held on to it for too long? Why, in the name of science, did no one seem to realise that here was something that scientists the rest of the world over would have given their right hands to get hold of and cherish with loving care? And all this time it was in Singapore, ours ... and we did not seem to care. Yes, it did cost money — $70,000 a year to be exact — to upkeep. Yes, it did take up 6,000 square metres of space. But it is part of our heritage, even if its cold scientific value was not appreciated ... and heritage is something that can never be replaced once it is lost. It is a shame that we placed ourselves in a position that saw us come so close to losing it. Which brings us to the fear that if this could happen to the Raffles Collection, could it not happen to some other facet of our heritage? Let the Raffles Collection episode be a warning; that there are things which mean more than cold cash and quick returns; that when it comes to what was once part of our lives or the lifestyle of our land of yesteryear, it is not becoming to stand up and say that it is old, that it is time for the new to take over, that looking back is not a farsighted policy. Our past did happen ... it is history, our history — and a lot of it is our heritage. It is our responsibility to see that it is preserved for the generations to come.[83]

Nanyang University had offered up the rooftop space on top of its library building to house the collection. After some renovations, Yang and her team took
some six months to move the Collection from Bukit Timah to its new home in
1980.

Safe at Last

Unfortunately, just six months after they had settled in, the collection had to
move again. ZRC staff was shocked and saddened by the dreadful news; they
were exhausted after moving the many thousands of specimens again. Moreover,
the frequent moving of the century-old, brittle specimens was extremely damag-
ing. The old Nanyang University had merged with the University of Singapore to
form the National University of Singapore (NUS) with its home in Kent Ridge.
The Nanyang campus was then given over to the new Nanyang Technological
Institute (NTI). Its administrators were anxious to reclaim the space for use as
its administrative office.

After news spread to the media that ZRC would have to make the fourth
move, NUS Deputy Registrar Ling Sing Wong told the press in November 1980
that NUS was thinking of moving the specimens 'because the collection is now
too far for specialists and undergraduates on the Bukit Timah campus to use
for research.'[84] This announcement — made without any consultation with the
Department of Zoology and certainly without consultation with Yang and her
team — was a particularly nasty blow to everyone. The report ended by stating
that an Australian University was interested in buying part of the collection,
but that 'many scientists are horrified at such a prospect of breaking up the
Collection.'[85]

In January 1981, Yang attended a meeting during which Mrs Sinclair read
out a letter from the American Museum of Natural History requesting to
purchase the collection and announced imperiously, 'The University is now
convinced that ZRC is an important collection and we will not dispose of it.'[86]
She also announced that a permanent home in one of the buildings at Kent
Ridge campus would be provided for the collection. Sinclair told the press that
the university had 'more than enough room at NTI for the zoological collection with 500 acres for an intake of only 800 students in 1982'. It was also
revealed that the NTI building committee had 'decided against shifting the
Institute's administration department to the Nanyang Library as originally
intended.'[87] Thereafter, plans were made to accommodate the entire collection
in the Science Library Building (S6) at Kent Ridge.

For seven years after the announcement (1980–87) the ZRC — except for the fishes — was maintained at the NTI campus and continued to serve the visiting researchers from overseas and the department staff and students. More specimens were added to the collection during this period through donations and collaborative work with visiting researchers. During this period, Yang and her team spent a lot of time repairing specimens which had been damaged by the moves and through poor storage.

A Kent Ridge Homecoming

In 1987, the Zoological Reference Collection moved to its 'permanent home' in three floors of the S6 building, which it shared with the Science Library of the National University of Singapore. Ironically, its new home was very close to the site of the old Romney Huts that it occupied from 1972 to 1978. It took more than a year to move, unpack and rearrange the entire collection in the new home. The fact that there was a separate Science Library building was itself remarkable especially since the science collection of books had hitherto been part of the Central Library's main collection. Credit for this must be given to Professor Gloria Lim, who was Dean of the Science Faculty from 1973 to 1977 and then again from 1979 to 1980. It was Lim who persuaded the University administration to build a separate Science Library, away from the main library. Initially, the Science Faculty was to have been located right next to the Engineering Faculty, and its proximity to the Central Library would have made a separate library for the science students unnecessary. However, when it was abruptly decided that the Science Faculty should be located at its current site, Lim argued 'very strongly and persuasively' for a separate library building to be built.[88] It thus became possible, within this new building, to provide for additional space that would be needed to house the ZRC.

For the first time since its creation, the ZRC was housed in a climate-controlled environment. The facility was air-conditioned 24 hours a day and maintained at between 22°C and 24°C with 55% to 60% relative humidity. The ZRC's new premises was opened on 31 October 1988 by Minister for Education, Dr Tony Tan. Also present at the launch were former Director of the Raffles Museum, Michael Tweedie and Eric Alfred, former Acting Director of the National Museum, Vice-Chancellor Lim Pin, Dean of Science Bernard Tan, and representatives of the British Council.[89]

Many zoologists, who had spent so many years worrying about the fate of the ZRC, were relieved and delighted to know it had finally found a permanent home. Gerlof Fokko Mees (1926–2013), Curator of the Rijksmuseum of Natural History in The Netherlands wrote, expressing his 'great satisfaction that the collection is now adequately housed', and that its scientific value is recognised by the University and by the Government of Singapore.'[90] Lord Medway, now Earl of Cranbrook, probably spoke for all of them when he wrote, expressing his gratitude to Yang:

> It is a personal triumph for yourself that the collection has survived its many moves and different homes these past years. When first the material was turned out of the old Raffles Museum building, I was not alone in thinking it lost forever. Only your persistence and determination has saved it through this period. Your modest demeanour hides a will of iron! I am so glad that you have succeeded in your aim.[91]

Life for Yang and her team became more settled after the move; it was a great respite from the stresses of first decade when they were harried from pillar to post in a bid to hold onto the collection. Yang's headaches were not completely over but the tide had by this time turned in the ZRC's favour and the Zoology Department was much more appreciative of the collection and supportive of Yang's efforts.

During its first decade back at Kent Ridge, the team spent most of its time sorting, identifying, arranging and cataloguing the collection and repairing damaged specimens and replacing most of the formalin used in the wet collection with alcohol since formalin is a known carcinogen. An increasing number of scientists consulted the collection, and more loans and exchanges of specimens were processed. Staff also actively participated in field work, faunal surveys and expeditions, and hence a substantial amount of new material was added.

The Raffles Museum of Biodiversity Research

By the time the collection moved to Kent Ridge, Roland Sharma had retired as Head of Department and was succeeded by Lam Toong Jin in 1981. He would remain Head of Zoology for an amazing 15 years and was an ardent supporter of the ZRC. Although he was not a taxonomist, Lam very quickly appreciated the value of the ZRC — especially its strengths in showcasing the biodiversity of the region — and resolved to support Yang and her team in all ways possible. Lam also had an excellent working relationship with Professor Koh Lip Lin, the Dean of the Faculty of Science. Koh, a chemistry professor from Nanyang University, had previously been Dean of Science at Nanyang University and was also very supportive of the ZRC. Koh got on particularly well with Lam whom he saw as a young man with plenty of drive and ambition.[92]

The first crisis Lam needed to resolve was that space limitation. Initially, the ZRC was only allocated two floors of the six-storey Science Library building but this was insufficient. For one thing, the collection had grown. When Yang took charge of the collection in 1972, there are some 126,000 specimens in the collection. By 1989, this had grown to 170,000 specimens. The other problem was that the space provided was bare and there were no funds to custom-make storage cabinets. Lam and Yang eventually found a Japanese company, Kongo Co Ltd, to design and install a compactor system suitable for the ZRC's use. As the system was 'incorporated' into the building structure, it was paid for out of the University's development fund, rather than from Lam's departmental budget.

Becoming the Raffles Museum of Biodiversity Research

In 1997, Lee Soo Ying, a professor of chemistry, succeeded Bernard Tan as Dean of the Science Faculty. Lee had known about the ZRC since the time he became Vice-Dean of the Faculty in 1993, and like Lam Toong Jin, was extremely proud of it and would always take guests to visit the collection. When he took over as Dean, Lee was anxious that this national treasure should be better utilised and publicised, and proposed that the Department of Zoology consider developing a series of outreach and education programmes to bring more people, especially school children, to the collection. It was, he said, 'like this huge box with so many wonderful things in it' and it was important that 'we open it up and share it with the public.'[93] Lee was fortunate in this enterprise in that he had the excellent support of Lam, who considered him a 'dynamic visionary'.[94]

One of the things Lee thought would help bring the ZRC a bigger audience was a public gallery. After speaking to Lam about this, Peter Ng — a dynamic young scholar who was now an Associate Professor in the Zoology Department — was tasked with preparing a paper to establish a Museum of Biodiversity Research. Ng sat down with Yang and drafted a comprehensive document entitled, 'Proposal for a Raffles Museum of Zoology: An International Research Institute for Southeast Asian Biodiversity and Establishment of a Public Display Gallery.'[95] The proposal, dated August 1997, stated that the 'primarily curatorial' mandate of the ZRC 'must be regarded as outdated' and that it was necessary to 'make museums academically stronger, more relevant for tertiary education as well as public-oriented to justify government funds.' Ng and Yang thus proposed the establishment of a 'public exhibition area' to meet the 'increasing interest shown by the public with regards to natural history and the ZRC over the last 10 years.' Once established, the ZRC would need to be renamed as something more in keeping with its new role. Ng and Yang proposed to call it the Raffles Museum of Zoology with a mission to enhance the Museum's research capability; the Collection's international research value; the use of the Museum in educating biology graduates as well as in secondary and pre-tertiary education; and to increase public awareness of Singapore's and Southeast Asia's natural heritage through the Museum. To do this, it was necessary to completely overhaul the museum's organisational structure, strengthen its research capability, establish a public display gallery, reorganise the facility, and appoint additional manpower to effect this plan.

Lee was very pleased with the proposal and lent ardent support to the scheme. Initially, some members of the department were not so keen on the proposed name, arguing that going back to the name 'Raffles' smacked too much of a colonial hangover. However, Ng and Lam were certain that the proposed name was timely and appropriate; after all, it was the historical name of the collection — everyone called it the Raffles Collection — and gave it instant recognition and a high profile.[96] Both Lee and Lam acted quickly. The proposal received strong support from the Vice-Chancellor, Lim Pin, and in 1998, the new facility was renamed the Raffles Museum of Biodiversity Research (RMBR).

It was now time to find someone to operationalize these proposals. As far as Lee and Lam were concerned, the only person for the job was the principal draftsman of the paper: Peter Ng. In his letter seeking approval to establish the Raffles Museum for Biodiversity Research and the appointment of Peter Ng as its first director, Lee Soo Ying wrote:

> The ZRC is part of the heritage of NUS, and dates back to the time of Sir Stamford Raffles. In the glory days of the 1930s and 1940s, the collection was an integral part of the Raffles Museum, one of the most famous research museums in Asia then. Today, we are ready to reassume that role and place ourselves at the forefront of biodiversity research again.[97]

As a pre-university student at Raffles Institution, Peter K.L. Ng (b 1960) became interested in crabs. Through Roland Sharma, he had connected with Leo Tan Wee Hin (b 1944) — who would later become his mentor — at the Department of Zoology. Ng was one of the department's most active and prolific scholars. He had, as a graduate student, helped overhaul the *Raffles*

Bulletin of Zoology, and was actively involved in many ZRC activities and in discovering new species and adding numerous specimens of crabs to the collection. Ng would visit the NTI campus in Jurong almost weekly to study the crab collection as well as to deposit new specimens he collected from Tuas, Johore Straits and the East Coast. Ng also met many visiting researchers and helped them identify crabs and other specimens and also helped in making logistical arrangements for visitors. After Ng joined the faculty at the Zoology Department, he continued to be actively involved in the activities of the ZRC and would often get his students to use the ZRC for collecting trips and in many other ways. As far as Lee and Lam was concerned, there was no better candidate for the Director's job. Yang, the ZRC's key custodian since 1972, agreed.

From RMBR to the Lee Kong Chian Natural History Museum

On 15 June 2001, Minister for Education Teo Chee Hean declared the RMBR's Public Gallery open. With that, many of the mounted specimens, which the ZRC recovered from the Science Centre in 1985, went on display for the first time in 30 years. Given its size, only a small portion of the ZRC could be displayed at any one time. The gallery featured rotating exhibits curated to introduce Southeast Asian biodiversity. In its first six months, the displays focused on ten topics: Singapore's biodiversity in Five Kingdoms; Tropical Habitats; Surprising Singapore; The Wonderful World of Crabs; Things People Eat; Conservation Issues; Biodiversity Research by staff and students; Education & Expeditions; Raffles the Naturalist; and New Discoveries in the Region. The tiny gallery — which was mainly visited and used by faculty and NUS students — attracted an average of 400 non-University related visitors a year.

One of the visitors to the museum in 2005 was Ambassador-at-Large, Tommy Koh, who was then Chairman of the National Heritage Board. Koh had known about the collection but was stunned to see so much of the century-old collection intact. He wrote to Vice-Chancellor Shih Choon Fong and persuaded him that with this important and historical Collection, the University should consider using it as the core for a natural history museum.[98] Koh's initiative was welcomed by the Science Faculty and in late 2005, with funding from the Faculty of Science and American entrepreneur Frank Levison (who was working with the University's Development Office) a study tour was organised. Five staff members of RMBR, including Peter Ng, left for America on a whirlwind tour to study the most successful natural history museums. They visited the California Academy of Sciences in San Francisco; the Biodiversity Institute and Natural History Museum at Kansas University; the Berkeley Natural History Museums; the Field Museum of Natural History in Chicago; the Smithsonian Institution; and the American Museum of Natural History in New York. The main object of this tour was to understand the financial models underpinning these museums. At the end of the trip, Peter Ng concluded that all successful natural history museums in America had three things in common: (a) good corporate governance; (b) a good endowment plan; and, most unexpectedly, (c) dinosaurs.[99] So, did Ng think NUS was ready for a natural history museum? Well, while NUS was certainly a well-governed establishment, the faculty certainly had no endowment for the museum and no dinosaurs. So, a new natural history museum? No way![100]

A Natural History Museum ... At Last

As a member of the Museum Roundtable, the Raffles Museum of Biodiversity Research took part in International Museum Day 2009. On 24 May, the museum opened its doors, only to be greeted by a huge crowd patiently waiting to be admitted. It was estimated that some 3,000 people showed up on that day. The response was completely overwhelming and the staff was completely exhausted but exhilarated by the response. This set off a giddy chain of events that would eventually culminate in the decision to establish Singapore's own Natural History Museum.

The first salvo was fired by Jaya Kumar Narayanan, a member of the public, who wrote to the *Straits Times* Forum Page highlighting the lack of space and the inaccessibility of the museum's location.[101] Victoria Vaughan of the *Straits Times* followed up with an article highlighting the museum's predicament,[102] in which she urged Peter Ng to speak bluntly.[103] Ng made an open call for a permanent natural history museum to be built, saying: 'We have an art museum, a civilisation museum, a heritage museum, but natural history is lodged in a corner of the university where no one can find it.'[104] Ng also revealed that over the last four years (2004–2008), he had been holding informal talks on the possibility of setting up a natural history museum with the Singapore Zoological Gardens, the Singapore Science Centre, and the National Parks Board but 'the zoo's commercial interest and the centre's education focus was thought to be in conflict with the RMBR's research agenda, and NParks already has its work cut out looking after plant specimens.'[105]

Ten days later, on the 14 June 2009, Tan Dawn Wei wrote a very influential full page article in the *Sunday Times* openly appealing: 'Let's Have a Natural History Museum for Singapore'.[106] Ng recalled:

> [We] were inundated with visitors — thousands of people crowding into a 200 sq m gallery sited behind a small building deep in the bowels of NUS. The visitors growled — tough to find, hard to get to, gallery too small, too little displayed — complaints galore. But there was one common denominator — they all loved the place and echoed Tommy Koh's hope: Bring back Singapore's natural history museum![107]

A few months before the fateful Museum Day, the university saw the return of Professor Leo Tan Wee Hin, who would eventually be a key player in the efforts to establish the natural history museum. Tan had joined the University of Singapore as a Senior Tutor in 1973 and taught at the Zoology Department from 1973 to 1986. In 1982, he was seconded to the Singapore Science Centre as its Director while concurrently teaching at the university. He left the University in 1986 to become the Centre's full-time Director and remained there till 1991. During his Directorship, he developed the Science Centre into one of the world's best. In 1991, Tan became Foundation Dean of the Faculty of Science at the newly-formed National Institute of Education (NIE). Three years later, he became Director of NIE and remained there till 2008 when he returned to the National University of Singapore as Professor and Director of Special Projects Dean's Office in the Faculty of Science.

The Mystery Donor

The newspaper articles and the public's positive response delighted the museum. However, if nothing further happened, that would have been the end of the matter — the little museum would simply continue chugging along quietly. However, a week after Tan Dawn Wei's article was published, 'a pleasant, non-descript gentleman turned up' and spoke to Tan Swee Hee (b 1971), a research officer at the museum, wanting to learn more about the museum. Tan brought this to the attention of his boss, Peter Ng.

This gentleman was a rather mysterious person who only initially left a mobile contact number, and was later known only as 'Anthony'. Ng had several more meetings with Anthony, who later said that he represented a group of anonymous donors and that they might be able to start the ball rolling in the fundraising efforts for the museum. Eventually, Anthony told Ng that the donor was prepared to give the university S$10 million to build the museum. Ng was dumbfounded but Anthony assured him that there would be no doubt as to the donor's genuineness and sincerity when all final arrangements were made. He was true to his word, naming a senior lawyer who was also a well-known alumnus and adjunct staff of the university.[108] Baffled, Ng conferred with Leo Tan — who had just returned to the Science Faculty. For Leo Tan, the donor was an important 'sign' that it was time to finally pitch for the building of a natural history museum. As a schoolboy, he had visited the old Raffles Museum many times and marvelled at its exhibits; as a young doctoral student, he spent many hours researching the ZRC that had been cast out of the museum. He made a solemn resolution to himself, 'If I could, one day, I'd like to restore the old Raffles Museum.'[109] Now, he had a chance to 'return the collection to the people of Singapore.'

With an offer of $10 million,[110] Leo Tan was confident that a new museum was within striking distance. He had seen how the Dentistry Faculty building — mainly of pre-fabricated construction — had been built for $17 million and surmised that that was all that was needed. Ng and Tan decided to approach the university's president Tan Chorh Chuan for advice on the matter.[111] 'That will do', he thought, as he went with Ng to meet the President:

> So we went to Chorh Chuan and we said, we can do it for $20 million. Assuming we have this $10 million, we need to only raise another $10 million, do we have your permission to commence fund-raising?[112]

The President's reply stunned Tan and Ng. After consulting with the university's building and estates people, he told them that S$10 million was not enough. They would need at least S$25 to S$30 million to build a respectable building. Leo Tan and Ng would therefore need to raise another S$25 million. The President added that he believed the museum to be a very worthwhile project and that he was prepared to set aside a piece of land at the soon-to-be-built University Town for it. However, he could not hold onto that piece of land for long. He gave the pair six months to raise the money![113] This timeline was imposed as the President could not hold onto the land any longer than that without giving due consideration to other demands on the space by other sectors of the university.[114] They accepted the terms and staggered out of the President's office around 7.00 pm that evening. Over dinner, they stared at each other and asked themselves, 'What have we just promised?' Tan recalled, 'We were just two scientists, who knew nothing about business or fund-raising, and here we were, having just committed to raising $35 million by the following June!'[115]

Raising the Funds

In his own inimitable and indomitable way, Tan set out to work to raise the remaining $25 million from various foundations, organisations and the public. It was a desperate race against time. To raise that much money so quickly, they needed as much publicity as possible. They approached Singapore Press Holdings who offered them a $5,000 donation but undertook to give the proposed museum the much-needed publicity.

By February 2010, they had only raised about $750,000. They needed a miracle. Over the next few months, they spoke to everyone they could think of. In April, Tan and Ng started their public campaign by each pledging $20,000 of their own towards the project.[116] They wrote 700 letters to friends and acquaintances asking them to donate to the cause. An external fund-raising consultant told Tan that this kind of 'spray and pray' strategy was foolish and often yielded very poor results. The final response, however, was overwhelming. They raised close to $1 million.[117] Even so, with S$10 million from the anonymous donor and just under S$1 million from the public, Tan and Ng were still S$24 million short and were in a bit of a quandary as the University's Development Office was already working with all the major foundations on gifts for the University Town project. Tan, whose network ranged far and wide, approached every notable he knew — ministers, businessmen, and even the President of Singapore. Another institution they approached was the Singapore Totalisator Board (Tote Board) as they were told the organization did not support scientific infrastructure or museums per se. It was a long shot, but their proposal was shortlisted but Ng had exactly 10 minutes to present and convince the Board to give the university the money for the museum. He prepared a very punchy and sharp set of slides and did his presentation within the allotted — his famous 10 minutes for $10 million speech — and was greeted with just a few polite questions. A few days later, the Board replied: Ng and Tan could add another S$10 million to the fund.[118]

The next S$15 million came through the Lee Foundation. Leo Tan had, at a private function, mentioned their fund-raising efforts to Dr Lee Seng Tee (b 1923), the second son of the legendary philanthropist, Lee Kong Chian. Dr Lee invited Tan and Ng to his Lee Pineapple office at the OCBC Building and asked for more details about the project. It did not take long for them to convince him of the importance of the museum and its immense heritage value for the country. Shortly after, Dr Lee pledged S$15 million to the project on behalf of the Lee Foundation.

Some weeks before the six month deadline, both Ng and Tan received another urgent call from the Lee Foundation, saying that Dr Lee would like to meet up with them again. This got them both very worried. Why would he call them at such short notice? Ng recalled:

> 'What now?' Was it going to be bad news? We walked in there and talked to him. It was very surreal. Dr Lee looked at us and said, 'We gave you S$15 million.' We said, 'Yes, thank you very much.' Then he said, 'Is S$15 million enough to make it world-class?' I tell you what, we'll round it up to S$25 million. Make it world-class, OK?' What was there to say? It was a very short meeting and as we walked out, I looked at Leo and asked, 'Leo, what just happened?' For the first time, I actually saw Leo looking a bit stunned.'[119]

When June came around, Tan and Ng had done 'the impossible'. They had raised $46 million for the museum in the immediate aftermath of the global financial meltdown of 2008. Most of the it had come from the Lee Foundation ($25 million); the Singapore Totalisator Board ($10 million) and the anonymous donor ($10 million), while the rest was made up of public donations.[120] With this huge sum in donations, the university was assured of a substantial matching grant from the Singapore Government.[121] University President Tan made the bold move of agreeing to allow the matching funds to be tied to the museum so that it could build up an independent endowment that would take care of part of the museum's operational costs. A day after Ng and Tan's meeting with Dr Lee Seng Tee, a note arrived, asking the university to consider naming the museum after Dr Lee Kong Chian. This was a request that was carefully considered and agreed upon for two reasons. First, the Lee Foundation made the most substantial donation, and second, Dr Lee Kong Chian was associated with the university as the first local Chancellor of the University of Singapore (predecessor of NUS). Ingredient Number 2 — an endowment plan — was now in place. In May 2014, Ng was appointed Director of the Lee Kong Chian Natural History Museum.[122]

One evening, just as the fundraising for the museum was coming to a close, NUS President Tan Chorh Chuan urgently sought out Ng and Tan. It was already past 6.00 pm, and they feared that something might have gone awry. Instead, they were met by an apologetic Tan who told them that the promised site was no longer available (it was given to NUS-Yale) but that he could now offer them the site of the soon-to-be vacated Estate Office. This was wonderful, since both Ng and Tan concluded that this was an even better site, given its greater accessibility and proximity to the University Cultural Centre, NUS Museum, and the Yong Siew Toh Conservatory of Music. Ironically, the Kent

Ridge Crescent location was the exact site the late Deputy Vice-Chancellor, Reginald Quahe had offered to house the collection back in 1977.

Work on the building began on 11 January 2013, and was completed in the first quarter of 2015. The 8,500 sq m, 7-storey 'green' building was designed by Mok Wei Wei of W Architects.[123] On 18 April 2015, the Lee Kong Chian Natural History Museum was declared open by President Tony Tan, the same dignitary who had declared open the ZRC's new home on Kent Ridge back in 1988.

References

[1] *Straits Times*, 20 Oct 1877, at 4.,

[2] 'The Year in 1884' *Straits Times* 2 Jan 1885, at 2.

[3] See 'Museum as aid to education', *Straits Times*, 6 February 1961, at 4.

[4] Ibid.

[5] 'Singapore's museum second to none in Malaysia', *Straits Times*, 30 June 1965, at 9.

[6] Ibid.

[7] See 'A Change for the National Museum', *Straits Times*, 14 December 1968, at 7.

[8] *Singapore Parliamentary Reports*, Vol. 28, 13 December 1968, col. 208.

[9] Ibid.

[10] Ibid.

[11] Eric R Alfred, 'Plan for the Reorganisation of the National Museum, Singapore as a Natural History Museum', Ministry of Culture, MC165-71 Pt 6, 29 Aug 1969.

[12] '$5m science centre to go up soon' *Straits Times*, 12 Mar 1970, at 5.

[13] 'S'pore plans get support from UNESCO' *Straits Times*, 25 Apr 1970, at 9.

[14] 'Plan for two divisions at $5.2 mil science centre' *Straits Times*, 8 Jul 1970, at 8.

[15] Ibid.

[16] See the Science Centre Act (Cap 286), Singapore Statutes. This was passed on 25 Sep 1970.

[17] 'Science Centre may attract 340,000 visitors each year', *Straits Times*, 23 July 1970, at 11.

[18] Ibid, at col. 142.

[19] 'Science Centre at Jurong', *Straits Times*, 3 December 1970, at 22.

[20] 'Culture Ministry will take over museum' *Straits Times*, 31 Mar 1972, at 12.

[21] 'Displays not meant to trace history of S'pore' *Straits Times*, 27 Dec 1972, at 9.

[22] Ibid.

23 'Schools science centre now at Alexandra' *Straits Times*, 13 Oct 1972, at 11.

24 'Culture Ministry will take over museum' *Straits Times*, 31 Mar 1972, at 12.

25 'Only the best people to run museum' *Straits Times*, 18 Nov 1972, at 14.

26 Interview with Eric R Alfred, 20 Jan 2014.

27 Interview with Bernard Tan, 3 Jun 2014.

28 Ibid.

29 Dixon to Pope, 9 Jul 1971, DF206/137, Museum of Natural History Archives, London.

30 'Type' or 'Holotype' specimens refer to the physical examples of an organism first used to describe the new species. There are also 'Syntypes' (one or more biological types listed in the description where no holotype has been designated) and 'Paratypes' (specimens of a type series other than the holotype). For example, the Zoological Reference Collection had, as at 2001, 48 'types' in its herpetological (amphibian) collection including 9 holotypes, 27 paratypes, and 12 syntypes. *See* Indraneil Das & Kelvin KP Lim, 'Catalogue of Herpetological Types in the Collection of the Raffles Museum of Biodiversity Research, National University of Singapore' (2001) 49(1) *The Raffles Bulletin of Zoology* 7–11.

31 This 'Lord Cranbrook' refers to David David Gathorne-Hardy (1900–1978), 4th Earl of Cranbrook who was Trustee of the British Museum between 1964 and 1973. His son, Dr Gathorne Gathorne-Hardy (b 1933) was the 'Lord Medway' referred to in this correspondence. Medway succeeded his father as the 5th Earl of Cranbrook in 1978. Medway served as Assistant at the Sarawak Museum between 1956 and 1958, and was later Senior Lecturer in Zoology at the University of Malaya in Kuala Lumpur from 1961 to 1971. *See* GWH Davison, Hoi Sen Yong & DR Wells, 'Cranbrook at Eighty: His Contributions So Far – Ornithologist, Mammalogist, Zooarchaeologist, Chartered Biolgist and Naturalist' (2013) *The Raffles Bulletin of Zoology* 1–7, Supplement No 29.

32 AP Coleman, 'Memorandum', 12 Aug 1971, DF206/137, Museum of Natural History Archives, London.

33 JP Harding (Keeper of Zoology) to Sir Frank Claringbull (Director, Museum of Natural History), 17 Aug 1971, DF206/137, Museum of Natural History Archives, London.

34 Berry to Lord Cranbrook, 22 Oct 1971, DF206/137, Museum of Natural History Archives, London.

35 'S'pore asks UNESCO aid' *Straits Times*, 4 Feb 1967, at 5.

36 Raoul Sèrene (1909–1980) was a famous French carcinologist who was UNESCO consultant. He was based in Singapore for many years. See Martyn ET Low, SH Tan and Peter KL Ng, 'The Raffles Bulletin, 1928–2009: Eight Decades of Brachyuran Crab Research (Crustacea: Decopoda) (2009) *The Raffles Bulletin of Zoology* 291–307, Supplement No 20.

37 Raoul Sèrene, 'What a Reference Collection Is', Ministry of Culture, MC 167/71 Pt 3, 18 Feb 1969.

38 Sèrene to Tham, Ministry of Culture, 2 Jan 1970, MC165-71 Pt 6, Ministry of Science & Technology, National Archives, Singapore.

39 Ibid.

40 Tham to Hoe, Ministry of Culture, 11 Apr 1970, MC165-71 Pt 6, Ministry of Science & Technology, National Archives, Singapore.

41 Hoe to Tham, 5 Mar 1971, MC165-71 Pt 6, Ministry of Science & Technology, National Archives, Singapore.

42 Tham to Quahe, 9 Mar 1971, MC165-71 Pt 6, Ministry of Science & Technology, National Archives, Singapore.

43 Hoe to Quahe, 30 Mar 1971, MC165-71 Pt 6, Ministry of Science & Technology, National Archives, Singapore.

44 Timothy P Barnard, 'The Raffles Museum and the Fate of Natural History in Singapore' in Timothy P Barnard (ed), *Nature Contained: Environmental Histories of Singapore* (Singapore: NUS Press, 2014) 184–211, at 202 [hereinafter 'Barnard'].

45 '$5m Property Takeover' *Straits Times* 1 Mar 1973, at 28.

46 When RMBC closed down in April 1978, the samples were shipped to the Marine Biological Center, Tokai University, Japan. Interview with Mrs CM Yang, 16 Jan 2014.

47 Interview with Mrs CM Yang, 16 Jan 2014.

48 Ibid.

49 The collection was made up of 15,000 mammals, 31,000 birds, 5,000 reptiles and frogs, 12,000 fishes, 10,000crabs, 18,000 molluscs, 10,000 insects, 25,000 other invertebrates. Interview with CM Yang, 16 Jan 2014.

50 Ibid.

51 Ibid.

52 Barnard, n 44 above, at 207.

53 FJ Ebling, 'Memorandum', July 1977, MC165-71 Pt 6, Ministry of Science & Technology, National Archives, Singapore.

54 Wells had obtained his PhD in 1966 at the School of Biological Sciences at the University of Malaya for his thesis, *Breeding Seasons: Gonad Cycles and Mount in Three Malayan Munias.*

55 Interview with David R Wells, 20 Jan 2014.

56 Ibid.

57 Wells to Short, 10 Sep 1977, on file with RMBR.

58 David W Snow (Sub-Dept of Ornithology, British Museum) to David Wells, 27 Sep 1977, DCF206/137, Museum of Natural History Archives, London.

59 Galbraith to Yang, 12 Dec 1977, DCF206/137, Museum of Natural History Archives, London.

60 Short to Yang, 21 Sep 1977, DCF206/137, Museum of Natural History Archives, London.

61 'Estimated Values of the Specimens of the Zoological Reference Collection' on file with RMBR.

62 Barnard, n 44 above, at 207.

63 Roland E Sharma, 'The Zoological Reference Collection: The Interim Years 1972–1980' in *Our Heritage: Zoological Reference Collection*, Official Opening Souvenir Brochure, 31 Oct 1988 (Singapore: National University of Singapore, Zoological Reference Collection 1988) 10–11, at 11 [hereinafter 'Sharma'].

64 Interview with Nancy Byramji, 31 May 2014.

65 Sharma, n 63 above, at 11.

66 Interview with CM Yang, 16 Jan 2014.

67 Sharma, n 63 above, at 11.

68 Yang to Short, 3 Mar 1979, on file with RMBR.

69 Interview with Peter KL Ng, 24 Oct 2014.

70 'Officials of nature society' *Straits Times*, 13 Jul 1977, at 7.

71 Interview with PN Avadhani, 19 Nov 2014.

72 Interview with CM Yang, 3 Nov 2014.

73 Nancy Byramji now goes by the pen name Aurora Hammonds.

74 Interview with CM Yang, 16 Jan 2014.

75 Nancy Byramji, 'Save our heritage: Priceless Raffles Collection may end in the dustbin unless $70,000 a year is found' *Straits Times*, 29 Apr 1979, at 1.

76 Ibid.

77 Interview with Nancy Byramji, 31 May 2014.

78 Nancy Byramji, 'Save our heritage: Priceless Raffles Collection may end in the dustbin unless $70,000 a year is found' *Straits Times*, 29 Apr 1979, at 1.

79 Ibid.

80 'How we can save our heritage' *Straits Times*, 7 May 1979, at 17; and Koh Yan Poh, 'Move to save century-old collection of specimens' *Straits Times*, 16 May 1979, at 8.

81 Interview with CM Yang, 16 Jan 2014.

82 'Talks to keep Raffles Collection intact' *Straits Times*, 13 Jul 1979, at 16.

83 'The heritage savers' *Sunday Times* 19 Aug 1979, at 12.

84 'No-home threat to priceless Raffles Collection' *Straits Times* 27 Nov 1980, at 13.

85 Ibid.

86 Interview with CM Yang, 5 Nov 2014.

[87] Sharma to Hooi, 15 Dec 1980, MC165-71 Pt 6, Ministry of Science & Technology, National Archives, Singapore.

[88] Bernard Tan to Kevin Tan, email dated 16 Sep 2016, on file with the author; see also Gloria Lim to Bernard Tan, email dated 2 Sep 2016, on file with the author.

[89] See *Our Heritage: Zoological Reference Collection*, Official Opening Souvenir Brochure, 31 Oct 1988 (Singapore: National University of Singapore, Zoological Reference Collection 1988).

[90] Mees to CM Yang, 19 Oct 1988, on file with RMBR.

[91] Cranbrook to CM Yang, 6 Aug 1988, on file with RMBR.

[92] Interview with Lam Toong Jin, 6 Nov 2014.

[93] Interview with Lee Soo Ying, 5 Nov 2014.

[94] Interview with Lam Toong Jin, 6 Nov 2014.

[95] On file with RMBR.

[96] Interview with Peter KL Ng, 24 Oct 2014.

[97] Lee Soo Ying to Lim Pin, 18 Sep 1998, on file with RMBR.

[98] Peter KL Ng, 'Destiny achieved: A journey of discovery' *Straits Times*, 25 Aug 2012, at D6.

[99] Interview with Peter Ng, 24 Oct 2014.

[100] Peter KL Ng, 'Destiny achieved: A journey of discovery' *Straits Times*, 25 Aug 2012, at D6.

[101] Jaya Kumar Narayanan, 'Museum needs more space, better access' *Straits Times*, 2 Jun 2009, at 18.

[102] Victoria Vaughan, 'Natural history needs more room' *Straits Times*, 4 Jun 2009, at B8.

[103] Interview with Peter Ng, 24 Oct 2014.

[104] Victoria Vaughan, 'Natural history needs more room' *Straits Times*, 4 Jun 2009, at B8.

[105] Ibid.

[106] Tan Dawn Wei, 'Let's have a natural history museum for Singapore' *Sunday Times*, 14 Jun 2009, at 26.

[107] Peter KL Ng, 'Destiny achieved: A journey of discovery' *Straits Times*, 25 Aug 2012, at D6.

[108] Interview with Peter Ng, 24 Oct 2014.

[109] Susan Long, 'Many of Science and Dreams' *Straits Times* 21 Mar 2014, at A21.

[110] Tan Dawn Wei, '$10m gift for natural history museum. Offer from unnamed donor boosts NUS bid to set up gallery for vast collection', *Sunday Times* 24 January 2010.

111 Interview with Leo Tan, 22 May 2014.

112 Ibid.

113 Interview with Peter Ng, 24 Oct 2014.

114 Ibid.

115 Interview with Leo Tan, 22 May 2014.

116 'Tan Dawn Wei, 'Museum in rush to raise $35m by June' *Straits Times*, 25 Apr 2010.

117 Interview with Leo Tan, 22 May 2014.

118 Interview with Peter Ng, 24 Oct 2014.

119 Interview with Peter Ng, 26 Oct 2014.

120 Victoria Vaughan, '$46m raised for natural history museum' *Straits Times*, 23 Jul 2010.

121 There was no matching grant for the Singapore Totalisator Board's S$10 million donation since it was considered a government financial institution.

122 Tan Dawn Wei, 'Crab expert named chief of new natural history museum' *Straits Times*, 7 May 2014.

123 Tan Dawn Wei, 'Work begins on new NUS msueum' *Straits Times*, 12 Jan 2013.

Chapter 29

The Scientific Society Movement in Singapore

Leo Tan Wee Hin, Andrew T S Wee and R Subramaniam

Abstract

Learned societies in the sciences have a long history of activism in the western world. However, it took quite some time for them to take root in Asia. This chapter traces the origins of the scientific society movement in Singapore and the genesis of the major societies in the sciences there. Two key aspects in the movement stand out — one is the emphasis on the 'invisible college' model and the other is the dependence on a voluntary cadre to drive the movement. It is argued that despite this model of governance and operation, with its attendant limitations and shortcomings, it is a framework which has been effective in the Singapore context, as can be seen from the track record of the various scientific societies in Singapore. An important development in the scientific society movement in the country, which does not seem to have occured in other countries, is the coming together of the various societies under the aegis of an umbrella organisation. Some reasons for this interesting development are advanced.

Introduction

Activism by interest groups is often a way to achieve certain objectives. Such interest groups have a long history of mushrooming for worthy causes in many countries. These groups often sprout up to cater to perceived deficiencies in societies that are not addressed effectively or not at all by the ruling dispensation or social structures. It is not something that can be done by an individual though often one person can be the driver for this activism.

Scientific activism is the focus of our interest in this chapter. Three key aspects of scientific activism stand out: there must be recognition by the scientific community that existing structures of governance do not adequately address some of their concerns, a rallying figure needs to emerge to champion the cause, and there needs to be a critical mass of like-minded people who are prepared to join this movement. The critical mass of people to support the cause need not be a large number — a handful of such people would be adequate, if the early history of scientific societies in the western world is an indication. The emergence of scientific societies thus has its basis in scientific activism. Here, we define scientific societies as learned associations formed by individuals to drive the cause of science in a country.

The principal objectives of this chapter are as follows:

- To trace the history of the scientific society movement in Singapore.
- To suggest reasons why the scientific society movement was relatively slow in taking root in Singapore as compared to that in the western world.
- To explore why an umbrella organisation was needed for the scientific societies in Singapore.
- To reiterate the 'invisible college' and voluntary cadre framework that operates for scientific societies in Singapore and its relevance to the developing world.
- To comment on the current state of affairs of scientific societies in the country.

It has to be noted that while there is a wealth of published literature on the scientific society movement in the western world, there is a paucity of scholarly publications in journals that focus on scientific societies in Asia. Some work on various aspects of scientific societies in the Singapore context seems to be an exception and have previously been reported. For example, Tan & Subramaniam (1999) briefly mentioned about the Singapore National Academy of Science in their article on the role of scientific societies in building better nations. There was some mention of the Singapore National Institute of Chemistry in their article on the role of chemical societies in boosting development (Tan & Subramaniam, 2000); and more detailed elaboration on the role of the Singapore National Academy of Science and its constituent societies was provided in another article (Tan & Subramaniam, 2009). The overall thrust of the present chapter, however, is significantly different from those which have been published though there is bound to be some elements of commonality owing to the nature of the topic being explored here.

Origins of Scientific Societies

The roots of group-based scientific activism can be traced back to the scientific society movement that started in the western world in the 17th century.

Table 1.   Major scientific societies set up in Europe and USA in early years.

Year of establishment	Name of scientific society	Country of origin
1660	Royal Society	UK
1666–1793	Academie Royale del Sciences	France
1700	Prussian Royal Academy of Sciences and Letters	Germany
1725	Polish Academy of Sciences	Poland
1728	Societatis Regrae Scientarium	Sweden
1739	Royal Swedish Academy of Sciences	Sweden
1742	Royal Danish Academy of Sciences	Denmark
1759	Bavarian Academy of Sciences	Germany
1783	Royal Society of Edinburg	Scotland
1785	Royal Irish Academy	Ireland
1817	New York Academy of Sciences	USA
1831	British Association for the Advancement of Science	UK
1848	American Association for the Advancement of Science	USA
1857	Norwegian Academy of Science & Letters	Norway
1863	US National Academy of Sciences	USA

We capture here salient aspects of the history of this movement so as to provide some context for the work reported in this chapter.

The formation of scientific societies in Europe coincided with the renaissance movement in the sciences that was spreading across Europe in the 17th century. The scientific intelligentsia spearheaded the formation of these scientific societies, which are independent by nature and comprise scholars elected for their contributions to their disciplines. While there are some doubts about the true origins of scientific societies, it was the Royal Society (UK) which made a success of it for others to emulate. Within a few years of its establishment in 1660, other academies started to form in other parts of Europe and the USA (Table 1).

By the 18th century, over 400 such learned societies in sciences have already been formed in Europe. The various academies and societies played a useful role in promoting science in their countries and focused also on scholarly pursuits such as publication of journals. However, it took quite some time for such august associations to be formed in Asia.

Establishment of Scientific Societies in Singapore

Despite the long history of the scientific society movement in the world, it took quite some time for scientific societies to be established in Singapore. In fact, the scientific society movement in Singapore is of relatively recent origins — that is, from the 20th century, and more accurately, from the 1970s.

Some reasons for their slow and late establishment could be as follows:

- Singapore, as a country, was founded only in 1819. Owing to its strategic location and mainly immigrant population, the emphasis for over a century was more on trade and commerce. There was relatively little emphasis on science and technology.
- The first university (University of Singapore) was established only in 1905, and for a number of decades, its focus was more on training and education, with relatively little emphasis on research and development. As a result, the critical mass of research-active scientists needed to support the evolution of scientific societies was not there in the earlier years. The situation appeared to be the same even after the establishment of the second university (Nanyang University) in 1956.
- There could be some influence from the brick-and-mortar model for the setting up of scientific societies, and this could have possibly delayed their establishment in Singapore.
- There was some degree of political instability in the early years of the 20th century — for example, continuation of colonial rule in the country, Second World War, Japanese occupation of Singapore, merger with Malaysia in 1963, and subsequent expulsion from Malaysia in 1965. These could have affected their establishment.
- There were very few scientists of international repute in the country who could drive the movement or galvanise others to join it in earlier years.
- Unlike social activism, which has a long history in Singapore, scientific activism had not taken deep root in the country for a long time.
- The success of the country in trade and commerce, as a result of its strategic location, may have sidelined science as an additional vehicle to tap on for socio-economic development.
- A core Singaporean identity took quite some time to evolve as many of the immigrant population continued to have linkages with their ancestral countries owing to a lack of sense of belonging to the country.

- There were no patrons of the sciences amongst the wealthy class who could support their establishment.
- The colonial masters who ruled the country did not see the relevance of catalysing the setting up of such societies here despite the importance and usefulness of the latter in their own country.
- There could be some recognition by the scientific community that existing structures of governance are adequate to cater for the needs of their community and that additional structures may not be necessary in the early years.
- A science culture has not taken deep root in Singapore in the early years. The term 'science culture' refers to the attitudes of the public towards science and their general understanding of it (Solomon, 1996). These attitudes take a long time to form, and systems and frameworks need to be in place to nurture these.

Some recognition of the relevance of scientific societies in the country began to emerge in the 1950s. It is of interest to note that the first scientific society to be set up in the country was not in the natural sciences but in mathematics, in 1952. It was only after 13 years that the second scientific society, the Science Teachers' Association of Singapore, was set up in 1965, the year of Singapore's independence. This was followed by the Singapore National Academy of Science in 1967. Those in the natural sciences emerged only in the 1970s (Chemistry in 1970, Physics in 1972 and Biology in 1975).

It is of interest to explore why the first three scientific societies set up in Singapore were not in the natural sciences. Mathematics as a discipline permits scholarly output with little or no funding, unlike generally those in the natural sciences, and the subject is of interest at the school level. Observations of established discipline-based scholarly societies in Europe suggests that these were formed after a tradition of scholarship has been established in a discipline, and this might well have been the case for mathematics in Singapore, even though we have not been able to obtain data to back this claim. Also, the critical mass of teachers in mathematics is already there for membership since mathematics is a compulsory subject at primary and secondary levels. In the case of science teachers, it was easy for them to set up their society as the critical mass was already there, especially as science (and its various disciplines) is a compulsory subject in primary and secondary schools. For the erstwhile Singapore National Academy of Science, its principal, mission objective at inception — promoting science and technology, was broad in scope and bereft of any disciplinary focus, so it made sense to promote science in general. It was thus easier for the limited number of scientists from the different scientific disciplines to come together under this academy and espouse the cause of science in general.

The success of the early scientific societies seems to have set the stage for the evolution of other learned societies in the sciences in later years. For example, the Singapore Institute of Statistics was formed in 1976, the Singapore Society for Microbiology was set up in 1983, the Singapore Biochemical Society was established in 1986, and the Materials Research Society-Singapore was formed in 1999. These learned societies can be considered to be more specialised societies in the sciences and their later evolution can be considered to be a logical sequel to serve more niche specialties and interest groups beyond the natural sciences.

Unique Model of Operation of Scientific Societies in Singapore

The established model of operation of scientific societies and scientific academies relies significantly on the setting up of premises and reliance on full-time staff to drive their operations, programmes and activities. While this is a viable model for emulation, it tends to inhibit their timely emergence since funding is needed to set up office and hire manpower — such funding is not easy to come by for a young nation which is trying to strike an identity of its own. The influence of this classical model of operation could have slowed down the establishment of scientific societies in Singapore.

A new model of operation was needed to jump-start their evolution. Though we do not have historical data to back up our claims, it seemed to the early pioneers that a possible way to replicate the scientific society movement in Singapore was to leverage on the 'invisible college' (Clarke, 1995) cum voluntary cadre model of operation. In this model of operation, there is no need to have premises and the office holders are all volunteers. A chief advantage of such a model is that it attracts only people who have passion and commitment for the cause to serve in the societies. That such an alternative model is viable can be seen from the fact that after the setting up of the first scientific society in Singapore in 1952, eight more societies were set up within a span of 35 years. This model of operation still exists in Singapore and is testament to the viability of this framework. In fact, Tan and Subramaniam (2002) have argued that a key reason for the slow establishment of scientific academies (and scientific societies) in the developing world is the scientific community's obsession with the brick and mortar model of operation and the belief that full-time staff are needed to support their operations. They have further suggested that the model of operation of scientific academies and scientific societies in Singapore is worthy of emulation by other developing countries, where start-up rates continue to be slow.

Key Scholarly Societies in the Sciences in Singapore

This section elaborates on the evolution of the key scientific societies in Singapore. Information about the societies was obtained from their websites, and paraphrased where necessary.

Singapore National Institute of Chemistry (SNIC)

SNIC was set up in 1970 with the express aim of representing the interests of the chemical profession in the country. In earlier years, their interests were taken care of by the local chapters of the Royal Institute of Chemistry (UK) — Malaya Section followed by the Singapore Section. With the increasing number of chemistry graduates produced by the two universities and with the rising demand for chemists in industry, the need for a local body that can better cater to the interests of the chemical profession as well as ensure higher standards of professional practice and ethical norms was felt. SNIC was thus born to address these challenges.

The scope of activities of SNIC has broadened over the years, and the major ones are now as follows:

- Organising of seminars, symposiums, talks, congresses, exhibitions and professional meetings.
- Conferring of gold medals and book prizes to university students who excel in Chemistry.
- Promotion of chemical education through talks on chemistry and chemistry education as well as via contests for secondary students.
- Fostering of interdisciplinary efforts in the promotion of science in general.

In earlier years, the Bulletin of the Singapore National Institute of Chemistry played an important role in providing an additional platform for the publication of research findings by the chemical community in Singapore. It was disbanded in the 1990s so as to amalgamate journals of the various constituent societies into a single flagship journal under the aegis of the Singapore National Academy of Science and that is published by an international publisher.

Singapore Institute of Biology (SiBiol)

The origins of SIBiol can be traced to the year 1974 when a group of biologists wished to promote the wider cause of the subject and its applications. The institute was officially registered in 1975.

Important programs of the institute include the organising of the Singapore Biology Olympiads, training of students for the International Biology Olympiads, election of Fellows and Honorary Fellows, and the SIBiol Prize for Early Career Biologists. In 2013, it established the online *Asian Youth Journal of Biology*, which has a focus on publishing papers written by students.

In earlier years, it published the *Singapore Journal of Biology*, which provided an additional platform for academics to publish papers. This journal has since closed down owing to the amalgamation of the journals of the various scientific societies into a single flagship journal of their parent body.

Institute of Physics Singapore (IPS)

IPS was established in 1972 to promote the wider cause of physics. Important programs of the institute include the organising of the Singapore Physics Olympiads, training of students for the International Physics Olympiads, election of Fellows and Honorary Fellows, conferring of IPS awards to recognise local physicists and those promoting physics, and IPS medals for distinguished contributions by physicists.

In earlier years, it published the *Singapore Journal of Physics*, which served as an additional vehicle for the publishing of papers by academics. It also published the *Physics Update*, a journal focusing on physics education. Both journals have stopped publication owing to its merger with the flagship journal of SNAS.

Science Teachers Association of Singapore (STAS)

STAS was set up in 1965 to serve the interests of science teachers in Singapore, with the express objectives of promoting the advancement of science education, and fostering educational, professional, social and cultural interests among its members. Important programs of the association include the co-organising of the long running Singapore Youth Science Festival with the Science Centre Singapore, the administration of the Primary Science Activities Club and Questa Club schemes with SAAS and SNAS, and the organising of the SAAS Outstanding Science Teacher Awards with SAAS. Since 2012, it has started organising the biennial Singapore International Science Teachers Conference.

Singapore Mathematical Society (SMS)

The Malayan Mathematical Society was set up in 1952 to serve the interests of the mathematical community in Malaya. Its name was changed to the Singapore Mathematical Society in 1967 so as to confine its focus to Singapore after its independence from Malaysia. Important programmes of the society include the organising of the Singapore Mathematical Olympiads, training of students for the International Mathematical Olympiads, and awarding of SMS Book Prizes to university students who graduate with first class honours.

It also publishes the long running Mathematical Medley, a magazine which publishes both research articles as well as educational articles.

Singapore Institute of Statistics (SIS)

SIS was established in 1976, with the principal objectives of promoting the advancement of statistics and its applications, encouraging the improvement of education in statistics and raising the status of the profession.

Its key programmes include the organising of workshops and seminars related to the wider cause of the discipline.

Singapore Society for Biochemistry and Molecular Biology (SSBMB)

Originally known as the Singapore Biochemical Society when it was founded in 1983, it had a name change to the Singapore Society for Biochemistry and Molecular Biology when its objectives were expanded in 1993. Its principal mission objective now is the promotion of the disciplines of biochemistry and molecular biology.

Important programs of SSBMB include the organising of the annual Young Scientists' Symposium for polytechnic students, awarding of prizes to the best honours student in biomedical science or molecular cell biology at the National University of Singapore and conferring of module prizes on polytechnic students who excel in the subjects. Regular international meetings and conferences are also organised by the society.

Singapore Society for Microbiology and Biotechnology (SSMB)

Formed in 1983 as the Singapore Society for Microbiology, it renamed itself as the Singapore Society for Microbiology and Biotechnology in 1997 as it wished to embrace biotechnology as well. Besides promoting its disciplinary foci, the society also has an interest in the regulatory mechanisms relating to these disciplines.

Key programmes organised by SSMB include workshops and seminars.

Singapore Association for the Advancement of Science (SAAS)

SAAS was established in 1976 with the principal objective of promoting science and technology in Singapore. Its key programmes are as follows:

- Young Scientist Badge scheme — this was established in 1982 and offers primary school students opportunities to engage in project-based investigations in 17 disciplines, as outlined in the respective activity cards. The 17 disciplines are as follows: Physics, Chemistry, Mathematics, Astronomy, Geology, Botany, Zoology, Ecology, Entomology, Meteorology, Environment, Ornithology, Information Technology, Food Science, Genetics, Energy and Water. Completion of the respective activity card earns the student a prestigious badge — for example, 'I am a Young Physicist' badge.
- Questa Club scheme — this is similar to the Young Scientist badge scheme but is for secondary school students and the investigations to be done are more open-ended rather than based on activity cards. Completion of the project earns the student a Questa badge. It was established in 1988.
- SAAS Outstanding Science Teacher Awards — this was launched in 1995 and serves to recognise outstanding science and mathematics teachers at the primary and secondary/junior college levels.
- SAAS Prizes — this is for the most outstanding trainee teacher in science and mathematics respectively graduating from the only teacher training institution in the country.

Materials Research Society-Singapore (MRS-S)

MRS-S was established in 1999 as a professional society to serve the interests of the evolving materials science community in Singapore.

Its key programmes include the organising of the biennial *International Conference on Materials for Advanced Technologies* and awarding of prizes to students who excel in their honours degree course at the National University of Singapore.

Chapter of Clinician Scientists (CCS)

CCS was established in 2012 as a chapter of the Academy of Medicine in Singapore. Its formation was predicated by the increasing role played by doctors who straddle both clinical work and research. A group of clinician-scientists provided the impetus for the formation of the chapter under the Academy of Medicine Singapore.

The chapter also serves to promote a distinct identity for clinician scientists, provide advice to the relevant agencies on matters relating to the training of clinician scientists, and be the academic voice of the Academy of Medicine. In 2015, it applied for affiliation to the Singapore National Academy of Science, of which it is now a constituent member. CCS organises workshops and seminars on a regular basis.

Singapore National Academy of Science (SNAS)

SNAS, in its later manifestation, was established as an umbrella organisation in 1976. The major scientific societies in the country are affiliated to it. Its principal objectives are the promotion of the advancement of science and technology and the representing of the scientific opinions of its Founder/Affiliate Members.

As an umbrella organisation, it ensures that its programmes and activities do not duplicate those of its constituent societies. Its focus is thus somewhat narrower. At the moment, its key programmes are as follows:

- Young Scientist Awards — the awards were established in 1997 and aims to recognise scientists of age 35 and below who have excelled in their research. Each award carries a cash prize of S$10,000 and a plaque. Up to four awards can be presented annually. The prizes are sponsored by the Agency for Science, Technology and Research but SNAS is responsible for its administration.
- SNAS Awards — these prizes are for university students who excel in the sciences in their final year examinations. Each of the awards carries a S$100 book voucher and a certificate.
- Proceedings of the Singapore National Academy of Science (COSMOS journal) — this is the flagship journal of SNAS and was re-launched in 2000 following the consolidation of the various journals published by its constituent societies. The focus now is on publishing special issues of interest in the sciences — this is to differentiate itself from other journals in the market. A sponsorship of S$100,000 was obtained from a leading foundation to launch this initiative.
- SNAS Fellowships — this was launched in 2011 to recognise outstanding scientists in the country. It carries a certificate of commendation. The SNAS Fellows are expected to advise and/or contribute to the government and other national organisations on various aspects of science, including research, teaching, science policy; as well as the promotion and public communication of science.

Table 2. Founding presidents and secretaries of constituent societies of SNAS.

Constituent society	President	Secretary
Institute of Physics Singapore	Hsu Loke Soo	Bernard Tan Tiong Gie
Singapore Biochemical Society	Ellen Wong Hee Aik	Pangajavalli Kanagasuntheram
Science Teachers Association of Singapore	Sng Yew Chong	Tan Choong Yan
Malayan Mathematical Society	Alexander Oppenheim	Richard Guy
Singapore Institute of Statistics	Saw Swee Hock	Yeo Gee Kin
Singapore Institute of Biology	Chua Sian Eng	Nga Been Hen
Singapore Society for Microbiology	Lim Kok Ann	Chan Yow Cheong
Singapore National Institute of Chemistry	Walter Rintoul	Henry Ong Wah Kim
Singapore Association for the Advancement of Science[#]	Ang Kok Peng	R. S. Bhathal
Materials Research Society — Singapore	Shih Choon Fong	Liew Yiew Wang
Chapter of Clinician Scientists	Tan Eng King	Lee Meng Har

[#] For SNAS, when it was first established in 1967, president was Thomas Harold Elliott and secretary was Chew Wee Lek. In 1976, it became SAAS.

In an effort to recognise the contributions of the early pioneers of the movement, we have compiled Table 2 to show the names of the founding president and secretary of each of the constituent societies. It has, however, to be recognized that other office bearers of the constituent societies at inception must also be regarded as pioneers of the movement. Of interest to note is that of the first four scientific societies to be established in Singapore, three of the founding presidents were Westerners.

Why an Umbrella Organisation for the Scientific Societies?

The conventional model for the operation of an academy of science hinges significantly on the setting up of premises, hiring of manpower for running its operations, allocation of funding by the government or other agencies, and the election of Fellows. However, the model used in Singapore is very much different. As explained earlier, the Singapore National Academy of Science is an umbrella organisation which represents the various scientific societies in the country.

As it is, the various scientific societies in Singapore are, in a way, an umbrella organisation that represent people working in different sectors to promote interest in their particular discipline. For example, the Singapore National Institute of Chemistry has members drawn from academia, industry, schools and the public sector. The idea of having an umbrella organisation to represent the various scientific societies in the country was a novel idea, and it does not seem to have been a model adopted in other countries.

The umbrella organisation formed to unite the existing scientific societies at that time was the Singapore National Academy of Science. Originally, it was one of the scientific societies in the country, having been established in 1967. When the need for an umbrella organisation was felt to further the cause of the scientific society movement in the country, it was felt that its name would be the most appropriate. The objectives of the erstwhile Singapore National Academy of Science was taken over by the newly formed Singapore Association for the Advancement of Science in 1976, with the former then transforming its objectives to represent those of an umbrella organisation. It appears that the terms 'national' and 'academy' in the name were the key considerations that went into selecting it as the umbrella organisation.

As there is very little archived information in this area, we feel that the following are some of the reasons that could possibly explain the emergence of an umbrella organisation in the Singapore context:

- An umbrella organisation with members (president and secretary) drawn from each of the scientific societies could better help to chart the progress of the overall scientific society movement in the country.
- It can present a collective voice to the establishment, where necessary, on issues of interest.
- Membership of a national academy, which is the highest body representing all scientists in the country, provides further clout to the various scientific societies.
- The academy has all along been headed by a distinguished and influential scientist, and this could be a reason for the various scientific societies to want to come under its aegis.

When the Singapore National Academy of Science became an umbrella organisation in 1976, there were only six societies affiliated to it — SAAS, SNIC, IPS, SMS, STAS and SIBiol. Over the years, a few more societies applied for affiliation to it, and were accepted as constituent societies. The Chapter of Clinician Scientists, which comes under the purview of the Academy of Medicine Singapore, applied for and was accepted as a constituent member of SNAS in 2015.

Table 3 shows the key office bearers of SNAS over the years.

Table 3. Key office bearers of the Singapore National Academy of Science till 2016.

President	1st Vice President	2nd Vice President	Secretary
Lee Chiaw Meng (1976–1978)	Chua Sian Eng (1976–1978)	Sim Keng Yeow (1976–1978)	Louis Chen Hsiao Yun (1976–1978)
Ang Kok Peng (1978–1993)	A.N Rao (1978–1998)	Loo Pui Wah (1978–1980)	R.S. Bhathal (1978–1981)
Leo Tan Wee Hin (1993–2013)	Ong Chong Kim (1998–2007)	Hsu Loke Soo (1980–1981)	Leo Tan Wee Hin (1981–1992)
Andrew Wee Thye Shen (2013–2016)	Bernard Tan Tiong Gie (2007–2013)	Huang Hsing Hua (1981)	Chia Woon Kim (1992–1994)
	Lim Tit Meng (2013–2016)	Lim Yung Kuo (1981–1992)	R. Subramaniam (1994–1996 (covering); 1996–2016)
		Tang Seung Mun (1992–1997)	
		Ong Chong Kim (1997–1998)	
		Lee Seng Luan (1998–2007)	
		Andrew Wee Thye Shen (2007–2013)	
		Ling San (2013–2016)	

Discussion

Consistent with the origins of the scientific society movement in the western world, it is noted that their evolution in Singapore was also spearheaded by a core group of individuals in the respective disciplines. Identification of a specific individual who rallied others for the cause in Singapore was difficult — it could be possibly because of the Asian ethos that group interests be subordinate to individual claims to glory that we were not able to find any archived information in this regard. It is likely that an individual must have catalysed the evolution of the respective learned societies in Singapore. In any case, a scientific society cannot be a one-man show as it is a group endeavor.

When examining the evolution of the scientific society movement in Singapore, it can be seen that there have been five distinct phases — emphasis on mathematics in the 1950s (SMS), general sciences in the 1960s (STAS and SAAS), natural sciences and statistics in the 1970s (SNIC, IPS, SIBiol and SIS), the more sophisticated sciences in the 1980s (SSBMB and SSMB), and the more technological sciences in the late 1990s and 2000s (MRS-S and CCS). We see this also as a consequence of the progressive maturation of the local science and technological scene and one that is consistent with the capacity building efforts in science that are taking place in the country.

A key aspect of the scientific society movement that has been observed by Tan and Subramaniam (2010) is the inclusion of science teachers in its movement. Besides their own association (STAS), many science teachers are also active in a number of other constituent societies. This augurs well for the movement as it enables reaching out to an important target segment — school students, as part of the outreach efforts of the respective constituent societies.

Another notable aspect of the scientific society movement in the country is the affiliation of these societies to an umbrella organisation. This development does not seem to have occurred in other countries and appears to be unique to Singapore. More importantly, the framework of operation of the Singapore National Academy of Science is very much different from other established academies such as the US National Academy of Sciences and the Academy of Sciences Malaysia. The latter academies have full-time staff manning their offices and receive regular funding from their governments for their wide range of programmes.

A key aspect in the establishment of an academy of science in many, if not all, countries where there is an academy of science, is the emphasis on the election of fellows. However, in the case of SNAS, it focused on building up a track record for nearly 35 years before it embarked on the election of fellows in 2011. Tan and Subramaniam (2009) note this as follows: "*Whilst the classical framework for the setting up of scientific academies has, as one of its core features, the election of scientists by peers, it was felt that an overriding priority for a newly industrializing state that has yet to attain the critical mass of scientists to be elected as Fellows, was the election of office-bearers to the academy who can help to drive the wider cause of the scientific enterprise in an honorary capacity. In effect, the framework offered the scientific community a space and voice to articulate their independent views, besides providing them an opportunity to contribute towards nation building efforts*". We feel that this framework has implications for developing countries.

The lack of government funding for the scientific society movement in Singapore can also be attributed to the largely top-down approach to research and development (R&D) planning in Singapore. In 1991, the National Science and Technology Board (NSTB) replaced the Science Council following the enactment of the National Science and Technology Board Act in November 1990. NSTB, later renamed as the Agency for Science, Technology and Research (A*STAR), was responsible for formulating and implementing long-term R&D strategies to transform Singapore from a newly industrialising state to a world-class, innovation-driven economy. Since 2006, the National Research Foundation

(NRF), a department within the Prime Minister's Office, sets the national direction for R&D by developing policies, plans and strategies for research, innovation and enterprise. While NRF has a scientific advisory board comprising internationally renowned scientists and industrialists, it does not directly consult the scientific societies in Singapore.

As a result of the lack of government funding, another key aspect of the scientific society movement in Singapore is the reliance on the 'invisible college' cum voluntary cadre framework of operation. This is a model that has worked in Singapore and has even been suggested as a framework for emulation by other countries in the developing world. Nevertheless, with the growth of a new generation of Singaporean scientists who have attained international recognition, SNAS could be on the cusp of change as it seeks government funding to enable it to play a bigger role in science policy formulation in Singapore.

With the scientific society movement in Singapore entering its 64th year, it is timely to reflect on its performance. We are of the view that the movement has been broadly effective and successful for the following reasons:

- Examination of the annual reports of the various scientific societies indicates that they have been active in organising a diversity of programmes to reach out to their target audiences. The impact of such actions on the target audiences cannot be underestimated.
- The movement continues to attract members of the scientific community to come forward to serve as office bearers. A good number of the office bearers are now from the younger generation. This talent pipeline augurs well for succession planning as well as continued vibrancy of the scientific society movement.
- The membership base of the scientific societies shows good diversity — for example, members come from universities, public sector organisations, research institutes and the private sector. It is fortuitous that even those beyond academia see relevance in being associated with a movement that is synonymous with the wider aspects of nation-building.
- Though there is no official support from the government in terms of provision of funds for running the operations of the scientific societies and academy, they seem to be held in high regard by the establishment — for example, ministers and other dignitaries continue to grace their functions and, on an *ad hoc* basis, government agencies have provided financial support for selected programs and events. In the early years of SNAS, the then President of the country, Dr Benjamin Sheares, and the then Minister for Health, Dr Toh Chin Chye, were its Patrons. (Dr Toh was previously Minister for Science & Technology and Vice Chancellor of the then University of Singapore).

- Philanthropic foundations often support some of the programs and activities of the scientific societies and academy with funds. A key reason for such support seems to be the recognition that initiatives driven by volunteers with no premises of their own and who are from the non-profit sector deserve some financial support.
- A number of scientific societies in the country have tie-ups with established scientific societies in the western world. For example, SNIC has forged links with the Royal Society of Chemistry that serve to better connect the chemical communities in the two countries. This would not have been possible if the scientific societies in Singapore had not chalked up an impressive track record.
- A number of well-established international conferences have come to Singapore owing to the efforts of the respective constituent societies. Some of these include the *4th Federation of Asian and Oceanian Biochemists Congress* in 1986, the *3rd International Union of Biochemistry and Molecular Biology Conference* in 1995, the *22nd Federation of Asian and Oceanian Biochemists and Molecular Biologists Conference* in 2011, the *Asian Chemical Congress* in 2013 and the upcoming *International Union of Microbiology Societies Congresses* in July 2017. Though it might be argued that this could be more because of Singapore's international branding as a MICE (Meetings, Incentives, Conventions and Exhibitions) destination, we should not overlook the fact that the ground work for bringing the conferences here is laid by the respective constituent societies, which went into the negotiations with an impressive track record.

The wide range of programmes and activities of SNAS and its constituent members over the years have also contributed in some measure to capacity building in the country. Though it is difficult to quantify this, we are of the opinion that it is modest but significant.

In particular, Singapore's impressive performance in the international Olympiads in physics, chemistry, biology and mathematics over the years can be traced significantly to the efforts of the respective constituent societies. At least for the past five Olympiads, the Singapore teams have never returned home without at least a few medals.

The scientific society movement in Singapore could, perhaps, have done more for the cause of science in the country. However, there are some limitations in this regard which cannot be overlooked. Firstly, the government has been doing a very good job of promoting science in multi-faceted ways to the people, so much so that scientific societies are hard pressed to identify niche areas which are not served or are underserved and where they can come in to make a difference. For example, the Science Centre Singapore has been playing a splendid role in promoting science and technology to students and the public over the years (Tan & Subramaniam, 1998); the Agency for Science, Technology and Research and the National Research Foundation disburse ample funds to universities and research institutes for the cause of research and development; and science and mathematics education in schools are internationally well regarded. Secondly, the scientific society movement still depends on a voluntary cadre to run its operations, unlike in other established academies and societies which have the help of secretariat staff — this necessarily restricts the scope and range of its operations Thirdly, the lack of premises puts a limit on what office-bearers can do in spite of all their good intentions. Fourthly, funding has often been an issue even though it has been circumvented to a significant extent through creative means as well as via reliance on the 'invisible college' cum voluntary cadre model of operation.

The scientific society movement continues to grow in strength in Singapore and the various societies are recognised as being key players in the local science and technology scene. Nevertheless, there is potential for the academy and scientific societies to contribute to scientific policy making as Singapore's R&D landscape matures, and diverse non-government views could make positive contributions. The next decade in SNAS's continued evolution could see some progress in this aspect.

Conclusion

Scientific societies in Singapore are late entrants to the scientific society movement that started in the western world in the 17th century. Despite a lag of about 300 years in joining the movement, the importance of the movement was recognised by the scientific community, with several societies being formed to cater to diverse interest groups in the sciences in the latter half of the 20th century. The various societies and SNAS have grown in strength over the years and are now active players in the local science and technology landscape. More importantly, they have evolved a model of operation that does not seem to be in operation in other countries — no premises and no full-time staff to run their operations, yet their annual reports show an impressive array of programmes and activities organised by their volunteer staff. It is a model that works and has even been suggested in various forums as a means to jump-start the establishment of scientific societies in developing countries.

Though there are some published studies on the scientific society movement in the Singapore context, the present chapter is significantly different. Aspects which have not been reported before but have been included here are the following: reasons for the slow and late establishment of scientific societies in Singapore, why the early scientific societies set up in Singapore were not in the natural sciences, why there is a need for an umbrella organisation for the scientific societies in Singapore; write-ups on most of the key scientific socie-ties in Singapore; and appraisal of the scientific society movement in Singapore from various angles. It is hoped that the present study will encourage further explorations of other aspects of the scientific society movement in Singapore as this seems to be a neglected area of research in Singapore.

Limitations

As mentioned earlier, there has been little effort by the scientific community in Singapore to focus on academic studies of the scientific society movement in the country. There is little archived information and data available in the public domain and research repositories besides annual reports and web-based resources. It is in this context that we have also drawn substantially from our combined experiences in the scientific society movement in Singapore over the years to provide some of the perspectives and insights presented in this chapter. Our focus has been on only the major scientific societies — that is, the Singapore National Academy of Science and the 11 constituent members affiliated to it. There are other related societies such as the Association of Mathematics Educators, Singapore Neuroscience Association, Nature Society, The Astronomical Society of Singapore, etc which we have not touched on as it is beyond the scope of this chapter to focus on a comprehensive coverage of the movement beyond its intended focus.

References

Clarke, R. (1985). *Science and Technology in World Development*, Oxford University Press, London.

Solomon, J. (1996). School science and the future of scientific culture, *Public Understanding of Science*, 5, 157–165.

Tan, W.H.L. & Subramaniam, R. (1998). Developing countries need to popularise sci-ence. *New Scientist*, 2139, 52.

Tan, W.H.L. & Subramaniam, R. (1999). Scientific societies build better nations. *Nature*, 399, 633.

Tan, W.H.L. & Subramaniam, R. (2000). Chemical societies boost development. *Chemistry & Industry*, 3, 91.

Tan, W.H.L. & Subramaniam, R. (2009). Role of scientific academies and scientific socie-ties in promoting science and technology: Experiences from Singapore. *International Journal of Technology Management*, 46, 38–50.

Chapter 30

Science in Singapore — The Next 50 Years

B T G Tan, H Lim and K K Phua

The essays and articles in this volume have given us a good picture of how science, and the scientific community have developed in Singapore over the 50 years since its dramatic emergence as an independent nation in 1965. From virtually nothing, an active and credible scientific research sector has been built up in academia, government and industry which has made its mark internationally in several key areas of research.

The RIE2020 Plan

The Singapore Government has just unveiled its next five-year plan for research funding as the Research, Innovation and Enterprise 2020 plan or RIE2020 plan.[1] The total sum set aside for RIE2020 is no less than S$19 billion, which is almost a ten-fold increase over the first such plan, the National Technology Plan in 1995, and dwarfs the sums set aside for research in the '60s and '70s.

Four major thrusts have been identified under RIE2020 to build on the progress achieved under RIE2015 (which cost S$15 billion) and to create greater value in Singapore from our investments in research, innovation and enterprise:

- Closer integration of strategies: to encourage multi-disciplinary and multi-stakeholder collaboration to allow greater efforts nationally towards achieving our research goals, and to invest strategically in both curiosity-driven and mission-oriented research.
- Stronger dynamic for renewal: a continued shift towards more competitive funding from 20% of public funding in RIE2015 to 40% in RIE2020 to support the best teams and ideas and more white space funding.
- Sharper focus on value creation: strengthen flow through from research to its eventual impact on society and economy.
- Better optimised RIE manpower: sustain a strong and innovative workforce in the private and public sector.

The greater emphasis in RIE2020 on curiosity-driven research and white space funding is an important development. In the past, much focus has been given to grant calls for thematic research in pre-defined areas. This is certainly an important aspect of research funding for areas which need immediate attention and development. However, it must be acknowledged that researchers who are striving for excellence and cutting-edge results in their fields should be given the freedom to explore avenues which are unpredicted or uncharted. That is, after all, the essence of the best research which probes deeply into not just known unknowns, but unknown unknowns.

Investigator-led Research

Many in the scientific community are also hoping that a greater proportion of the research funding will be allocated for investigator-led research which is often the most effective way of funding the best and most productive researchers in academia and in research institutions.[2] Such funding is based on the excellence of the investigator's track record and allows the principal investigator greater leeway in the use of research funding (not forgetting that the strict governance of such funds must always be imposed) in accordance with the progress and direction of their research. This means that funds may have to be granted in smaller and more palatable sizes much more like, for example, the R01 grants of the US National Institutes of Health or NIH.[3]

It is always easier for research funding agencies to give large grants which, from the agencies' point of view, can only be justified by ambitious research objectives requiring a large team of researchers. Characteristically, such grants are given to research groups which the agencies consider to be multi-disciplinary teams drawn from several different entities. These entities therefore have to collaborate with each other to formulate such grant applications which usually ask for multi-million dollar sums or larger.

Indeed, agencies often specify that applicants must form multi-disciplinary teams in which collaboration between at least two or three groups in different areas is seen by the agencies as a major positive factor for grant approvals. There are indeed research initiatives where a multi-disciplinary approach is a key part of the research methodology, but to view a multi-disciplinary approach as essential to every research project is not always helpful or productive.

A large proportion of the scientific community's best research is still initiated by individual principal investigators working deep within their disciplines. Such researchers generally do know the need to be aware of work outside of their area which might impinge on their research. This awareness is an essential tool of skilled researchers who do not need to be shoehorned into a multi-disciplinary framework by research policy makers who have little direct experience in research but who are captivated by buzzwords like "multi-disciplinary" and "collaboration".

Indeed, collaboration with research colleagues, whether in your own institution or in outside institutions, and whether in your own field or in other fields, is the lifeblood of good science today. However, collaboration with another individual is an extremely personal part of the research process, and requires the right conditions and the right personal chemistry. It cannot be mandated and forced on researchers from the outside but should arise naturally and with mutual respect between potential collaborators. The lure of funding can tempt researchers into seeking collaboration, but this may not always lead to truly productive partnerships or fruitful results.

Basic and Applied Research

Research policy makers also need to understand the importance of basic research and its relationship to applied research as well as to real-world applications in industry and society.[4,5] Political leaders must of course be accountable to society for the use of public funds. Both policy makers and the public naturally best understand this justification in economic terms. Public funds should rightly be used to advance the nation's economy by improving its industrial and commercial sectors and its social services. It is thus unsurprising that policy makers and the public often mandate that funds set aside for research should be used towards this objective.

But is this best achieved by the funding of only applied and industry-based research? The answer would seem to be in the positive, as this kind of research would appear to lead more directly to the progress of industry and the direct improvement of the public's living standards. Hence the call to mainly support applied research and industrial research seems logical, as such downstream research would seem to lead most directly to new products and processes which would have a positive impact on the economy and society.

However, in reality, the belief that there is a linear chain from basic research to applied research to industrial research and on to product and process development is highly simplistic. The evolution and creation of new products and processes is highly complex and cannot always be reduced to a linear chain of events, and is also very dissimilar for different sectors like the biomedical/pharmaceutical industry and the electronics/IT industry. It is fallacious to think that one could always benefit from "applied" research without spending resources on "basic" research, as though one can separate a portion of lean meat from a portion of fat in a pork chop.

A research activity which is labelled "basic" in one laboratory can indeed be indistinguishable from another research activity labelled "applied" in another laboratory. For example, a physicist in one laboratory measuring the microwave dielectric constant of a new material may believe that he is doing basic research on the properties of that material, while another physicist engaged in an identical experiment elsewhere may believe that she is performing applied research leading to an improved microwave phase shifter.

What is important is that each scientist employs similar scientific principles in designing their experiments and in producing, recording and interpreting their results. The same standards of scientific excellence and integrity should and must be applied to their experiments and indeed to all scientific experimentation as well as to theoretical research, model building and verification.

Gravitational Waves and Go

Two recent important scientific and technological events have highlighted the extremely close and complex relationships between basic research, applied research and technology. First, the astounding discovery of gravitational waves in February 2016 was an extremely important event for scientists because it was a direct experimental confirmation of Einstein's General Theory of Relativity and belonged firmly in the field of basic research.[6]

These gravitational waves originated from a binary black hole system which was rotating and eventually merging into a larger black hole. The gravitational waves were detected by LIGO, the Laser Interferometer Gravitational-Wave Observatory located in Louisiana and Washington states in the US, a research project which was initiated over 40 years ago.

LIGO uses the most advanced detection techniques in laser interferometry to observe the incredibly weak perturbations in the space-time fabric reaching Earth propagated by these waves over huge interstellar distances. Laser interferometry is an exquisitely sensitive distance measuring technology which is based on the laser, itself certainly a creation of basic research. It is often the case that such delicate tools which were designed for scientific work will eventually make their way into the real world of industrial applications.

The other recent notable event was the Go (a board game also known as *weiqi* or *baduk*) match between the 18-time Go World champion, Lee Sedol, and the AlphaGo computer programme created by the Google DeepMind team.[7] AlphaGo is perhaps one of the most advanced manifestations of current artificial intelligence technology known as Deep Learning. Go is much more complex than Chess, and hence not amenable to "brute force" computing approaches such as were used by IBM's Deep Blue programme, which was able to defeat Gary Kasparov.

Using a combination of deep neural networks and tree search,[8] AlphaGo was able to defeat Lee Sedol by a margin of four games to one. However, the fact that Lee was able to defeat AlphaGo in one game indicates that while AlphaGo has attained and possibly exceeded the highest levels of human Go ability, it still has some weaknesses which will undoubtedly be ironed out in the future. While game playing programmes like AlphaGo are certainly remarkable technological achievements, they are firmly based on basic research in mathematics and theoretical neuroscience.

Pokemon Go and Joseph Schooling

Indeed, the all-encompassing and ubiquitous reach of both basic and applied science is embodied in both the very recent Pokemon Go craze and Singapore's first Olympic Gold Medal won by swimmer Joseph Schooling at the Rio Olympics. Pokemon Go, which is the first widespread popular application of mixed (or augmented) reality, is rooted in the complex mathematics of GPS navigation, computer modelling and graphics.[14] The application of hydrodynamics and biomechanics to the improvement of swimming stroke techniques and the design of swimwear for drag minimization are now integral to performance improvement in high level competitive swimming.[15,16]

Recognising and Funding Good Research

If one experiment is deemed "basic" research, and another deemed "applied" research, so be it! It is more important to allow researchers in both experiments the flexibility and freedom not to adhere to a preplanned fixed schedule, but to vary and even to radically change direction in accordance with their gut feel, when they sense that the ongoing results of their research warrant such changes. After all, true research is a step into the unknown whose results cannot possibly be entirely predicted, certainly not all of the time.

Building up a solid foundation of good basic research is the soundest basis for any national research programme, and this in itself can be deemed a public good. As all applied research depends on a foundation of basic research, a nation that wants to benefit economically by pursuing only applied research would not only be pursuing an ethically questionable course, but also an unwise one as the full benefits of the exploitation of applied research often require a thorough understanding of the basic research underpinning it. This understanding is undoubtedly best gained by active and serious participation in the process of basic research.

If we allow, as we should, a good proportion of our research funding to go to basic or "undirected" research, as opposed to applied or "directed" research, on what basis should such funds be allocated, if there is no obvious direct economic benefit which would be the outcome? The most logical procedure would be to fund research which attains the highest possible standards, producing results widely recognised as first-rate and impactful. How would one identify research proposals with such potential? An easy answer would be to scrutinise the track records of the potential Principal Investigators or PIs. This might mean looking for publication track records with papers in the most highly cited journals like *Nature* or *Science*, and this is indeed a common approach.

This may mean, however, that talented researchers without such track records would never get any funding. Important as high impact journals like *Nature* and *Science* may be, they certainly do not have a monopoly on good research results.[9] Research managers and department chairs must be able to recognise researchers with the right qualities who have good research ideas and are deserving of funding, and not just rely on journal branding. Unfortunately, too many universities and research institutions are losing the capability of identifying good research and instead rely heavily on journal and institutional branding for this purpose. There is plenty of first-rate science still being done in less high-profile places — you just need to be able to recognise it when you see it!

Science Education

Most nations, and particularly those in the early stages of economic development, believe that scientific and technical education is one of the main keys to a nation's economic progress, as much of industry and our societal infrastructure, such as health, public utilities and transportation, is based on science. A solid grounding in physics, for example, is one of the cornerstones of a credible engineering education.[5] At the school level, a sound basis must be laid in mathematics and the basic sciences which will allow students to benefit from training for professions and vocations based on science and technology.

One continuing trend is that bright students may not now choose science and technology-based courses such as engineering as their first choice, as they now perceive that such courses are too mathematical or quantitative in nature and thus deemed to be difficult. Furthermore, the jobs market for these courses does not provide adequate financial recompense, especially when compared to jobs which are in the banking and financial industries.

Hence it is widely recognised that there is a need to encourage more young people to be interested in what has now come to be called the STEM vocations, i.e., in Science, Technology, Engineering and Mathematics. Some countries have come up against a deficit of graduates trained in STEM vocations and a surfeit of graduates in business and finance. This is of current concern in Singapore, particularly with respect to the engineering professions and vocations which are vital to the maintenance and development of the nation's core technical expertise in areas like transportation, health, manufacturing and defence. More seriously, this deficit may see a complete hollowing out of our nation's technical and technological skills and expertise if not addressed adequately.[10]

Indeed, the nation will always require well-trained graduates in the various branches of science and engineering to maintain this core technical expertise and to prepare it for the technological and scientific changes and challenges which are bound to arise in the future. The major challenges which the world faces this century — in energy, climate change, pollution, food and water resources — will require scientific and technological revolutions which can only be accomplished by future well-prepared cohorts of scientists and engineers who must attack these challenges in the coming decades.

Public Awareness of Science

Perhaps just as important as having well-trained scientists and engineers is the need for the general public, the media and policy makers to have a good understanding of basic scientific concepts. They do not need to have the detailed understanding of science which professional scientists or researchers with a tertiary level training in their disciplines would have, but they need to be able to grasp the basic scientific concepts and ideas which underpin our daily lives and the fundamental infrastructure of society.

This is because the major challenges which the world will have to face in the coming decades are going to be ameliorated mainly by the application of technological solutions based on scientific principles. For many of these challenges, difficult decisions and choices will have to be made as to which courses of action should be taken. Scientists and engineers must be able to accurately present the hard choices and options whenever a decision has to be made, using sound and up-to-date scientific and technical data and information.

What is important is that these decisions should not be made just by the scientists and engineers involved in the technical details of the options presented, but must be made by society as a whole, since the results and effects of these decisions will affect everyone, not just the scientists and engineers. It is the job of the scientists and engineers to present the options available accurately and clearly, so that good decisions will be made by society, whose individual members must have a basic understanding of the scientific issues involved.

Hence it is imperative that the individuals who make up society and who will make these decisions collectively should have a good understanding of the basic concepts of science. If science is effectively taught at the primary school level and if students grasp its basic concepts firmly, then this would be more than sufficient for the next generation of citizens and decision makers to be able to make well-considered decisions on key national policy options which involve scientific and technological concepts.

For example, the climate change issue will require tough decisions to be made in the next few decades in topics such as energy generation and energy conservation. This will require a good understanding of how global climate works and the advantages and disadvantages of each type of energy source. One important energy source is nuclear energy, which allows energy generation without any carbon dioxide production, but which has other issues such as waste disposal and radiation dangers in the wake of nuclear accidents. The 2011 Fukushima nuclear incident in Japan revealed that there is widespread misinformation and much ignorance amongst the general public, as well as the media and policy makers, about the dangers of nuclear radiation and its effects.[11]

Such misinformation is difficult to dispel, especially if generations of citizens have been ignorant or misinformed about the effects of nuclear radiation. The danger is that long-term decisions about nuclear energy (or any other type of energy, for that matter) will be made based on misinformation and ignorance, when they should be made on the basis of accurate, unbiased and scientifically-based information about its societal impacts. I certainly would not like to see our nation make such (or any) long term decisions which are coloured by a lack of proper information or understanding about the basic scientific facts of each option.

In the case of nuclear radiation, about which large sectors of the public are particularly ignorant and misinformed, there is an urgent need to ensure that the next generations of citizens are well-informed enough to make decisions about nuclear energy and other key scientifically-based issues which are rational and logical. One effective way of ensuring this is to insert a few simple sentences about nuclear radiation into the primary science syllabus so that students will understand that such radiation is a perfectly natural phenomenon like water or fire, and like them is dangerous only in excessive quantities.

The Singapore Nuclear Research and Safety Institute (SNRSI)

The Pre-Feasibility Study conducted in 2010 by the Ministry of Trade and Industry on Nuclear Energy concluded that "nuclear energy technologies presently available are not yet suitable for deployment in Singapore".[12] The study however stated that Singapore needed to strengthen its capabilities in understanding nuclear science and technology. Hence, research in relevant areas of nuclear science and technology would be supported, and a pool of scientists and experts would be trained.

Therefore it was perfectly logical and appropriate that on 23 April 2014, the National Research Foundation announced the formation of the Singapore Nuclear Research and Safety Institute (SNRSI) under the leadership of Lim Hock of the NUS Physics Department.[13] SNRSI will be a national resource hosted by NUS, focuses on research and developing capabilities in nuclear safety, science and engineering. SNRSI's role will be to monitor the development of nuclear energy technologies, and support research in nuclear science and engineering, objectives espoused by the Pre-Feasibility Study.

SNRSI, as the newest science-based research institute in NUS and Singapore, is thus an affirmation of the nation's resolve to be well-prepared for the many challenges which lie ahead the rest of this century, and to ensure that our scientific and technological capabilities will continue to be vigorously developed for this purpose. As we look forward to the next 50 years in the nation's history, we are confident that science in Singapore will continue to be a key factor in our ability to weather future crises and challenges and to build a better society for our children and grandchildren.

References

[1] National Research Foundation. National Research Foundation, Prime Minister's Office, Singapore. [Online]. www.nrf.gov.sg/research/rie2020

[2] House of Commons Science and Technology, *The impact of spending cuts in science and scientific research: Sixth Report of Session 2009–10, Volume II*. London: The Stationery Office Limited, 2010.

3 National Institutes of Health. National Institutes of Health. [Online]. https://grants. nih.gov/grants/funding/r01.htm

4 Steve Hyman. Society for Neuroscience. [Online]. https://www.sfn.org/ News-and-Calendar/Neuroscience-Quarterly/Fall-2015/Message-From-the-President

5 K.K. Phua, "加强基础科学研究的投入," *Lianhe Zaobao*, February 2, 2016.

6 B.P. Abbott *et al*, "Observation of Gravitational Waves from a Binary Black Hole Merger," *Physical Review Letters*, vol. 116, no. 061102, pp. 061102-1 to 061102-16, 12 February 2016.

7 "Man falls to machine in final Go game," *The Straits Times*, March 16, 2016.

8 David Silver *et al*, "Mastering the game of Go with deep neural networks and tree search," *Nature*, vol. 529, pp. 484–489, 28 January 2016.

9 Stephen G. Lisberger, "Sound the Alarm: Fraud in Neuroscience," *Cerebrum*, May 2013.

10 Han Fook Kwang, "Shrinking pool of engineers poses national risk," *The Sunday Times*, July 12, 2015.

11 George Johnson. The New York Times. [Online]. http://www.nytimes.com/ 2015/09/22/science/when-radiation-isnt-the-real-risk.html?_r=0

12 Ministry of Trade and Industry, "Factsheet: Nuclear Energy Pre-Feasibility Study," 15 October 2012.

13 National University of Singapore. NUS News. [Online]. http://news.nus.edu.sg/ highlights/7618-singapore-nuclear-research-and-safety-initiative-to-be-hosted-at-nu

14 University of Toronto. U of T News. [Online]. Pokemon GO craze shows that augmented reality is hitting its stride, U of T experts say. https://www.utoronto.ca/ news/pokemon-go

15 Top End Sports. [Online] Biomechanics and swimming. http://www.topendsports. com/sport/swimming/science-biomechanics.htm

16 Hon Jing Yi, "Science propelled Schooling to victory in Rio," Today, December 1, 2016.

Index

www.ingramcontent.com/pod-product-compliance
Lightning Source LLC
Chambersburg PA
CBHW080127270326
41926CB00021B/4384